EARTH, AIR, FIRE, WATER

EARTH, AIR, FIRE, WATER

Humanistic Studies of the Environment

Edited by

JILL KER CONWAY,
KENNETH KENISTON,
AND LEO MARX

University of Massachusetts Press
AMHERST

Copyright © 1999 by
The University of Massachusetts Press
All rights reserved

Printed in the United States of America
LC 99-16970
ISBN 1-55849-220-8 (cloth); 221-6 (paper)

Designed by Dennis Anderson
Set in Janson Text by Graphic Composition, Inc.
Printed and bound by Sheridan Books, Inc.

Library of Congress Cataloging-in-Publication Data

Earth, air, fire, water : humanistic studies of the environment /
    edited by Jill Ker Conway, Kenneth Keniston, and Leo Marx.
        p.   cm.
    Includes bibliographical references.
        ISBN 1-55849-220-8 (cloth : alk. paper). — ISBN 1-55849-221-6
    (pbk. : alk. paper)
        1. Environmental degradation—Social aspects.   2. Nature—Effect
    of human beings on.   I. Conway, Jill K., 1934– .   II. Keniston, Kenneth.
    III. Marx, Leo, 1919– .
    GE140.E18   1999
    304.2′8—dc21   99-16970
                        CIP

British Library Cataloguing in Publication data are available.

This book is published with the support and cooperation of the
University of Massachusetts Boston.

तस्माद्वा एतस्मादात्मन आकाश: संभूत: ।

आकाशाद्वायु: । वायोरग्नि: । अग्नेरापः । अद्भ्यः पृथिवी ।

पृथिव्या ओषधयः ।

तैत्तिरीयोपनिषद् २-१

From that Brahman, which is the Self, was produced space.
From space emerged air. From air was born fire.
From fire was created water. From water sprang up earth.
From earth were born the herbs.

Taittiriyopanishad 2-1

Source: *Eight Upanishads*, translated by Swami Gambhirananda, Advaita Ashrama,
Calcutta (1977), 287

Contents

Preface

THIS VOLUME grew out of the four-year Workshop on Humanistic Studies of the Environment that was sponsored by the John D. and Catherine T. MacArthur Foundation at the Massachusetts Institute of Technology. Information about the workshop, the participants, the speakers, and the students supported by the project may be found in the Appendix. We are grateful for the understanding and support of the MacArthur Foundation and its officers.

JKC, KK, LM

EARTH, AIR, FIRE, WATER

JILL KER CONWAY,
KENNETH KENNISTON, AND LEO MARX

The New Environmentalisms

FIFTY YEARS after the atomic bomb was dropped on Hiroshima, the conviction of environmental crisis to which it gave rise has intensified. The first use of a nuclear weapon in 1945 made humanity aware that it had acquired the power to inflict irremediable damage on the biosphere, a destructive power that might even lead to human self-extinction. As it turned out, in fact, Hiroshima was only the first in a series of events that seemed to portend an ecological apocalypse.

In the aftermath of Hiroshima, the intellectual results of this mounting anxiety were immediate, profound, and lasting. In the academy the first members of what would become a large and steadily growing international cohort of scholars—most of them scientists—began to work on problems of nuclear contamination. In subsequent years the range of major ecological problems was enlarged by the discovery of such new (or hitherto undetected) hazards as the potential "nuclear winter" phenomenon, global climate change, the depletion of the ozone layer, and the accelerating rate of species extinction. With each discovery an alarm was sounded, and the worldwide fear of an impending ecological disaster intensified. By now that fear has been extended to the damaging effects of many everyday technologies, and we see harm lurking in such innocuous sites as the local garden shop, with its lawn fertilizers and gas-powered mowers, and the supermarket, with its array of detergents and chemically improved meats and vegetables.

In response to these fears, a set of new environmentalisms has emerged—movements, arguments, and analyses that target the newly identified environmental problems of the late twentieth century. To be sure, men and women were concerned with preserving their environment long before Hiroshima. But in the last five decades, initiated by the use and testing of nuclear weapons; impelled by books like Rachel Carson's *Silent Spring*; embodied in local groups like the Love Canal activists; highlighted by disasters such as Chernobyl, Bhopal, and Three Mile Island; and armed with regulatory power through governmental bodies such as the Environmental Protection Agency, the new environmentalisms have acquired unprecedented public support and political importance. Despite their major differences, these environmentalisms share a concern with today's apparently unprecedented and accelerating rate of environmental degradation.

The prevailing assumption both within and without the academy has been that for self-evident reasons, it is scientists who bear the major intellectual responsibility for coping with this degradation. When we think of the forms of environmental decline calling for most urgent attention—"eroding soils, shrinking forests, deteriorating rangelands, expanding deserts, acid rain, drained aquifers, stratospheric ozone depletion, the buildup of greenhouse gases, air pollution, poisoned water supplies, and the loss of biological diversity"—it seems only logical that scientists should be the people mobilized to tackle these problems.[1]

It also seems obvious that the human sciences, the term we use to embrace the humanities and humanistic (nonquantified) social sciences, have little to contribute to our understanding of these threats to the biosphere. Until recently, humanists themselves accepted this popular assumption. What can Homer tell us about nuclear winter? How can students of language help halt the destruction of forests? Surely only scientific experts are capable of discovering a hazard like the greenhouse effect. And clearly only scientists can monitor it accurately and thus perhaps devise effective remedial measures. Where else should we look but to scientific expertise for the resolution of problems resulting from the interaction between the peoples of modern societies and nonhuman nature?

And yet it is the seemingly self-evident nature of this response that should give us pause. As cultural historians have often demonstrated, the more obviously self-evident a human response to change seems, the more likely it is to embody an unconscious, or largely unconsidered, reflex of the prevailing collective mentality. This is not to imply that all such "common sense" responses are skewed or misleading, but they often are, and in the case of environmental degradation there are good reasons for skepticism about the humanists' failure to engage with the problem. Notice, for example, the heavy burden of ideological assumption carried by the heavily scientific, technological names we routinely use to designate environmental problems. (Few behaviors are more revealing of cultural bias than naming practices.) Each of the labels mentioned—*eroding soils, shrinking forests, acid rain*—designates an environmental problem by naming its chief biophysical symptom. Missing entirely are the simple, short everyday words by which people actually refer to their biophysical world: *earth, air, fire, water.* The scientific labels convey no hint of human agency. Indeed, they seem to convey that such forms of environmental deterioration are spontaneously occurring, "natural" (i.e., nonhuman) biophysical processes. Such a designation places the entire process of environmental deterioration within the realm of expertise of scientists who study natural phenomena.

Once we examine these names critically, we realize that they are highly misleading, because, although they locate such phenomena as acid rain and soil erosion in the biophysical realm, not one of these phenomena is wholly attributable to the operation of natural (nonhuman) processes. Each, in fact, has its origin in human behavior, in complex socioeconomic practices with long histories. So, although it is not impossible, it is highly unlikely that any of them could be corrected or compensated for by a simple technological fix. In fact, these nature- and science-oriented names mask the fact that many such forms of damage to the environment cannot be ameliorated or corrected without extensive long-term changes in social behavior, such as prevailing beliefs about and attitudes toward the interaction of humanity with nature. Amelioration does not require exclusively scientific knowledge, but rather changes based upon law and public policy, institutional structures and practices, habits of consumption, and countless other facets of daily life.

Thus if we are to understand and devise effective solutions for today's environmental threats, we must locate them within their larger historical, societal, and cultural setting. Only when placed in this context will they be recognized for what they are: immediate, short-term, partial manifestations of the increasingly heavy burden that modern urban industrial societies place upon the finite capacities and resources of the biosphere. The root problem of this demand is human, not physical, not natural—although, of course, scientists, engineers, and other technical experts can help us chart its dimensions. Once we have framed the issues in this way, we can see that many, perhaps most, of our most pressing current environmental problems come from systemic socioeconomic and cultural causes and for this reason their solutions lie far beyond the reach of scientific or technical knowledge. Thus to answer an earlier, seemingly rhetorical question, all the disciplines that elucidate human behavior and the functioning of social and cultural systems are essential for understanding environmental issues and devising effective approaches to their amelioration.

This book, then, is an effort at correcting our deceptive nomenclature by locating ecological problems in the behavior of human beings—in the human institutions, beliefs, and practices that mediate between humankind and that obscure, beautiful, nonhuman world we call "nature." It opens with a section devoted to the elements and the way humans have understood them in past times. It continues with a section devoted to social institutions and the ways we can learn from current and past efforts to understand the interaction between people and nature. The concluding section analyzes the culture of modernity and the ways the human imagination has changed in response to the arrival of modern technology, for it is this change that has contributed most significantly to our distancing ourselves from the natural phenomena we now consider to be the exclusive concern of scientists.

As a framework for the examples of humanistic studies of environmental thinking that make up these three sections, we here lay out some major concerns that any humanist proposing to work on environmental subjects will encounter. These arise from critical oppositions inherent in current thinking about humans and their interaction with nature.

The "Constructed" versus the "Real" Environment

One of the first questions confronting humanists who work on environmental problems is, What constitutes reliable knowledge of the natural world? Or, put differently, the problem is knowing how to steer a reasonable course between two equally extreme viewpoints: naïve positivism (or realism) and all-embracing social constructionism (or the assertion that what we call "nature" is merely a figment of our cultural imagination).

The positivistic position assumes that reliable, unmediated knowledge of "nature" or the "environment" is obtainable by means of direct sense perception and that it may then simply be added to the cumulative findings of science. Those findings are assumed to constitute a true picture of the world. This picture is not considered problematic or seriously influenced—unless based on erroneous data—by the unique outlook, history, or culture of the observer. Nature, environment, and the world are a transparently accessible domain of incontrovertible fact.

In contrast those who hold the constructionist view regard what we call "reality" as in actuality a kind of narrative, or "text," that we construct about our surroundings. Such narratives are in some measure unique to each individual, and they invariably are the distinctive products of particular historical contexts, cultures, and social groups with particular interests—especially national, economic, class, racial, or gender interests. The notion of the "environment," or "nature," as a transhuman reality disappears, replaced by a variety of interpretative lenses through which individuals convince themselves (falsely, of course) that they are seeing something beyond—not reflective of—their subjectivity or the distinctive positions they occupy in their social and cultural settings. In particular, radical social constructionists deny the hegemony of scientific knowledge as the only truly reliable—or, as they say, "privileged"—conception of the world. Science is thus merely one among many lenses on the world, a lens with no justifiable claim as a source of superior knowledge. To the constructionists, the humanists' task is to understand, analyze, and deconstruct discourse about nature and our environmental dilemma, and in the process to challenge the illusion that we have access to the ostensibly "real," knowable environment.

To state these two positions in this extreme, caricatured form is to underscore the latent contradiction that often makes itself felt in humanistic inquiry into environmental issues. Concepts such as "environment," "nature," and "wilderness" are often assumed to be constructs whose contours are defined less by "objective" reality than by the interests, history, and presuppositions of the observer. Thus, many recent humanistic studies of the "environmental crisis" have been studies of writings about the crisis, or of definitions and so-called "constructions" of the crisis, rather than studies of the ways that human beings, through their culturally and historically influenced behaviors, help aggravate or ameliorate the condition of the biophysical world that surrounds them.

The familiar parable of the blind men trying to describe the elephant, each insisting that a leg, the trunk, or a tusk is the whole of the beast, is a useful analogy for our own thinking. We agree with the social constructionists who insist that the world—and especially large interpretive concepts about the world, such as "environment," "wilderness," and "nature"—is invariably seen from a particular vantage point and through a particular lens constituted by history, culture, and individual idiosyncrasy. There are indeed many natures, environments, ecologies, and wildernesses, as scholars insist.[2] But the parable of the elephant derives its ultimate meaning precisely from the fact that there *is* an elephant, a real elephant, that each blind man only partially describes.

As human beings and adherents of a culture, therefore, we have no way of seeing other than through the lens of our own culture, history, and personality. But the fact that we each see the world from a distinct context and a unique perspective in no way denies the world's existence; on the contrary, only if there *is* a world to be seen through our different lenses does the act of perception make any sense. Analogously, arguments over the meanings of "nature" or "wilderness" in no way deny the existence of a nonhuman biophysical reality over whose characteristics we may argue. In fact, the existing nonhuman biophysical reality constitutes a large part of what people perceive, and what they disagree about.

We share, then, the belief of most natural scientists that the "environmental crisis" is real; that it is global as well as local; and that science gives us an especially reliable and useful, though not unique, way of understanding the crisis. But of course the natural sciences make no claim to a deep or sophisticated understanding of the dimensions of life that derive from human behavior, culture, personality, social organization, or history. Quite the contrary: the sciences most engaged in the study of the environment are mute when it comes to the human (or "anthropogenic") sources of recent environmental problems. Thus computer models of the impact of greenhouse gases on global climate often include projections of the increases in carbon dioxide emissions likely to result from human activities over the course of the next century. But the question of why or whether humans are seen as likely to increase carbon dioxide emissions is not one that atmospheric scientists address. To explore that question, the methods of humanists and social scientists are needed. Several years after a major international effort to integrate scientific studies of the global environment (the International Geosphere-Biosphere Programme: A Study of Global Change) was organized, a Human Factors group was finally established, as if in belated recognition that, after all, humans' activities are at the root of virtually all the world's most pressing environmental problems.

Another reason to doubt the exclusive authority of the scientific viewpoint is that scientists rarely achieve unanimity on environmental issues. They can differ among themselves as much as nonscientists do about the meanings, implications, causes, and remedies of environmental problems. Scientific knowledge of the environment is generally new and hence contested: little of it is established, textbook knowledge. Like all frontier knowledge in science, knowledge of the environment is therefore peculiarly susceptible to conflicting interpretations, alternative forecasts, and disputed remedies.[3] In analyzing these conflicts, it is important, though by no means sufficient, to acknowledge their cultural origins—their roots in differing perceptions, politics, interpretations, interests, and histories. As with all contested, frontier knowledge in science, moreover, continued exploration and lively debate also are needed, for these alone can transform contested knowledge into scientifically established (if always open to reexamination) textbook truths.

Perhaps more than most humanists we acknowledge the importance of scientific findings and accept their legitimate claim to special, if limited, authority and usefulness. And at the same time, we here stress the obligation of humanists to study the ways that human beings actually interact with—not merely talk about—nonhuman nature. Humanists and humanistically inclined social scientists have a double task. First, they can (and do) contribute to an understanding of environmental discourse—the ways that ideas (including scientific ideas) about nature embody extrascientific interests and presuppositions; the historical origins and shifting meanings of central concepts (e.g., "nature," "environment," and "wilderness"); and the role of the socioeconomic and political context, culture, ideology, and history in forming the lenses through which we perceive and interpret the biophysical world.

But humanists and their social scientist partners have a second and often neglected task: to study the precise ways that culturally and psychologically patterned behavior contributes to the despoliation of the environment and the possibility or impossibility of alleviating it. It is important, for example, to understand the steady, worldwide growth of consumerism, its changing character over time and across cultural boundaries, and its relationship to today's well-nigh universal quest— even in the richest nations whose populations' basic needs have long since been satiated—for a continuously rising standard of living. Similarly, it is important to understand why some people are politically mobilized—and others are not—against perceived environmental problems, be they global in scope (chlorofluorocarbon [CFC] emissions) or local (water pollution, deforestation, or the exhaustion of arable land).[4]

Varieties of Environmental Experience

In carrying out any such analysis, humanists must recognize the instability and ambiguity of the term environmentalism. Almost no one professes anything but goodwill toward the environment or its protection, yet few social movements elicit greater hostility than—or embody such deep divisions and bitter controversies as—the diffuse collection of ideas and groups labeled "the environmental movement." The "environmentalism" of the National Rifle Association and of sports trophy hunters is no less passionate than that of deep ecologists and the tree-hugging members of Earth First! To be sure, mainstream environmentalists regard the "environmentalism" of international paper companies or the nuclear power industry as self-interested, exploitive, and manipulative. Although none of the authors in this volume endorses the views of those corporations, we are reluctant simply to charge them with hypocrisy, but see them as embracing a different conception of the environment, based on different historical time spans, different interests, and different assumptions about the essential relationship between humanity and nature. One of the essential tasks of the humanist, therefore, is to disentangle the multiple meanings of "environmentalism."

Although the classifications that follow are tentative, we think them a useful introduction to the essays in this book. They highlight that there are many varieties of environmentalism, many sets of attitudes, values, and beliefs subsumed under the omnibus term "environmentalism."

Ecocentrism versus Anthropocentrism

Nowadays, environmental thinking is widely assumed to be polarized between two opposed, probably irreconcilable doctrines: ecocentrism and anthropocentrism. Ecocentrism is a moral philosophy whose exponents, a vocal minority of environmentalists, are dedicated to changing radically the way humans think about their relations with nature. They look upon mainstream environmentalists as weak compromisers who inveigh against the despoliation of the environment but in practice are all too accommodating to the despoilers. Such compromising is predictable, the ecocentrists contend, because reform environmentalists and despoilers, whatever their differences, are indistinguishable in one crucial respect: both assume that our chief reason for protecting the environment is its usefulness to ourselves, to human beings. From an ecocentric viewpoint nothing we could possibly do to arrest the accelerating devastation of the global ecosystem would be more effective than to rid ourselves of the complacent illusion that nature exists to serve humanity. "No intellectual vice is more crippling," writes the Harvard sociobiologist and ardent ecocentrist E. O. Wilson, "than defiantly self-indulgent anthropocentrism."[5]

The radical transformation of human consciousness envisaged by Wilson and his fellow ecocentrists, which they see as a belated accommodation to the inescapable dictates of biological reality, would be as profound as that which followed the discoveries of Copernicus, Newton, and Darwin. It entails acceptance of the far-reaching implications they draw from an unarguable fact of nature, namely, that *Homo sapiens* is only one of the myriad, intrinsically valuable, interdependent species on Earth, and their more arguable conclusion that we therefore have no right to reduce the diversity of life or assess the worth of other forms of life (or even, for that matter, inanimate parts of nature) merely on the basis of their value to ourselves. To satisfy our basic needs, of course, humans might continue to kill some animals, consume plants, and use nature in various other ways, but these and all other human activities should henceforth be restricted by the ruling imperatives of ecocentrism: to live lightly on the earth, restrict the scope of technological innovation and intervention, and treat all forms of life and all parts of the cosmos with reverence, responsibility, and care.

The intellectual genealogy of the ecocentric doctrine leads back to the religious origins of contemporary attitudes toward the nonhuman environment. The ecocentric lineage may be traced, by way of the Norwegian philosopher Arne Naess, to modern nature writers including Rachel Carson, Aldo Leopold, and John Muir; poets and novelists such as Robinson Jeffers, Gary Snyder, D. H. Lawrence, and Thomas Hardy; the great Romantics Rousseau, Coleridge, Wordsworth, Blake, and Goethe; and, especially for their shaping influence on American attitudes toward nature, the prominent Transcendentalists Ralph Waldo Emerson and Henry Thoreau. Almost without exception, these writers accorded the natural environment a reverence of the kind and intensity their forebears had reserved for divinity.

Emerson and Thoreau, in particular, were pivotal in effecting the transition in America between predominantly theological and predominantly secular views of nature. They played a role analogous in many ways to that played by Coleridge, Carlyle, and Wordsworth in England; Rousseau in France; and Goethe in Germany. The religious roots of Emerson's and Thoreau's environmental thinking seem more obvious, however. They patently were the heirs of Jonathan Edwards, the greatest philosopher produced by New England Calvinism, and of three or four generations of Puritan thinkers who preceded him. Although they adopted a less explicitly religious language to discuss human interactions with the environment, that discursive change was somewhat misleading, for it disguised the degree of underlying continuity between their ideas and those of their religious precursors.

Thus Emerson, a descendent of a long line of New England ministers, began his career as a Unitarian pastor, and he never stopped thinking of nature as—to invoke his formulation in the seminal book *Nature* (1836)—"the present expositor of the divine mind." His mature philoso-

phy was a somewhat idiosyncratic amalgam of Anglo-German Romanticism (much of it indirectly borrowed from the eighteenth-century
German *Naturphilsophen*), post-Kantian idealism (above all, Schiller's version), and his hereditary Yankee protestantism.

Thoreau, fourteen years younger, began his career as Emerson's disci-
ple; at first he adopted most of the Transcendentalist doctrine, but he
soon took a more independent course. He became a knowledgeable
woodsman and amateur naturalist, and he developed a distinctive literary
style based on the exact observation and depiction of natural facts. The
purest examples of his brilliant nature writing are to be found in his immense *Journal*, but his most popular and influential work, *Walden* (1854),
also conveys a passionate aversion to the dominance of society by an
acquisitive commercial ethos that issues in a well-nigh systematic degradation of the environment. In his nature writing, Thoreau exemplified a
pragmatic yet worshipful attitude toward nonhuman nature that now has
made him the patron saint of ecocentrism.

Unlike ecocentrists, who emphasize the attributes humans share with
other species, anthropocentrists hold that we humans have a unique responsibility as stewards of the environment. That responsibility derives
in part from religious doctrine, such as the biblical injunction (in Genesis) "to replenish the earth, and subdue it, and have dominion over . . .
every living thing that moveth upon the earth," and in part from humanity's manifestly distinctive capacities—intellectual, moral, technological—to manage the resources of Earth. The concept of "resource management" is a hallmark of the anthropocentric relationship with the
environment. Environmentalists of that utilitarian persuasion remind us
that most species that ever existed are extinct; that the history of nature
is marked by unceasing change; and that though each species modifies
its habitat in some degree, the extent to which humanity's modification
of its global habitat exceeds that of all other species amounts to orders
of magnitude. To the charge that anthropocentrism represents an arrogant, self-serving presumption of human superiority, the anthropocentrists respond by charging the ecocentrists with what appears to be
an even more arrogant refusal to accept the responsibility, for which
Homo sapiens is the uniquely qualified species, to oversee the maintenance of a life-enhancing ecosphere.

We are presenting the dichotomy between ecocentric and anthropocentric environmentalisms in its sharpest, most melodramatic form.
To be sure, each of these extreme viewpoints has its adherents, but they
constitute a small minority. Most active environmentalists, as well as
most members of the general public who advocate the protection of the

environment, almost certainly hold opinions of a measured, pragmatic, utilitarian—or anthropocentric—tenor. However, as unrealistic or impractical as the severe ecocentric code of environmental probity may seem, it nonetheless provides a challenging long-term goal of harmonious accommodation to nonhuman nature and the unillusioned recognition of certain unmodifiable, bedrock imperatives of human survival. The value of ecocentrism, like other visionary or utopian doctrines, is to generate long-term aspirations—to educate desire.

Apocalypticism versus Gradualism

A parallel, closely related, spectrum of opinion along which environmentalists differ is defined by the degree of urgency they bring to their proposals. Ecocentrists in general tend toward a more extreme, even apocalyptic, sense of urgency, whereas anthropocentrists are more likely to advocate a temporizing, gradualist agenda. They consider it more prudent and effective in the long run to make haste slowly.

At the apocalyptic extreme is the view that the environmental "crisis" has already reached catastrophic or near-catastrophic proportions: we currently risk the destruction of the habitat of humankind and of most species through actions already taken or imminent. Typical culprits are global warming; the proliferation of toxic chemicals; the population explosion; the pollution of air, water, and earth; and the accelerating rate of species extinction. In this apocalyptic view, the carrying capacity and recuperative powers of the planet have been exceeded or are about to be exceeded. Barring massive immediate changes in human behavior, irreversible and catastrophic destruction—including the death of billions of human beings and the possible extinction of life on the planet—will result.

This apocalyptic view is typically accompanied by calls for farreaching changes in the way we live, organize our institutions, and view the world. Apocalyptic environmentalism is analogous to—and indeed often has historical roots in—millennial religious movements, with their inherited notions of imminent destruction and their calls for dramatic and total reform, repentance, and spiritual reawakening. Indeed, the modern sense of an oncoming ecological apocalypse owes a great deal to the ancient Christian tradition of millennial evangelism and fundamentalism. In the United States, where today's "deep ecology" and ecocentric doctrines draw heavily on the writings of the New England Transcendentalists, especially Emerson and Thoreau, there is a direct line of descent from the eschatological tenor of the Puritan churches (via John Muir and the Sierra Club, for example) to ecological apocalypticism. In eighteenth-century Western thought, moreover, there was a widespread tendency to transfer qualities previously reserved for divinity to an abstract, post-Newtonian concept of nature. Thus the despoliation of the environment has come to have close affinities with the kinds of mortal sin that merit severe divine punishment.

At the opposite extreme is the gradualist view, especially common among scientists, politicians, and industry representatives, that we should take no rash action, since we do not know enough yet. Gradualists stress the admitted uncertainty of many scientists who work on ecological problems, and they are concerned about the possibly harmful effects of action taken prematurely, in the absence of certain knowledge. They are less impressed by the rapidity than by the slowness of changes in the state of the environment, and consequently they stress the ways recent human, political, and economic actions already have achieved improvements. Thus, for example, they point to the positive results of the environmental protection laws, or international agreements, adopted in the last twenty years by the industrialized nations. Above all, gradualists stress the hazards of taking action in the absence of truly reliable knowledge.

It is easy to attribute self-interest to gradualism when it is adopted by spokespersons for corporations and other institutions called upon to adopt innovations that are economically or humanly costly. But this view is also held by many who have no self-serving economic or political interest in deferring action. They insist on the inadequacy of existing models of environmental change; the uncertainties of ecological knowledge and theory; and, most important, the human, economic, and social costs of taking the radical measures advocated by the environmentalists of the apocalyptic cast of mind. Whatever the environmental toll of the pesticides, tube wells, herbicides, and "artificial" fertilizers associated with the Green Revolution, for example, their abolition would dramatically diminish the world's food supply. This might be ecologically sound from a long-term point of view, but in the short term it would produce massive food shortages and might well result in the death from starvation of millions, even billions, of people. Gradualists contend that as yet we have no sure evidence of irreversible environmental damage and that remedial or preventive action should await a knowledge of its consequences.

Materialism versus Idealism

Another divide between environmentalists separates those who believe that environmental problems are in essence material or technological problems from those who regard them as essentially problems of consciousness, values, or beliefs. For the latter, the environmental dilemma is largely ideological, spiritual, aesthetic, cultural, or psychological in character. In contrast, at the materialist extreme are those who assume that history is generally a record of continuous, cumulative progress and see contemporary environmental problems as the result of inadequate and poorly conceived technologies, such as polluting energy sources, unsafe nuclear reactors, toxic organophosphates, inadequately reprocessed industrial wastes, and automobiles with excessively damaging exhausts. For them, the central environmental problem resides in inadequate or antiquated technologies, or in methods of intervention in the environment developed without adequate knowledge of their potential results. They stress the malign impact of the law of unintended consequences.

Almost invariably, then, gradualists contrive and optimistically endorse technological solutions. The "green technology" movement, with its emphasis on "reducing the waste stream," devising "cleaner forms of energy production," building "fail-safe third generation nuclear reactors," or creating non- or low-polluting methods of transportation, typifies the views of those who conceive of both the problem and the solution as technological. As a president of the Massachusetts Institute of Technology once put it, "The answer to bad technologies is not no technologies, but good technologies."[6]

At the other extreme are those who view the ultimate sources of environmental problems as essentially moral, spiritual, aesthetic, ideological, or cultural in character. Our relations with nature do not originate in tangible, material circumstances so much as in the beliefs, values, and meanings of which whole ways of life—entire cultures—are constituted. "'Tis said," Emerson once remarked, "that the views of nature held by any people determine all their institutions."[7] Thus the assumption that nature exists to serve humankind is decisive. It manifests itself in the culturally shaped and instilled desire for a standard of living far beyond that necessary for the maintenance of life and health; in the advent of consumerism propelled by a powerful advertising industry whose purpose is to create needs for new products that the population never knew it needed (VCRs, high-definition television sets, automatic bread makers, computers, etc.); and, most important, in a materialist mentality that places the satisfaction of material needs, particularly acquisitive and consumerist needs, ahead of nonmaterial aesthetic, moral, or spiritual satisfactions. These manifestations of the assumption that the purpose of nature is to serve humanity are seen as primary causes of the environmental crisis.

The solution, accordingly, lies not in better scrubbers, more ellective catalytic converters, or safer nuclear reactors, but rather in a massive transformation of human aspirations to a willingness to dispense with superfluities and embrace a life of voluntary simplicity. This would entail a radical change of values: a relinquishment of the pursuit of a steadily rising level of consumption (standard of living) in favor, as society's chief economic goal, of equitable sufficiency. Instead of a commitment to limitless growth, the primary aim of an ecocentric economy would be to dispense with superfluities and concentrate on providing the truly necessary material goods to all the world's people. In this view, fulfillment would be identified with the achievement of satisfying human relationships, the life of the mind and spirit, and the attainment of a more harmonious coexistence with nature. In short, the nonmaterial aspects of life would be given priority over the anticipated benefits of increasing human control of nonhuman nature. The call, then, is for transformation of the collective consciousness, renunciation of today's pervasive consumerism, and abandonment of the obsession with technological and economic "progress" that dominates the lives of people in virtually all contemporary societies.

Primitivism versus Presentism

Another critical distinction between environmentalists is related to their evaluation of the mindsets, outlooks, and practices of "primitive" (i.e., premodern) or non-Western peoples. Often associated with an ecocentric and millennial outlook, the "primitivists" see premodern and non-Western societies as an important source of ideas and practices that could help solve contemporary environmental problems. The outlook of premodern societies is often characterized as animistic, as not drawing decisive distinctions between humankind and the rest of nature, as committed to "living lightly on the land," and above all as showing a loving respect and concern for all living things. Some primitivists look with special admiration on the spiritual reverence with which Native American tribes are said to have regarded animal or vegetable totems, and others encourage the re-creation of premodern rituals or the deliberate search for wilderness experiences as a means of recovering a direct relation with nature.

One variant of primitivism looks less to premodern societies than to non-Western societies, and in particular those that are not influenced by the Abrahamic tradition of God-given dominion over nature, that is, not influenced by Judaism, Christianity, or Islam. In societies like India or Japan, it is said, even today people have a more reverential, more "ecological" attitude toward the biophysical world. One Japanese observer claims, for example, "Nature is at once a blessing and friend to the Japanese people. . . . People in Western cultures, on the other hand, view nature as an object and, often, as an entity set in opposition to humankind."[8]

At the opposite pole are those who question the relevance of premodern and non-Western attitudes to contemporary environmental problems or deny the claim that these attitudes are truly environmental in any useful contemporary sense. Some critics of primitivism point out that premodern societies have often despoiled and even destroyed their environments and argue that many previous civilizations have collapsed because of self-created ecological disasters. Others question whether non-Western societies like Japan are truly environmentally oriented in any comprehensive way. For example, one student of Japanese environmental attitudes argues that the Japanese "reverence for nature" is in fact a "highly restricted" attitude, "confined to particular species or individual animals, frequently admired in a context emphasizing control, manipulation or contrivance.[9]

Most important, those who reject the views we are calling primitivist believe that contemporary environmental problems are sui generis— unlike those faced by any previous civilization. They attribute today's problems chiefly to the enormous expansion in human understanding of, control of, and power over the environment brought about by the scientific, technological, and industrial changes of the last two centuries. Modern societies have the technological power to destroy their environment and perhaps, indeed, to cause irremediable damage to the global ecosystem, whereas previous societies usually did not. Having "wilderness experiences" on plastic rafts roaring down rapids created by the timed release of water from an upstream hydroelectric plant—such experiences may replenish the spirits of those who can afford them, but they do not truly speak to the major contemporary environmental problems, all of which involve complex sociotechnological systems. And it is simply not clear to critics how simple reverence for nature or premodern rituals, even if they *did* characterize premodern and non-Western societies, can help us deal with contemporary problems such as global warming, acid rain, ozone depletion, and toxic chemicals.

Worldview versus Issue

Another contrast between environmentalisms is that which separates environmentalism viewed as the fulcrum of an embracing, comprehensive philosophy of life, society, and politics and the preservation of the environment as simply one important value among other possibly equally or more important objectives.

The contention that environmentalism is or should be central to an all-inclusive philosophy of life and social organization is closely associated with millennial, spiritual, and global perspectives. The essential claim, as with ecocentrism, is that a drastic reorientation of all existing outlooks is required, such that the first criterion of every individual action, social policy, or political act should be its bearing on the preservation and enhancement of the environment. At the individual level, environmentalism therefore means adopting lifestyles characterized by voluntary simplicity; at the social level, it requires a redesign of social institutions to enhance those that preserve the environment and eliminate those that degrade it; at a political level, it means reorganizing policy and politics, and perhaps even redefining political boundaries, so as to promote environmental preservation. So seen, environmentalism is an overriding philosophy, sometimes described as a "new worldview," which must supplant consumerist, capitalist, socialist, individualist, or other allegedly environment-destroying outlooks.

The alternative view sees environmental preservation as only one among many important social values, for example, social justice, peace, economic development, human rights, and the fulfillment of individual ambitions. Proponents of this view deny that ecological principles constitute an adequate base for an entire philosophy and note that there are environmentalists of every political stripe from the reactionary right to the radical left. Other values, such as equity and individual liberty, may at times conflict with and deservedly override environmental values. Reverence or care for nature in itself tells us little about how we should organize our daily lives, social institutions, and political affairs. In the northern industrial societies, to be sure, environmentalism is today usually associated with a left-wing point of view, but in the 1920s and 1930s, some ardent environmentalists were ultraconservatives or fascists who saw nature worship as part of an embracing rejection of contemporary industrial society and a return to values of blood and brotherhood. Similar alliances between environmentalism and ultraconservatism are seen today in Russia, where some environmentalists, dubbed "ecofascists," combine a reverence for the vast, unspoiled Russian taiga with anti-Semitism, anti-industrialism, xenophobia, opposition to democracy, and the call for a return to a command-and-control economy. In short, the defense of the environment provides inadequate guidance as to how to organize life, society, or the polity: for that, we need additional goals and values. However important, protection of the environment does not in itself constitute the basis for a comprehensive worldview.

Global versus Local Perspective

Another distinction among environmental movements is between those that adopt a global perspective and those that adopt a local one. Global environmentalists, who have emerged as a powerful force in recent decades, stress the worldwide despoliation of nature. The objects of their concern are transnational, indeed planetary. They began, in the era of nuclear weapons testing, by stressing the dangerous spread of radioactivity around the world and then moved on to concerns over acid rain; CFC contamination; the diminution of biological diversity and stability as a result of human activities; the menace of overpopulation; the global threat produced by overfishing and modern agricultural methods; species extinction; and, perhaps most important in the late 1990s, the threat posed by global warming and related changes in the global climate.

Such global changes, it is argued, threaten to end—or already have ended—the concept of nature as an accessible realm free of human intervention.[10] By now the very sky above is polluted by CFCs, ozone, and greenhouse gases created by human activity. Nothing in our corner of the cosmos is left uncontaminated by human interventions. The fragile layer of earth, air, and water that sustains human activity on the surface of Earth is threatened, and its protection must be given the highest priority for remedial action. Globalists applaud the Montreal agreement to ban CFCs; they urge reduction in the emission of carbon dioxide, especially by the industrial nations; they worry about the increase in other greenhouse gas releases in the industrializing nations. Most of those who express such global anxieties are not—at least not yet—personally affected by the trends that alarm them, but they have informed intellectual, idealistic, and scientific reasons for concern about the future of the planet.

The concerns of local environmental movements are very different: they habitually focus on a particular problem in a particular locale and involve those immediately affected by the problem. Thus the so-called "toxics movements," usually led by women concerned for the welfare of their families, are directed against specific local dangers. These movements, in most cases focussed in the scope of their concern to a single local problem, are a worldwide phenomenon, as characteristic of India and Kenya as of the United States and Norway. Epitomized in the

United States by the activist residents of Love Canal, they direct attention to, say, a dam in India that is being built to support industrial development and alleviate the national shortage of electric power but also threatens the living space of tens of thousands of villagers; a toxic waste dump, often located in a community of poor and unempowered minority citizens; the proposed location of a nuclear plant near a downwind village; the industrial pollution of what had been until recently a pristine lake in Siberia. Thousands of such local movements of resistance to local despoliations have arisen on every continent: to some observers they constitute today's most energetic and promising form of environmental action. They have suggestive common attributes: they are usually led by women; they typically mobilize individuals not previously active in environmental movements; they often activate those who are dispossessed, propertyless, or politically inert; with a few notable exceptions, they resist affiliation with larger, national groups; and they tend to disband once their local objectives have been achieved.[11] The chief point, in any case, is that these movements devote their energies to coping with concrete, visible, palpable local problems.

Ecofeminists versus Material Feminists

One of the more striking dichotomies in environmental outlooks is found within the feminist movement. At one extreme are ecofeminists, who base their view of the nature of and remedies for environmental degradation upon an essentialist construction of male and female temperaments, in which men seek power over nature and women protect and revere the Earth and its fecundity. At the opposite extreme are material feminists, who argue that in specific circumstances, particularly in Third World countries, the undermining of inherited gender roles and rights, often through mistaken transposition of Western gender ideologies, has resulted in mismanagement of land and water resources and the production of cash crops in place of traditional food staples, mobilizing women because they are most materially affected by these changes.

Ecofeminists clearly fit within the millennial, spiritual renewal spectrum of environmental thought, since they argue that the planet will be destroyed by male aspirations to technological power over nature and by the male quest for ever more powerful nuclear and biological weapons. As a counterbalance to this assumed male drive they propose return to worship of the mother goddess and revived reverence for the Earth and the fertility of nature. In this sense, ecofeminists seek to convert humankind through a spiritual revival based upon worship of the feminine principle, on pacifism, and on a return to a prehistorical, simple agricultural society.

Material feminists, on the other hand, see some successes in the efforts to preserve women's control of use of land in parts of South Asia and Africa and in the education of development agencies about women's role as the primary food producers in much of Asia and Africa. Their programs seek political solutions through which rights of land use can be converted to female-owned property, the harvesting of forests can be carried on respecting traditional women's knowledge of forestry, and government plans for transforming land tenure systems can recognize female as well as male rights within village societies. They also favor agricultural education schemes targeted at women food producers, rather than at males who do not till the soil.

Ecofeminist ideas are usually global and ecocentric, while material feminists are concerned with specific local issues and fine-grained studies of why women's food-producing role has been ignored in development projects in specific regions. Although highly critical of gender hierarchies, material feminists do not essentialize male and female temperaments, and they are not opposed to technology, provided women have equal access to its use and equal voice in its control.[12]

North and South: Conflict versus Community

Almost from the beginning of environmental debate, the differences and parallels between the interests of the "North"—the highly industrialized nations—and those of the "South"—the poor, less developed, or developing nations—have been discussed. A major divide in debates about the relationship between economic development and environment is the degree to which conflict is stressed as opposed to community of interest between North and South.

The conflictual analysis emphasizes that the industrialized nations of the North, above all the United States, are principal contributors to worldwide pollution, and especially to those processes we label "global change." Per capita outputs of almost every known manmade pollutant are highest in the United States and in other industrialized nations. Southern nations, in contrast, with their low per capita incomes, greater reliance on agriculture, and low energy outputs, produce less global pollution both on a per capita basis and on an aggregate basis, even though the South constitutes 75 to 80 percent of the world's population.

Given the commitment of the South to economic development, environmental conflict with the North seems inevitable to many. For example, were the nations of the South to reach the same levels of per capita environmental degradation as the North, the Earth's carrying capacity might well be exceeded with catastrophic results. It is claimed that China and India alone, which together contain one-third of the world's population, have the capacity to overwhelm the planet's environment should they reach the levels of per capita pollution that characterize the United States.

If this analysis is accepted, two quite different conclusions can be drawn: that the nations of the South must limit or strictly control their economic development, or that the nations of the North must radically reduce their own level of environmental damage to make ecological room for increased development and pollution from the South. To the nations of the North, then, the ideal solution might be to try to slow the development of the southern nations or to insist on their use of complex (and expensive) environmental technologies such as scrubbers, "green" production facilities, and low-polluting energy sources. To the nations of the South, in contrast, the obvious answer is for countries like the United States to reduce dramatically their own levels of environmental degradation.

Emphasizing the conflict between North and South usually entails the further assumption that the wealthy nations are those most concerned with environmental preservation, whereas the poor ones are concerned chiefly with economic development. Only when a high level of economic development has been reached, it is assumed, are people likely to adopt "postindustrial" values such as environmentalism. In the impoverished nations, environmental concerns must take a back seat to issues of subsistence and economic growth.

An alternative perspective stresses instead the areas of similarity and potential collaboration on environmental issues between North and South. It emphasizes that most environmental problems are global in nature, and so are their solutions. Loss of biodiversity; destruction of forest cover; global warming; degradation of soil; salination of arable land; depression of water tables; depletion of the ozone layer; acid rain; and poisoning of land, animals, and people through intensive use of pesticides all affect the developing nations at least as much as they do the industrialized world.

Underscoring the global nature of environmental problems, poll studies show that individual attitudes of environmental concern bear no relationship to the level of economic development of the nations studied. For example, more Filipinos and Nigerians say they are personally concerned about the environment than do Americans. As the authors of one study conclude, "Conventional wisdom is wrong about the existence of major differences and levels of environmental concern between citizens of rich and poor nations."[13] In short, the notion that concern with the environment is a postindustrial characteristic of rich people or rich nations is simply incorrect.

A final argument supporting the community of North and South is the similarity of both the arguments and the movements organized around the environment in both parts of the world. Wherever they are tolerated by political authorities, as in India, citizens' movements to protect the environment in developing nations are extraordinarily like those in, say, the United States or northern Europe. The structure of discourse and debate about the environment, the conflicts within environmental movements, and the arguments over the most efficacious means of protecting the environment in Latin America, Africa, and Southern Asia differ little from those in Scandinavia, Australia, or the United States.

Wise Use versus Forever Wild

The contradiction between the "wise use" and "forever wild" attitudes toward nature has given rise to political controversy in the United States for at least a century. A variant of the anthropocentric/ecocentric dichotomy, its political ramifications are exemplified by the Hetch Hetchy River controversy in Yosemite, California, in the late nineteenth century. At that time, engineers working for the city of San Francisco, whose aim was to dam the river as a new source of city water, came into sharp conflict with John Muir and his allies, all militant preservationists.[14] The arguments of the dam builders anticipated the "wise use" doctrine—today most often advocated by the forestry service, lumber companies, ranchers, hunters, and other land owners—which holds that nature is a reservoir of energy and other raw materials for human use. (A corollary of the doctrine holds that property rights entitle landowners to compensation for any economic losses incurred as a result of environmental regulations.) People are entitled to use natural resources by, for example, judiciously "harvesting" trees at reasonable intervals; "culling" flocks of wild animals for human consumption; "taming" wilderness areas to prevent flooding; and "controlling" undesirable species such as wolves, coyotes, bears, and jaguars. The goal is to render the natural environment productive, pleasant, and agreeable for human use. If a species, such as wolves, poisonous spiders, scorpions, or rattlesnakes, requires elimination, and if that can be shown to benefit humankind, then it may be done. If clearcutting proves to be the most efficacious long-run mode of harvesting timber, then nonmaterial, aesthetic, or sentimental considerations—and, in some cases, rules for the protection of endangered species—should be subordinated to the material needs of the population.

At the other extreme is the "forever wild" or "wilderness" preservation outlook. It is exemplified by the deed of Baxter State Park surrounding Mount Katahdin in Maine and in the "nature preserve" movement in the former Soviet Union. Here, what remains of the unspoiled biophysical environment, far from being regarded as a source of material resources, is seen as a sacred or quasi-sacred place with an inherent claim to inviolacy. Lovers of wilderness regard the natural landscape as a source of spiritual and aesthetic nourishment, but only if it is left in its pristine, untouched, or "wild" state. For people without faith in a supernatural divinity, the unspoiled reaches of the natural world, which existed prior to the evolution of humanity and presumably will outlast humanity, constitute the only remaining locus of transcendence. The Russian nature preserves are an extreme example: they are substantial areas of "wilderness" from which the entire population (other than attendants and working scientists) is wholly excluded. In the United States today, those who wish to prevent "harvesting" of forests, mining of minerals, grazing of cattle, or the encroachment of tourists driving motor homes on public lands almost invariably embrace some variant of the "forever wild" view.

In recent years, however, the concept of wilderness has come under sharp postmodernist attack as a typically deceptive social construction. After all, the vast areas of North America that arriving white European settlers called "wilderness" had for millenia been home to millions of Native Americans. It is easy to demonstrate that what Americans call wilderness, especially when it refers to areas of our national forests and national parks, is an elaborately constructed cultural artifact. Recently, the environmental historian William Cronon offended many ardent adherents of the "forever wild" school by arguing that we should dispense entirely with the misleading, indefensible space-oriented concept of "wilderness"—wilderness as a topographical entity—and transfer our allegiance to the spatially neutral concept of "wildness." Wildness, as identified with aspects of life unmodified by human intervention, can exist anywhere, indeed everywhere. It is inherent in our own being. Thus, Cronon suggests, a bird in a city, say a migrating warbler in the Ramble area of New York's Central Park, is as wild as it would be anywhere else. Wildness is not restricted by space. Recall that Thoreau's famous dictum, motto of the Sierra Club, is "In Wildness [not Wilderness] is the preservation of the World."[15] Thoreau, like other nineteenth-century American writers, thought of "wildness" as an attribute of *Homo sapiens* as well as other animal species. In any case, many recent debates in the United States about the use of public lands, endangered species, and environmental regulations generally have involved aspects of the conflict between adherents of "wise use" and "forever wild."

Government Intervention versus Market Changes

Another recurring distinction in environmental debates, finally, is between state-interventionist and individualist, or market-based approaches. In essence, this opposition turns on the issue of which agency (or tactic) is most effective in resolving environmental problems. From an interventionist vantage, isolated individual human actions, however sincere, are of little avail in an advanced, complex, highly institutionalized, tightly organized urban industrial society. Even if 100 percent of the population recycled all household wastes, they argue, it would have almost no impact on the major sources of environmental degradation, which are industrial, military, and governmental. Barry Commoner argues that the most notable successes of environmental policy have entailed the simple prohibition by public authorities of the use of toxic substances like DDT, lead in gasoline, and CFCs.[16] The results, as measured by the diminution of toxicity, have been immediate, dramatic, and progressive. The general principle is that intervention by official (governmental) mandate—regulation—is usually the best means of improving environmental quality.

The opposing view is that individuals acting because of changed economic incentives in a free market can in the long run best effect a reduction in environmental degradation. Rejecting direct governmental regulation as bureaucratic, inefficient, and easily circumvented, proponents of "free market" environmental measures propose instead such indirect market interventions as taxes on environmentally undesirable behaviors or products, the use of sellable "pollution rights" to encourage industrial conservation of resources, and efforts to "internalize externalities" by market mechanisms that obligate organizations and individuals who do environmental damage to pay the real (long-term) costs of repairing the harms they do. At the extreme, free market environmentalists may even argue that, in the end, all environmental problems will be solved simply by the automatic mechanisms of the market. For example, as oil supplies are exhausted, the price of oil will rise so steeply that individuals and firms will be obliged to find other energy sources and conserve oil.

Government action is warranted only to enforce, reinforce, or strengthen market mechanisms, and not to intervene directly through regulation, standards, and requirements that are difficult to enforce.

Affinities between Positions

It is obvious that there are natural affinities or likely groupings between the positions we have separated above. For example, ecocentrists tend to emphasize the spiritual over the technical nature of environmental problems; view environmentalism as an aspect of an all-embracing worldview; and see environmental problems in a global, millennial perspective. Conversely, those who believe that environmental problems are largely technological in nature tend to be gradualists; see environmentalism as one issue among many, rather than a complete philosophy; and stress the uniqueness of contemporary environmental problems. Like other cultural values and political outlooks, environmental attitudes tend to come in "packages," or clusters of associated ideas.

It seems pointless (and misleading) for us to try to identify any one viewpoint, or any one cluster of ideas, as "true" environmentalism, with the rest, presumably, being "false." As humanists, however, we believe that environmental programs involving *exclusively* technological solutions are limited and ultimately inadequate. We insist on the need for enhanced comprehension of the extratechnological—human, cultural, psychological, political, and religious—dimensions of any effective inquiry aimed at instituting better measures for arresting the deterioration of the global environment.

Furthermore, we believe that many of the views we have referred to as dichotomous are in fact not as incompatible as we (and others) have implied. Thus there are issues to which the extreme ideas of the apocalyptic environmentalists reasonably apply and which require immediate action if irreversible damage is to be avoided. The banning of CFCs, which evidently contribute to the long-term destruction of the upper ozone layer, is a case in point. In contrast, there are other issues where a prudent gradualism makes sense: for example, issues involving the causes and remedies of global warming. In this case, present knowledge is limited, and existing models do not enable us to predict catastrophe if we fail to take immediate, costly action, even though prudence would nonetheless seem to justify a serious international effort to reduce the emission of greenhouse gases. Nor do we view innovations in technology as necessarily incompatible with preserving the spiritual benefits of our relations with nature. On the contrary, the well-being of the environment seems to involve importantly *both* changes in the values that issue in rampant consumerism—including a willingness on the part of the rich nations to alter their behavior with a view to reducing inordinate levels of environmental pollution—and, at the same time, changes in technology that will permit them to do so and will permit other nations to realize their aspirations for a more adequate standard of life without overloading the planet's fragile environmental balance.

In one area, however, we do take sides: while we appreciate and understand the ultimate, long-term educative value of the ecocentric doctrine, we believe that it is untenable. Or, rather, we believe it is much less tenable than the anthropocentric view that stresses the material and political needs of humankind. To be sure, conflict between the human species and other species can and should be reduced and, if possible, avoided. In the end, though, we believe that the ultimate justification for environmental preservation, far from inhering in the absolute and equal rights of all species, is humanity's moral obligation to its own kind. Moreover, without a reasonable improvement in the degree of equity in the conditions of human life, no resolution of our environmental problems is conceivable. Anthropocentrism, as we would endorse it, does not provide a rationale for ravaging nature to satisfy the trivial needs of human beings; rather, it entails preserving the environment, protecting, nursing, shepherding, and husbanding it precisely because we, as human beings, so desperately require a flourishing global landscape.

The Humanities and the Environment: What Is to Be Done?

In our view, many aspects of contemporary environmental thought involve issues of major concern to humanists. The scholar of the humanities has disciplinary training to elucidate the millennial and apocalyptic nature of much environmental writing, the covertly authoritarian assumptions behind some plans to coerce changes in consumption, the uninformed idealization of traditional cultures and their environmental practices, and the essentialist view of gender differences enshrined in ecofeminism. All of these views of human history, expectations about the future, and wholesale rejections of contemporary science and technology touch on deep themes in the modern and postmodern consciousness.

Environmental thought today also raises issues once thought settled in the age of the Enlightenment. Is there such a thing as progress? What is the moral standing of animals, plants, forests, groundwater? Are we to face a Malthusian future in which the human population will outrun resources? Does consumerism touch deep structures in the human psyche, such that we cannot imagine a cultural era based upon rational voluntary restraints on consumption? Are North/South concerns about environmental issues really so different? Analysis of the patterns of thought represented in Indian environmentalism, for instance, shows the same dichotomies we have identified for the West. These should alert us to the possibility that thought about humans and nature as cultural categories may be more global than our current focus on ethnicity and cultural difference allows.

Despite the importance of such questions, efforts to engage humanists in systematic work on environmental issues have been only partially successful. We thus have asked ourselves whether there are ways in which the professional training of humanists and the ends toward which they direct their work might be reformulated to bring the human/nonhuman environmental relationship into sharper focus.

We see this question as important partly because of the postmodern attack on the ideas of the Enlightenment, which is one way the professional training and ethos of humanists have been altered, often negatively, vis-à-vis environmental issues. One of the consequences of postmodernism lies in its defining a broad range of questions or intellectual territories as outside the sphere of the humanities, that is, as not part of the humanist task to explore what it means to be human. Among these questions are a number central to the understanding of contemporary environmentalism.

For example, the preparations for our workshop involved a search for an art historian who could explicate how the history of representations of nature might illuminate the nonverbal and emotional changes that have accompanied environmental degradation. Artists began to paint landscape only after land defined as private property became the norm. And the history of art shows us how nature gradually became a backdrop for human beings in early modern times. But we were able to find no tradition of seeking to understand what that change means in terms of the human/nonhuman relationship.

We also searched in vain for an economist or historian of ideas who could help us understand just when and why humans became defined as and encouraged to be insatiable consumers.[17] Our workshops helped us to see that in the wealthy modern societies, consumption is as powerful a cultural activity as production and the "masses," Marxist theory notwithstanding, exercise aesthetic judgments and sensibilities as consumers. But much humanistic thought has been based on the demeaning notion of "mass society" as devoid of aesthetic concerns, a point of view shaped in part by European émigrés' encounter with fascism as a mass phenomenon. Professional training that contested these received ideas from a variety of cultural perspectives would be a valuable preparation for teaching and research in the humanities today.

The contemporary study of ethics does indeed address issues raised by the need to constrain or redirect consumption in the interests of intergenerational environmental equity. But we found that much remains to be done to move such concerns into the everyday language of the humanities. We believe that they need to be as much discussed as, say, the impact of the machine on the human imagination or the alienation of the landless poor following the closing of the commons.

The discipline of history has in recent years shown a growing concern with the study of events that occur outside a human timescale: for example, the impact of climate on changes in vegetation, the rise of sea levels, and other natural phenomena. But the standard professional training of historians as yet places little systematic emphasis on the understanding of such macroenvironmental events, leaving nature as much of a backdrop to most historians as it was to Renaissance artists. Moreover, while there are now many and controversial accounts of the relationship between the expansion and the exhaustion of resource bases of ancient empires, those themes are rarely treated as standard in the professional preparation of historians who study the contemporary era.

The humanities and the social sciences converge in the study of myth, but here, too, we found little systematic study of apocalyptic imagery in contemporary environmental thought, and even less analysis of those mythologized "traditional societies" that are often invoked to instruct late-twentieth-century men and women about how to live in supposed harmony with nature.

The final section of this volume deals with the problem of modernity, a problem that calls for systematic inquiry in all humanistic disciplines concerned with environmental issues. Postmodernist theory has made many contributions. It is a useful corrective to the frequent modernist rejection of technology. Postmodernism also contains an invaluable commentary on imperialism and its cultural rationalizations, embodiments, and consequences. It rightly insists upon the breakdown of barriers between the organic and the engineered, barriers that were central to the modernist mentality.

Nonetheless there remain many crucial environmental issues to be investigated by postmodernist thinkers. Should environmentalism abandon totally the Enlightenment concern with human reason? Is the eighteenth-century stress on religious toleration irrelevant to human experience in Bosnia and Kosovo today? While it is undoubtedly true to note that war crimes are defined by the victors, are there not some universal notions of human rights that should inform our responses to the local and tribal conflicts of today, to the degradation or exhaustion of natural resources, or to the abuses of power seen in modern commercial imperialism? Central among these questions are concerns for the rights of women and men to use common land and forests and to retain some balance between rural and industrial life. Though these issues are usually defined as economic, having to do with development policies, they are also humanistic, having to do with human/nonhuman environmental relations in the context of contemporary politics.

Recent years have shown a steady movement by humanists toward sustained analysis of environmental issues. Many of the authors represented in this volume have been leaders in that movement. Yet this work reminds us that much remains to be done: the humanities and the humanistic social sciences have barely scratched the surface of sustained inquiry into environmental issues. Environmental questions, we believe, must be more central to the concerns of humanists, preoccupied with the most fundamental questions of human existence. A humanistic training that neglects environmental issues sets the humanities at the margins, rather than at the center of modern concerns. To the skeptic who questions the relevance of the humanities to environmental issues, we commend these essays as examples of the fruitful linkage of the humanities and the environment.

NOTES

1. Lester K. Brown, ed., *State of the World, 1990: A World Watch Institute Report on Progress toward a Sustainable Society* (New York: W. W. Norton, 1990), 10.

2. For a collection of provocative essays, most of them exemplifying this viewpoint, by scholars in many humanistic disciplines, see William Cronon, ed., *Uncommon Ground: Toward Reinventing Nature* (New York: W. W. Norton, 1995).

3. See S. Jasanoff, *Science at the Bar: Law, Science, and Technology in America* (Cambridge, MA: Harvard University Press, 1995).

4. The perspective here developed accords closely with that of Riley Dunlap, especially in R. Dunlap and W. Catton, "Struggling with Human Exemptionalism: The Rise, Decline, and Revitalization of Environmental Sociology," *The American Sociologist* 25 (Spring 1994): 5–30.

5. E. O. Wilson, *On Human Nature* (Cambridge, MA: Harvard University Press, 1978), 17.

6. Jerome Weisner, personal remark to Kenneth Keniston, 1977.

7. Ralph Waldo Emerson, *English Traits,* in *Works,* vol. 2 (Philadelphia: John D. Morris & Co., 1906), 46.

8. Quoted in S. Kellert, *Journal of Social Issues* 49 (1993): 53–69.

9. Riley E. Dunlap, George H. Gallup Jr., and Alec M. Gallup, "Of Global Concern: Results of the Health of the Planet Survey," *Environment* 9 (November 1993): 35–36.

10. See, for example, Bill McKibben, *The End of Nature* (New York: Random House, 1989).

11. For a further discussion, see the chapter by Barbara Epstein in this volume.

12. See Vandana Shiva and Maria Mies, *Ecofeminism* (London: Zed Books, 1993), and Bina Agarwal, *A Field of One's Own: Gender and Land Rights in South Asia* (New York: Cambridge University Press, 1994).

13. Dunlap, Gallup Jr., and Gallup, "Of Global Concern."

14. Michael Smith, *Pacific Visions: California Scientists and the Environment, 1850–1915* (New Haven: Yale University Press, 1987), 171–185.

15. See William Cronon, "The Trouble with Wilderness; or, Getting Back to the Wrong Nature," in *Uncommon Ground: Toward Reinventing Nature,* ed. William Cronon (New York: W. W. Norton, 1995), 69–90. For a series of critical responses to this argument, including a reprint of the essay and a response by the author, see *Environmental History,* 1 (January 1995): 7–55.

16 Barry Commoner, "The Successes of Environmentalism," lecture given at the Soviet Academy of Sciences, October 3, 1991.

17. On the related issue of the reconceptualization of exchange transactions in terms of "the market," see Karl Polanyi, *The Great Transformation* (Boston: Beacon Press, 1980).

I.   HISTORICAL STUDIES

INTRODUCTION

HUMANISTS BEGAN systematic study of the history of the environment some thirty years ago. They began with intellectual history—charting the development of ideas about the biosphere and about human interactions with it—a history that had a dramatic marking point in the explosions at Hiroshima and Nagasaki. The discipline of ecology, concerned with the study of the interaction of population groups in relation to given habitats, gained prominence in the 1960s and early 1970s, prompting the development of a narrative describing human alterations of natural ecosystems. That history drew its data from new capabilities in the scientific analysis of pollen deposits in ponds, better carbon dating, and the immemorial practice of counting the rings on trees. Once we began to chart the course of human alterations of natural ecosystems, much history written with the idea of progress as its organizing theme began to be rewritten. Thus, even in my graduate school days, we were taught to marvel at the positive contribution of the European religious orders to the rise of the European economy because of their success in draining swamps and bringing "marginal" land into production. Now we wonder what the impact of those practices was upon bird life, and how that ecological disruption played out in the extinction of species.

Certainly we see Europe in the year Columbus set sail as a continent virtually denuded of its once thick timber cover, a society running out of fuel and building materials, and therefore looking outward for new sources of supply—something perhaps more mind expanding than the core ideas of the Renaissance. When we realize that Europeans once slashed and burned their way through the entire woodstock of the European continent, we see contemporary Brazil somewhat differently.

Moreover, when we lose our European myopia and begin to look at the transformations of the biosphere wrought by Mayan civilization, or by the civilization that built Angkor, we begin to place our own times in a more complex context and to abandon the simplistic idealization of traditional peoples, which is one of the cherished dogmas of New Age environmental thought.

When historians like Stephen Pyne began writing in the 1980s about such primal factors as the history of human use of fire, this New Age fantasy became even more suspect, as Pyne charted the use of fire in prehistory to enhance land fertility and provide better opportunities for hunting and trapping small animals. Thus hunter-gatherer peoples began altering the biosphere even before the introduction of agriculture. In this time perspective, the apocalyptic theme that runs through much contemporary environmental thought becomes muted while the environmental consequences of human behavior take the foreground as part of an ongoing process of great antiquity. As we bring a global perspective to bear on this human history we begin to comprehend how much of Western attitudes to the natural world have been shaped by the experience of a Europe denuded of its woodlands—for instance, our fear of forest fire and our export of temperate climate concerns to arid areas and the tropics.

What can we learn from studying the history of climate? First, we have to adapt the timescale on which we examine trends of change, because climate cycles are much longer than human life cycles. In his essay, "Climate and History: Lessons from the Great Plains," Donald Worster tells us that despite the much vaunted human capacity for adaptation, we do poorly in adapting to a climate not fully understood. Climate, in Worster's terms, is non-Newtonian, perpetual random motion that we may never predict successfully. There has been a lengthy history of hopeful or despairing human extrapolation of minor blips in weather cycles into long-term trends, but in reality we simply don't understand changing weather.

Moreover, if his Great Plains example is susceptible of generalization, technologies aimed at controlling or modifying nature don't succeed. They don't provide the opportunity for successful adaptation, because they aim at short-term results and ignore the long time frame of climate patterns. Worster gives us many examples to prove his point: deep well irrigation on the High Plains; the impact of irrigation on soil fertility recorded in many different ages and cultures; and the irony of crop growers' being paid not to produce, as in Great Plains agriculture since the New Deal. His view of climate history makes him reject the efficacy of technological innovation as an enabler in overcoming the limits set by climate and soil, even though new technological fixes do appear to foster the expansion of commercial agriculture.

Worster thinks we can't engineer a successful adaptation to climate and agriculture on marginal soils using the cultural and economic systems of private property and the regulation of crop production by markets. Climate cycles are too long and market cycles too short to allow culture to face up to the reality that nonadaptive agriculture may offer short-term returns but spell longer-term disaster.

Worster's detailed analysis of the effort to develop irrigated agriculture on the relatively arid Great Plains is just one incident in the vast planetary history John Richards describes in "Only a World Perspective Is Significant: Settlement Frontiers and Property Rights in Early Modern World History." Richards extends his timescale to five hundred years, still only a short moment in climatic terms. He sees the history of humans on the planet as one of human transformation of the environment, first by the introduction of domesticated animals on grasslands and savannahs, and then through the clearing of forests and jungles and the so-called reclamation of marshes for the introduction of plough cultivation.

Until the threat of climate change led historians to adopt a global perspective on the history of human settlement, Richards notes that historians ignored the continuity of frontier patterns throughout human history. The frontier process was and is a global one based upon state power, access to capital, intensive agriculture connected to commodity markets, and the existence of surplus population in some segments of the globe. The entire process of frontier settlement is characterized by the cumulative commercialization of the soil and the definition of the forest as an enemy to be cleared.

Richards sees this process occurring in three early modern and modern phases: from 1500 to 1800 in Eurasia and Africa; from 1800 to 1930 through European colonial expansion; and from the 1930s to the present in the humid tropics, in areas such as the Brazilian rainforest. The frontier process in each phase was similar, no matter what the occupying culture. So, if we take Richards seriously we should be wary of the tendency to blame environmental degradation on the Christian cosmology alone. Moreover, the emerging modern state—whether Russian or Chinese or Mughal India—responded to population pressures by pushing back indigenous peoples and introducing peasant smallholders who engaged in intensive agriculture. In the same era, 1500 to 1700, Finnish pioneers developed the techniques of forest clearing and log-cabin building that were to be transferred to North America in the nineteenth century and to tropical forests in the twentieth. In each era and culture, the forest was seen as an enemy to be moved out of the way of settlement supported by a global trade in forest and agricultural products.

For Richards, the Great Depression of the 1930s marked the end of frontier expansion, an event partially compensated for by renewed expansion in tropical areas after the 1939–45 War. Changing attitudes to forests and the ecological networks they sustain may moderate this expansion, but, Richards warns us, at the cost of major cultural transformation. Up until the present time, he points out, most human societies have had access to abundant, seemingly underutilized land and its accompanying low-cost resources. Of course, that access has been based on a fundamental misunderstanding of the uses of land by indigenous peoples. Rooted in misperception, frontier myths, based upon such widespread access, have nevertheless played a profound part in shaping cultural attitudes, not only in the West, but in non-Western societies as well. Since Richards doubts that technological innovation can provide successful adaptation to the constraints of climate and land resources, he argues that the dynamism encouraged by frontier mythologies may be on the wane, to be replaced by constrained resources. We may well ask whether this is the case or whether the frontiers of cellular biology and artificial intelligence may excite similar expansive mythology—now deployed about a fabricated rather than a natural environment.

Pyne concerns himself with the air we breathe and its relationship to fire in his essay, "Consumed by Either Fire or Fire: A Prolegomenon Approach to Anthropogenic Fire." The time frame in which he casts the subject extends even further than Richards's analysis of frontier movements since 1500. Using a time frame even longer than Richards's, Pyne wants his readers to see fire as naturally occurring long before early human beings learned to use it as their principal tool for interaction with nature. That interaction concerned the burning of grasses and forest underbrush for hunting, to fertilize the soil, and to promote the easy movement of hunter-gatherers.

Pyne constructs the history of humans' use of fire to shape the environment, from primitive slash and burn agriculture, which followed the use of fire to facilitate hunting, up to the invention of modern technologies based on burning fossil fuels. A major turning point in that history was the eighteenth-century creation of forestry, a discipline devoted to engineering the forest and controlling spontaneous fire, which was seen as endangering the artificial woods, no longer "nature" but a carefully husbanded "natural resource." One of his most striking observations is that the modern forester, because of a determination to control spontaneous fire and prevent the burning of stubble to fertilize the soil, has left "the Earth with less free-burning fire now than when Columbus sailed."

What has changed is the substitution for agricultural or forest fire of industrial fire, a substitute fire that burns without reference to the living environment. The net result, Pyne argues, is hard to calibrate, because industrial demands could destroy forests through logging, and yet the substitute technology of industrial fire no longer required the burning of woodlands for fuel—a demand of the earlier fire technology that had stripped early modern Europe of its woods by 1500 and made the air around its cities gray with wood smoke.

Now that we have moved fire technology from consuming the woods to fossil fuels, Pyne tells us, we measure health by the quality of air, no longer by the level of humus in the soil, which was the indicator of well-being in an agricultural society. Industrial fire burns quite apart from the capacity of the biota to renew itself, while at the same time we attempt to limit spontaneously occurring fire, which historically was a source of regeneration and renewal of plants and soils. No one knows the planet's carrying capacity for burning, but it is by no means clear that there is more fire on the earth now than in past times. Pyne leaves us with the question of whether fire is changing the world climate or merely changing world opinion about our culture's relationship to fire. Such questions can be raised only by using the timescale Pyne has adopted for his analysis.

In "As the World Runs out of Breath: Metaphorical Perspectives on the Heavens and the Atmosphere in the Ancient World," Gregory Nagy returns to the ancient Greco-Roman world to analyze the way a premodern Western people thought about humans and their relation to the atmosphere, the sky, the heavens—the element that today's industrial society sees as the real indicator of our health. For contemporary Western women and men that element is part of nature and can never be fabricated. But for the ancient Greeks, Nagy tells us, the artificial could include the natural, a mode of thought with significant implications for the ancient concepts of nature and the atmosphere. Things made by the gods (fabricated) could also be part of nature, such as the divine west wind (the breezes of the Zephyros), which blows to renew and reanimate Menelaos in the *Odyssey* and, by extension of the myth, all humans. There is much to be learned from the Greek organic and artificial models of the universe. They are interchangeable—categories defined in terms of one another—just as we are learning today to see the concept of nature as an artifact, the product of human culture. Only when we have internalized this concept—that culture "makes" what we see as nature—can we recognize the interchangeability of terms, a reversal that is a precondition for holistic visions of humans and their environment.

JKC

GREGORY NAGY

As the World Runs out of Breath

Metaphorical Perspectives on the Heavens and the Atmosphere in the Ancient World

THIS ESSAY puts to the test a perspective on the relationship between nature and humanity. According to this perspective, human beings now have it in their power not only to change nature but to end it. In *The End of Nature*, Bill McKibben (1989) explains his title:

> By the end of nature I do not mean the end of the world. The rain will still fall and the sun shine, though differently than before. When I say "nature," I mean a certain set of human ideas about the world and our place in it. But the death of those ideas begins with concrete changes in the reality around us—changes that scientists can measure and enumerate. More and more frequently, these changes will clash with our perceptions, until, finally, our sense of nature as eternal and separate is washed away, and we will see all too clearly what we have done. (8)

I concentrate on one particular aspect of nature, the atmosphere. The idea that humanity has contaminated, and forever changed, the entire atmospheric envelope of earth is in the forefront of McKibben's apocalyptic vision:

> We have not ended rainfall or sunlight; in fact, rainfall and sunlight may become more important forces in our lives. It is too early to tell exactly how much harder the wind will blow, how much hotter the sun will shine. That is for the future. But the *meaning* of the wind, the sun, the rain—of nature—has already changed. Yes, the wind still blows—but no longer from some other sphere, some inhuman place. (48)

The wording here implies that our world, this sphere of ours, is somehow animate, even human, affected as it is—and forever altered—by our own flawed humanity.

In this essay, I test this implication by comparing it with the worldviews of ancient Greek society, examining how these views do or do not embrace the concept of atmosphere, or sky, and how they relate to that society's image of human existence. I also touch on the cumulative experience of all societies, as expressed through their traditions, in all

37

their countless varieties.[1] For these purposes, the search is for images that are not modern, images either antedating or bypassing the patterns of thought ordinarily associated with the current "Western world." My perspective here relies in part on the discipline of anthropology, though my profession is the classics and most of my examples are taken from the ancient Greco-Roman world.

I am specifically concerned not with the earliest aspects of the history of science in Greece and Rome, what we know as history of ancient science, but rather with patterns of thought inherited even by the earliest scientists, patterns that a social anthropologist would describe as myths. Let us think here of myth in anthropological terms, not from the standpoint of modern usage, in which *myth* is conventionally equated with something other than the truth. For a working definition, we may adopt Walter Burkert's formulation: myth *within* any given traditional society is "a traditional narrative that is used as a designation of reality. Myth is *applied* narrative. Myth describes a meaningful and important reality that applies to the aggregate, going beyond the individual" (29).[2] It is important to observe, in the same breath, that different aggregates of people may have different and potentially conflicting truth values, so that the myths of different societies, or even different subsets within one society, may likewise differ among themselves.[3] Accordingly, we can expect a great diversity in the mythological images of the universe in general and the atmosphere in particular, and an even greater diversity of philosophical explanations rooted in myth. Still, despite all the mythological divergences, we may discern some patterns of convergence in theme, to which I now turn.

Throughout this essay, I concentrate on McKibben's (1989) point that the atmosphere of our world becomes animate, even human, once we fully comprehend how human error has changed it permanently to our own detriment. Even now, McKibben argues, we resist comprehension: "In our minds, nature suffers from a terrible case of acne, or even skin cancer—but our faith in its essential strength remains, for the damage always seems local"(58). What surprises us now about our universe is that it may not be independent of, after all. We think we have known all along that we depended on our universe, but now we are told that it depends on us in turn. When we say "it," of course, we are thinking of the universe as we have come to know it, as we now depend on it. It is our own human realization of codependency with the universe that makes us see it as animate, as part of the human sphere:

An idea, a relationship, can go extinct, just like an animal or a plant. The idea in this case is "nature," the separate and wild province, the world apart from man to which he adapted, under whose rules he was born and died. In the past, we spoiled and polluted parts of nature, inflicted environmental "damage." But that was like stabbing a man with toothpicks: though it hurt, annoyed, degraded, it did not touch vital organs, block the path of the lymph or blood. We never thought that we had wrecked nature. Deep down, we never really thought we could: it was too big and too old; its forces—the wind, the rain, the sun—were too strong, too elemental. (McKibben 48)

The human error against the universe—if we globalize the ongoing neglect and abuse of nature—is seen here as an injury inflicted on an animate being. It follows that such an injury can even be seen as a crime that calls for the punishment of the offenders, that is, all humanity, where the punishment is exacted by an injured universe that can now manifest the characteristics of an animate being. Such an image is actually formalized in the myths of some traditional societies. In a type of narrative known to anthropologists as the "revolt of objects" myth, human error can turn even the inanimate objects of everyday life into living things that join forces with a reanimated universe in taking revenge on humanity. In a traditional prophecy of the Zuñi, a people living in the southwestern United States, the vengeful "revolt of the objects" against humanity is correlated with an apocalyptic mass asphyxiation inflicted by an elemental atmosphere:

> Maybe when the people have outdone themselves,
> then maybe, the stars will fall upon the land,
> or drops of water will rain upon the earth.
> Or our father the sun will not rise to start the day.
> Then our possessions will turn into beasts and devour us whole.
> If not, there will be an odor from gasses which will fall from
> the air we breathe, and the end for us all shall come.
>
> (Zuñi 1972, 3)[4]

Let us explore further the notion of an animate universe. Of course, such a notion is but one of many different mythological constructs. An even more basic notion is that of the world as a *structure*. Moreover, the *order* of this structure may alternate, either at random or in a given pattern, with *disorder*. For my premier example, I cite the ancient Greek word *kosmos*, which can denote simultaneously

• the structure of the universe, or cosmos (Xenophon *Memorabilia* 1.1.11; cf. Empedocles *fr*. 134 line 5);

- the structure of society, in particular one's own native society (Herodotus 1.65.4),
- the structure of the myth that tells about the universe and society (Pindar *fr.* 194; cf. *Odyssey* vii 489).

The cosmos as structure may be visualized specifically as an *organic* structure.[5] According to the Greek philosopher Democritus, a human being is the microcosm (*mikros kosmos*) of the cosmos (Democritus *fr.* 134). Moreover, according to the earlier Greek philosopher Anaximenes, "just as our *psūkhē*, being *aēr*, holds us [i.e., our individual bodies] together, so does wind or *aēr* enclose the whole cosmos" (Anaximenes *fr.* 2; cf. Lloyd 1966, 235).[6] The *psūkhē* is to be understood here as the 'breath of life' (by extension, the 'soul') in an individual human being,[7] while the *aēr* is the 'air' of the atmosphere (Kirk, Raven, and Schofield 1983, 146). Although a scientist like Aristotle criticizes the belief that *psūkhē* pervades the whole universe (*De anima* 411a7ff) or that the elements are themselves alive (a11ff), he nevertheless resorts to biological models in explaining celestial dynamics while at the same time explicitly rejecting the notion that the four sublunary elements are actually alive (Lloyd 1966, 258, 264).[8]

Aristotle recognized that the analogy between celestial movements and those of living things is just that, an analogy, and we, too, may concede that the principle of vitalism in analogies of macrocosm/microcosm "nearly always led to the proposal of crude and grotesque theories on the structure of man and of the universe" (Lloyd 1966, 267).[9] Still, what is merely analogy in the realm of scientific investigation may become metaphor in the realm of myth.

We have just observed the metaphor of a living organism, and we earlier noted in passing another metaphor for the universe, that of a structured society, even a political state.[10] To repeat, the Greek word *kosmos* can designate either cosmos or society, and the *kosmos* as society may be envisioned in all sorts of specific forms, such as the constitution of a state, which is in ancient Greek terms the sum total of a state's customary laws: this was the case in ancient Sparta (Herodotus 1.65.4). Besides these two metaphors for the cosmos, I need to mention a third that is comparably widespread in ancient Greek myth: the universe as an artifact made by an artisan.[11] A derivative of *kosmos*, borrowed into English as the word *cosmetic*, is a reflection of this third metaphor.[12]

I stress that these three metaphorical models for the universe as (a) political state, (b) living organism, or (c) artifact made by an artisan turn

out to be the basis for a dazzlingly wide variety of scientific and philosophical perspectives. Each model is pervasively attested in the whole
spectrum of Greek artistic, philosophical, and scientific thought, with
clear resonances even in the most highly developed conceptual realms of
Plato and Aristotle. The case of Aristotle is particularly interesting, since
he is on record as condemning the use of metaphor in reasoning (Lloyd
1966, 302, 361).

Of these three metaphors for the universe, the third—artifact—turns
out to be pivotal for understanding the second, especially with reference
to ancient concepts of the atmosphere. To compare the cosmos to an
artifact is to think of the universe as a structure that is manmade, artificial rather than organic and natural. I use the term *manmade* only for the
time being: it turns out to be misleading from the viewpoint of the ancient Greeks, which will become clear as the discussion proceeds.

We see here the makings of an important question: is the universe
itself to be viewed as an artificial or as a natural structure? For us the
answer seems simple. The universe and nature are coextensive, and
therefore the cosmos is surely natural, not artificial. For Aristotle, too,
the cosmos is a thing of nature, making it easier even for him to slip into
analogies of the living organism—although he does indeed reject the
notion that the four sublunary elements are actually alive.

In ancient Greek myth, however, the idea that the universe is artificial
exists as an acceptable alternative to the idea that it is natural, inasmuch
as the metaphorical model of the universe as artifact exists as an alternative to that of the universe as living organism. Or at least, let us say that
such a set of alternatives was manageable in the case of the four elements
of earth, water, fire, and *aēr* 'air'. But then there is Aristotle's fifth element, the *aithēr* 'ether, ethereal realm' (Lloyd 1966, 268). In this case, as
we shall see, *aithēr* cannot conceivably be viewed as natural.

Going back in time beyond Aristotle, we find a vital distinction between *aēr* and *aithēr* in ancient Greek mythmaking, which is essential to
the evolving concept of the atmosphere. Whereas the *aēr* can be equated
with the air we breathe, in all its atmospheric reality of observable winds
and storms, the *aithēr* is that vast immeasurable stretch of space that separates our mortal existence from the immortal existence of the gods.
Even in cases of divine intervention, the *aēr* is the element of immediacy,
and the *aithēr*, of remoteness. To become invisible, one is enveloped in
aēr by the intervening god, so that one momentarily vanishes into "thin
air" *through the god's presence.*[13] The god's intervention is thus immediate.
In contrast, the luminous *aithēr* highlights the limitless gulf that sepa-

rates immortals from mortals. True, whenever a divinity like Aphrodite wishes to intervene in human affairs, she can plummet through the immeasurable space of the *aithēr* in the twinkling of an eye, an occurrence that is vividly pictured as happening then and there in Sappho's immortal poem known as *Prayer to Aphrodite*.[14] But even this instantaneous presence only accentuates the incomprehensibly vast distance that regularly separates the gods from mortals. That immeasurable space up there and out there is the *aithēr*.[15]

The immortal realm, kept apart from our mortal realm, was envisaged as artificial *precisely because it was immortal*. So, too, the *aithēr* that kept it apart. That is why only the gods could cross this space.[16] It is an irony that this vast space, which was thought to exist beyond the atmosphere, is now included in *our* concept of atmosphere. There is no need any more to keep the *aithēr* separate from the *aēr*, since the *aithēr* is now part of our own existence. "The gods," Sam Bass Warner has wryly predicted, "will soon come tumbling down through the ozone hole."[17]

If indeed whatever is mortal was understood as natural and whatever is immortal, as artificial, the difference is in the exclusion of death by artificiality, as symbolized by, for example, the image of the recurring Golden Age, during which even mortals can become immortalized after death: such immortalization is consistently symbolized by the Artificial as opposed to the Natural (Nagy 1979, 151–210). One key word is *aph-thitos*, literally meaning 'unwilting' or 'unfailing' and applicable either to the life force of one who is immortal or to any artifact made in the realm of the immortals (Nagy 1979, 175–190). It becomes clear from the Greek usage, then, that it is ill advised to equate their ancient notion of the *artificial* with our modern notion of *man-made*. If anything, the *artificial* can be equated primarily with the *god-made*: it is the gods who are credited in myth with the very invention of the artificial, the technological, whatever is culture itself, and it is ultimately from the gods that mortals received culture—or stole it.

The question remains; Does the Artificial exclude the Natural in ancient Greek thinking, as it clearly does in our own modern usage? As we shall see, the Artificial in fact *includes* the Natural, even if immortality excludes mortality *in terms of* the Artificial. Such a thought pattern of inclusion has a bearing, as we shall also see, on the ancient concept of nature in general and the atmosphere in particular.

In order to grasp this thought pattern, let us review what linguists describe as *unmarked* and *marked* categories of an opposition. These terms are defined as follows by Roman Jakobson: "The general meaning of a marked category states the presence of a certain (whether positive or negative) property A; the general meaning of the corresponding un-marked category states nothing about the presence of A" (Jakobson 1984, 47; also Jakobson 1939; Waugh 1982). The unmarked category is the general category, which can include the marked category, whereas the reverse situation cannot hold. For example, in an opposition of the English words *long* and *short*, the unmarked member of the opposition is *long* because it can be used not only as the opposite of *short* ("I am read-ing a long essay, not a short one") but also as the general category ("How long is this essay?"). In an opposition of *interesting* and *boring*, the un-marked member is *interesting:* if we say "A is more interesting than B," we do not necessarily mean that both A and B are interesting or that B is boring, whereas to say "A is more boring than B" presupposes that both A and B are boring. To ask a question such as "How interesting is this?" does not commit the questioner, unlike asking "How boring is this?" To ask "How short?" is to presuppose shortness, whereas asking "How long?" does not presuppose either shortness or length. Thus the marked member, as in the case of *short*, is defined in terms of the un-marked member, in this case, *long*, and not the other way around.

To cite another example: in an opposition of *day* and *night*, the un-marked member is *day*, which serves not only as the opposite of *night* ("It is daytime, not nighttime") but also as the general category ("There are seven days in the week"). The unmarked member is inclusive, in that the marked member can be an aspect of the unmarked. It can be exclu-sive, however, if it negates the marked member, as when we say "It is not *night*; it is *day*." The negation of the marked by the unmarked has been called the *minus interpretation* of the unmarked (for example, "day, not night"), as distinct from the *zero interpretation* (for example, "day"); the assertion of the marked member is the *plus interpretation* (for example, "night" or "night, not day"). The term *plus interpretation* designates not "positive" but "marked, either negatively or positively." The zero inter-pretation of the unmarked member includes, as an overarching prin-ciple, both the minus interpretation of the unmarked member and the plus interpretation of the marked member. The opposition of *long* and *short*, for example, is a matter of *length*.

Further, the opposition of unmarked *order* and marked *disorder* is a matter of overall order (Nagy 1990a, 8). The cosmos is a structure that contains both structure and lack of structure, order and disorder. The default category is the containing category: you default from nonstructure to structure, from disorder to order. In other words, structure *includes* nonstructure, order *includes* disorder.

Let us apply these considerations to the opposition of Natural and Artificial, examining the proposition that, for the ancient Greeks, the cosmos is a world that combines the Natural and the Artificial. Following the lead of social anthropologists, we may substitute the terms *nature* and *culture*.[18] From here on I use the word *culture* to designate whatever is perceived by a given society as the product of craftsmanship or technology.

In the cosmos of a wide variety of mythmaking traditions, culture explicitly *includes* nature. Thus the cosmos, embracing all the things of nature, can be seen as the handiwork of a divine craftsman or artisan or Demiurge (*dēmiourgos* is a Greek word for 'artisan') (Glacken 1967, 14). Here we see the zero interpretation of the unmarked member, where culture is unmarked while nature is marked. But culture can also exclude nature when it switches roles from default category to defining category. Here we see the minus interpretation of the unmarked member ("long, not short"). The traditional Iranian idea of Paradise, for example, reflected in the Greek borrowing *paradeisos* from the Iranian word *pairidaēza*, meaning 'walled-off enclosure' and designating the royal park of the Persian King, is the image of an enclosure that *excludes* or fences off the hostile and unchecked forces of nature (Giamatti 1966, 11).

We see an ironic inversion of this pattern of thought in the state of our world today. Today's wildlife nature preserve is an enclosure that supposedly *excludes* or fences off the hostile and unchecked forces of culture, of a runaway technology. Leo Marx relates an anecdote from his recent trip to Moscow, "where a young scholar characterized Soviet nature preserves as 'sacred spaces' whose existence helps people to cope with the messiness of the 'profane' world."[19] In this latter-day worldview, culture is still unmarked and nature is marked, but the old positive and negative values assigned to culture and nature, respectively, have been inverted. Even the proportions are inverted. On the one hand, the idea of Paradise implies a selective foreground of order against a random background of disorder, with the outposts of cultivated land or gardens highlighted against vast stretches of hostile wilderness. On the other hand, the idea of a wildlife sanctuary implies the survival of isolated pockets of nature amid the overwhelming encroachments of the denatured world that surrounds them.

The idea—or ideal—of the wildlife sanctuary brings us back to the point made by McKibben (1989), concerning our stubborn "faith" that the essential strength of nature remains, "for the damage always seems local" (58). The problem is, once we shift our perspective from localized considerations of spaces on earth to the actual atmosphere that envelops the earth, the basis of that faith is destroyed:

> The idea of nature will not survive the new global pollution—the carbon dioxide and the chlorofluorocarbons and the like. This new rupture with nature is different not only in scope but also in kind from the salmon tins in an English stream. We have changed the atmosphere, and thus we are changing the weather. *By changing the weather, we make every spot on earth man-made and artificial.* We have deprived nature of its independence, and that is fatal to its meaning. Nature's independence *is* its meaning; without it there is nothing but us. (McKibben 58)[20]

According to this argument, the reality of what has happened with the atmosphere has canceled the very distinction between nature and culture. To put it another way, there is no more nature to be excluded by culture.

There is perhaps some hope left, however, in contemplating the ancient models of Paradise as a microcosm reflecting a macrocosm. Besides the minus interpretation of Paradise, in which culture excludes nature, there is also a zero interpretation, in that the ideal garden can *include*, literally enclose, the forces of nature. We may say that Paradise excludes the "hostile" aspects of nature but includes its "benign" aspects. Much the same can be said of the Garden of Eden in the Hebrew Bible, perpetuated also in Christian and Islamic traditions (Meisami 1985, 231). Inside the enclosure, the gardener as artisan can make order out of chaos:

> The earthly garden functions both as an object of man's contemplation and as a setting for important human activities; it differs from a natural landscape by virtue of being an artifact, constructed according to design (a fact no less true of literary gardens than of real ones) as well as by the frequent opposition of the garden world to the wilderness beyond it (gloomy forest or desert waste), the abode of forces hostile to man and to order. In proportion as its design and constituent elements are seen as reflecting principles of cosmic order and beauty, the garden itself becomes an ideal place wherein such principles may be observed. (229–230)

In Islamic traditions, as influenced by the earlier Iranian models, the earthly garden both reflects and prefigures Paradise (Meisami 1985, 231). The traditional Iranian garden "was designed to imitate not only Paradise, but also the conception of the cosmos" (231). In other words, Paradise is a microcosm that organizes the macrocosm. Moreover, the

containment of nature by culture in a garden becomes the ultimate Islamic model for an ordered universe:

> The medieval Islamic garden was a *hortus conclusus*, walled off and protected
> from the outside world; within, its design was rigidly formal, and its inner
> space was filled with those *elements that man finds most pleasing in nature*. Its
> essential feature included running water (perhaps the most important ele-
> ment) and a pool to reflect the beauties of sky and garden; trees of various
> sorts, some to provide shade merely, and others to produce fruits; flowers,
> colorful and sweet-smelling; grass, *usually growing wild* under the trees; birds
> to fill the garden with song; the whole cooled by a pleasant breeze. The garden
> might include a raised hillock at the center, reminiscent of the mountain at
> the center of the universe in cosmological descriptions, and often surmounted
> by a pavilion or palace. (231)[21]

The detail of the pleasant breeze that cools the garden is a common
feature of classical Persian poetry, as we see in the description, in the
poem *Haft Paykar* by the twelfth-century poet Nezāmi, of the royal park
where the king Bahrām Gur celebrates the New Year ritual of springtime. The same poem juxtaposes such positive images of a garden with
the archetypally negative image of the Garden of Eram, which God destroyed with a violent blast of wind (*Koran* 89:6–7) (cf. Meisami 1985,
234). The air from within the ideal garden is benign, while the air from
without threatens to destroy it.

In the *Discourses* of the thirteenth-century Persian poet Rumi, the
coming of springtime in the ideal garden is "resurrection," signaled by
the West Wind: "on the resurrection day when the zephyr blows, all
things will melt away" (Schimmel 1980, 69; cf. Meisami 1985, 243).
About two thousand years earlier, in ancient Greece, the verses describing the ultimate garden of Elysium were taking shape in the Homeric
Odyssey, and here, too, it happens to be the Zephyros, the West Wind,
that signals the reanimation of mortals:

> It is not fated for you, sky-nurtured Menelaos,
> to die in horse-grazing Argos and meet your fate that way.
> But you will be taken to the Meadow of Elysium and the Edges of the Earth
> by the immortals, where golden-haired Rhadamanthys abides,
> and where life is most easy for humans.
> There is no snow, no snowstorm or rainstorm,
> but the breezes of the Zephyros, ever blowing with their clear sound,
> are sent forth by River Okeanos in order to reanimate humans.
>
> (iv 561–568)[22]

In the language of myth, the cosmos that includes nature is not only
the macrocosmos. It is also the microcosmos of humanity, of the individ-

ual human organism. The breezes of the Zephyros are pictured as literally reanimating humans, as if breath and wind belonged to a cosmic continuum. The organic human reality of breathing, as reflected in the Latin word *animus* 'spirit', is in fact cognate with the cosmic reality of the winds in the atmosphere, as reflected in the Greek word *anemos*, meaning 'wind' (See Nagy 1990b, chap. 4). In the plural, *anemoi* also means 'spirits of ancestors' (Nagy 1990b, 116). Latin *animus* 'spirit' and Greek *anemos* 'wind' are derived from a verb that still survives in Sanskrit as *aniti* 'blow', just as Greek *psūkhē* is derived from a verb that not only survives in Greek as *psūkhō* 'blow' but even designates the blowing of winds in *Iliad* XX 440. In sum, the linguistic heritage shared by the Latin and Greek languages reveals a thought pattern that equates the breath of life with the atmosphere, the winds of the world or universe.

In the language of myth, the atmosphere as conveyed by Greek *anemos* literally *animates* the breath, and vice versa—to use the appropriate English derivative of Latin *anima*, by-form of *animus*. The Greek word *menos*, meaning 'power', is derived from the same root *men-*, as in Latin *mens, mentis* 'mind': *menos* is used in the language of Homer to designate both the 'power' of the wind (as in *Odyssey* xix 440) and the 'power' literally *breathed* into the hero by a god (as in *Iliad* X 482). This *menos* or 'power' is, etymologically, a *reminding* executed by divinity. The god *reminds* the mortal to breathe, *reminds* the wind to blow.[23] We may compare the breath that the God of *Genesis* infuses into human creation (2:7), and the mighty wind that swept over the waters at the very beginning of cosmic creation (1:2), even before God intones, "Let there be light" (1:3).

What inferences can we draw for our own world? Explicitly in myths and implicitly in the etymological history of Greek words like *kosmos*, we have seen a variety of formulations pointing to the vital interdependency of body, society, and cosmos. We have also seen an interchange of subordination between macrocosm and microcosm, nature and culture, and this idea of interchangeability comes together in the metaphor of the breeze that blows in the garden to reanimate humanity, so that humans may yet take time out to draw a deep breath. Yet, time may run out. The universe has existed without humanity, and may well do so again, lasting beyond the time allotted for human habitation.

While humanity depends on the universe, the immortality that myths attribute to this perfect artifact of a universe depends in turn on our humanity, even our mortality. So also with the mythical image of an organic universe: it depends on our own physiological identity. And it also depends on what we have done to make the universe different, however infinitesimally, from what it had been before humanity. That infinitesimally small difference, if it becomes the difference between life on earth or death, still means the whole world to us as humanity: it is our *organic* world, because we depend on it and it depends on us.

The organic and the artificial models of the universe are interchangeable, just as the environment is interdependent with both the physiological and the cultural essence of humankind in all its many dimensions, including the affective, the cognitive, the conceptual, and even the technological. As we contemplate our newfound technological power to transform for all time the container of our environment, the atmosphere, it is vital to relearn and rethink the now antiquated patterns of human consciousness that identified the breath of life with the winds of nature. These patterns may help to reorganize a holistic vision of humanity and environment.

In the language of myth, then, how are we to answer the question, what is happening to our universe, the only world we have? What is blowing in the wind? The answer, my friend, is that the world may be running out of breath.

NOTES

1. I propose to use the concept of *tradition* or *traditional* with a focus on the perception of tradition by the given society in which the given tradition operates, not on any perception by the outside observer who is looking in, as it were, on that tradition. While a given tradition may be perceived in absolute terms within a given society, it can be analyzed in relative terms by the outside observer using empirical criteria: what may seem ancient and immutable to members of a given society can in fact be contemporary and ever-changing from the standpoint of empiricist observation. It is from this perspective that I have used the word *tradition* in Nagy (1990a, 57–61, 70–72; cf. also 349, 411).

2. My translation (and emphasis), with slight modifications, of Burkert (1979).

3. I offer some examples in Nagy (1990b, 43–47).

4. Quoted by Quilter (1990, 60) with reference to the "revolt of the objects" theme at line 5. In the iconography of the Moche region of Peru, the revolting objects grow or get arms, legs, and heads (Quilter, 54). In a myth from the Xingu region, "the sun is killed with a cudgel which then transforms itself into a snake while the blood of the sun turns into spiders, snakes, centipedes, and other noxious creatures" (60). I hasten to add that such details about "revolting objects" may have different functions in different myths. Moreover, the narratives themselves may refer mythically to the past, prophetically to the future, or even, cyclically, to both (cf. 61).

5. A survey of examples is given in Lloyd (1966, 232–272).

6. The Pythagoreans argued that *pneuma* 'breath' and the void are inhaled by the universe (Aristotle *Physics* 4.213b).

7. Later in the discussion, we shall consider the derivation of the noun *psūkhē* from the verb *psūkhō*, meaning 'blow'.

8. On the Empedoclean model of four elements, see Kirk, Raven, and Schofield (1983, 286) and Glacken (1967, 9–10).

9. On analogy and induction, see Lloyd (172).

10. A survey of examples is given in Lloyd (210–232).

11. A survey of examples is given in Lloyd (272–294).

12. On *kosmos* as the 'arrangement' of beautiful adornment (as in *Iliad* XIV 184), see Nagy (1990a, 145n45).

13. I use the expression "thin air" to convey the sensation of the hero Achilles when he strikes at Hektor in battle, only to be striking at the "deep *aēr*" where formerly his enemy had stood (*Iliad* XX 446); the presence of Hektor had been spirited away by the god Apollo, who "enveloped" his protégé "with much *aēr*" (XX 444). In such situations, the *aēr* is invisible, translatable as 'thin air', in that it substitutes for what had been there and cannot be seen there any more. In other situations, the *aēr* is visible, translatable as 'mist', in that it blocks the view of the sun itself: at the spot on the battlefield where the Achaeans are struggling with the Trojans over the corpse of Patroklos, "you would think that there was no sun or no moon any more" (XVII 366–367) as all the warriors are enveloped in *aēr* (368), while elsewhere on the battle-field the struggle of Achaeans and Trojans continues in full sunlight (371–372), "underneath the *aithēr*" (371). Then the narrative switches back from the clear periphery to the mist-filled center point (*en mesōi*, 375) of attention, the spot on the battle-ground where the struggle over the corpse of Patroklos continues and where the warriors are struggling in "battle and *aēr*" (376). Later, an Achaean warrior who is standing at a distance from this spot and yearns to catch sight of his fellow warriors who are struggling there prays to Zeus that he save his comrades from the *aēr* and that the god make brightness or *aithrē* (646; abstract noun derived from *aithēr*) so that they may become visible again (646).

14. Original Greek and translation appear in Campbell (1982, 55).

15. To say that the highest fir tree on the highest peak of Mount Ida stretches "through the *aēr*" to "reach the *aithēr*" (*Iliad* XIV 288) is hyperbole: the top of the tree reaches the sky, as it were.

16. Besides the example from Sappho, already discussed, I cite the analogous passages in Homeric poetry, such as *Iliad* XIX 351 (cf. Worthen 1988, 16–17).

17. Sam Bass Warner, recorded from a *viva voce* remark, 16 October 1991.

18. This terminology is fully developed in Lévi-Strauss (1964); see also Glacken (1967).

19. Leo Marx, recorded from a *viva voce* remark, 16 October 1991.

20. Emphasis, except for *is*, is mine.

21. Emphasis is mine.

22. My translation. On the verb *anapsūkhein*, which I translate as "reanimate," see Nagy (1979, 167). The River Okeanos is the fresh-water cosmic river encircling the universe.

23. For further discussion, with Indic parallels, see Nagy (1990b,114).

REFERENCES

Burkert, W. 1979. Mythisches Denken. In *Philosophie und Mythos*. Edited by H. Poser, 16–39. Berlin / New York.

Campbell, D. A., trans. 1982. *Greek Lyric I: Sappho and Alcaeus*. Cambridge, Mass.

Giamatti, A. B. 1966. *The Earthly Paradise and the Renaissance Epic*. Princeton, N.J.

Glacken, C. J. 1967. *Traces on the Rhodian Shore: Nature and Culture in Western Thought from Ancient Times to the End of the Eighteenth Century*. Berkeley and Los Angeles.

Jakobson, R. 1939. Signe zéro. *Mélanges de linguistique, offerts à Charles Bally*, 143–152. Geneva. Reprinted in Jakobson 1984:151–160.

———. 1984. *Russian and Slavic Grammar: Studies 1931–1981*. Edited by M. Halle and L. R. Waugh. The Hague.

Kirk, G. S., J. E. Raven, and M. Schofield. 1983. *The Presocratic Philosophers: A Critical History with a Selection of Texts*. Cambridge.

Lévi-Strauss, C. 1964. *Le cru et le cuit*. Paris. Translated 1969 as *The Raw and the Cooked*, by J. and D. Weightman. New York.

Lloyd, G. E. R. 1966. *Polarity and Analogy: Two Types of Argumentation in Early Greek Thought*. Cambridge.

McKibben, W. 1989. *The End of Nature*. New York.

Meisami, J. S. 1985. Allegorical Gardens in Persia. *International Journal of Middle East Studies*. 17:229–260.

Nagy, G. 1979. *The Best of the Achaeans: Concepts of the Hero in Archaic Greek Poetry*. Baltimore.

———. 1990a. *Pindar's Homer: The Lyric Possession of an Epic Past*. Baltimore.

———. 1990b. *Greek Mythology and Poetics*. Ithaca, N.Y. Reprinted 1992, with corrections.

Quilter, J. 1990. The Moche Revolt of the Objects, *Latin American Antiquity*. 1:42–65.

Schimmel, A. 1980. *The Triumphal Sun: A Study of the Life and Works of Mowlana Jalaloddin Rumi*. London.

Waugh, L. R. 1982. Marked and Unmarked: A Choice between Unequals in Semiotic Structure. *Semiotica*. 38:299–318.

Worthen, T. D. 1988. The Idea of "Sky" in Archaic Greek Poetry. *Glotta*. 66:1–19.

Zuñi. 1972. *The Zuñis: Self Portrayals by the Zuñi People*. Translated by A. Quam. Albuquerque, N.M.

Donald Worster

Climate and History

Lessons from the Great Plains

THE GREAT PLAINS of North America offer a revealing place to study the challenges of rural communities trying to cope with climate and that chaotically changing envelope of gases we call the atmosphere. Despite the seeming monotony of flat, immutable land meeting big, unchanging sky, the plains are in fact the most volatile place on the North American continent. Their complexity lies not in land forms but in climate. Nowhere else do Americans confront such extremes of cold and hot or such rapid oscillations around the crucial point that divides wet from dry.

The old pioneer song brags about how steady and cheerful is our home on the range, "where the skies are not cloudy all day," "where seldom is heard a discouraging word," and "where the deer and the antelope play." But we know that the man who wrote that song must have been heavily editing his data. The song leaves out the flies, the scummy waterholes, and the hard winters that the animals regularly endured. Today, the deer that have survived the trauma of invading white civilization are probably choking on tractor dust or getting drenched by center-pivot irrigation systems. We have to admit, too, that dark clouds hang over the plains from time to time, and now and then they carry no water—they're full of dust. And as for no discouraging words, in fact the plains have heard many such words over the past century: words full of human grief, words expressing anger, surprise, and disillusionment, words once considered unfit to utter in a public place.

Over the next hundred years the words may get more discouraging than ever, if many climatologists are right. Planet Earth is beginning to warm up, they believe, and the plains are going to get more than their share of the warmth. The Intergovernmental Panel on Climate Change, a large, distinguished group of scientists assembled by the World Meteorological Organization and the United Nations Environment Programme, has calculated that the global mean temperature will rise three degrees Centigrade before the end of the twenty-first century, if no radical changes occur in technology and human behavior and we continue on our present course of polluting the atmosphere freely with carbon.

Those three little degrees would constitute a greater atmospheric change than we have seen in the past ten thousand years, and it would come far more rapidly than any remotely comparable change in the history of civilization.[1]

The panel goes on to predict that "temperature increases in . . . central North America" will be "higher than the global mean, accompanied on average by reduced summer precipitation and soil moisture."[2] They are referring directly to the Great Plains, along with the upper Mississippi Valley—essentially all the states from Indiana west to Colorado. Though no one can be quite sure what the specific regional effects of global warming will be, most qualified observers agree with that prediction. In effect, the area will very likely be returning to and exceeding the conditions that existed during the mid-Holocene Altithermal (or Hypsithermal) Period of some five thousand to eight thousand years ago, when mean temperatures in the middle latitudes were 1.5 to 2.5 degrees Centigrade higher than today and the present U.S. corn belt was a dry prairie and the western wheat belt a near desert.[3] Other scientists expect to see a similar trend toward increasing aridity showing up across the already arid Southwest and into California. In the future of this broad midwestern and western area, the consensus seems to be, we may expect to experience, beginning about the years 2005 to 2020, a warmer climate than we are used to and, beginning about the year 2030, to enjoy less rainfall during the growing season and less soil moisture for raising our crops.[4]

How little do we understand the complicated forces, natural and now manmade, roiling the sky overhead. Who would have thought in the days of the sodbuster fighting off grasshoppers or of the calvaryman dodging Indian arrows that *carbon* would one day be the greatest threat to rural security? But then the sodbuster did not imagine that his or her descendants and fellow countrymen would one day burn natural gas, coal, and oil, some of it derived from underfoot on the plains, or that automobiles, tractors, combines, and irrigation pumps would one day be in widespread use, requiring those fuels in massive amounts.

The warming trend predicted by the scientists is the result mainly of extracting carbon buried deeply in the ground and putting it up into the air in the form of carbon dioxide. The nation's economy does that all the time; it could not do otherwise and survive. Likewise, the rural plains economy is based on burning fossil fuels at a ferocious rate. A typical plains farmer burns, directly and indirectly, thousands of gallons of gasoline a year. Winter wheat production, for example, requires the energy equivalent of about 0.6 barrel per acre, as do sorghum and soybean production. Irrigated sorghum requires 4.0 barrels per acre, irrigated corn 5.7 per acre. Each of those barrels has the energy equivalence of over 47 gallons of gasoline. Thus, on a thousand-acre dryland wheat farm, the farmer may be burning 28,200 gallons of gasoline to produce a single annual crop, not to mention the energy needed to ship that wheat to a grain elevator, a flour mill, a bakery, a supermarket, a restaurant, or a kitchen. The farmer's ancestors did not foresee such an energy-intensive way of life, but he finds it completely natural and is loath to give it up. In practicing that modern agriculture, however, he is unwittingly helping to create a desert in his grandchildren's future.[5]

So the words frequently heard around tomorrow's home on the range will be: "Hotter and drier." "No rain in sight." "Not much rain last year, either, or the year before." "The ground has no plant cover, the wheat has not taken root, the wind is starting to blow the topsoil away."

Those were precisely the conditions prevailing in the infamous Dust Bowl years of the 1930s, when 100 million acres suffered from serious wind erosion and in some years nearly a billion tons of dirt blew from farm to farm, and even blew east toward New York and the Atlantic Ocean. "The prairie," observed a reader of the *Dallas Farm News* in 1939, "once the home of the deer, buffalo and antelope, is now the home of the Dust Bowl and the WPA." In some counties one in every three farmers drew a relief check from the Works Progress Administration (WPA) or other federal agencies.[6] Drought produced crop failure, crop failure produced erosion, and erosion produced poverty. What the scientists are now predicting is nothing less than the return of some of those Dust Bowl conditions, though on a more gradual, permanent basis.[7]

The future may see an inexorable desiccation, not a temporary drought of two or eight or even twenty years. The winter wheat belt, now located in the central and southern plains, may gradually shift northeastward into Iowa, Minnesota, and Canada. That may seem like good news if you own land or a grain elevator there, but wheat is a less profitable crop per acre than corn, and the shifting of the wheat belt would mean the replacement of the corn belt and an overall decline in prosperity. Watching the Chihuahuan desert creep toward them would not be a pleasant sight for most plains farmers or for the rest of the country or anybody overseas who depends on us for food. The shift of the breadbasket northward will mean lower overall production of wheat in the United States. Americans will likely still have enough to eat, but we will have less and less to export to our major markets abroad, including the Soviet Union and Japan. In 1990–91 we shipped 1.1 billion bushels of wheat, or about half the total harvest, overseas. That level will probably not be maintained. As Karl Butzer writes, "the United States, now the world's major food exporter, will barely remain self-sufficient, while the Soviet Union will be heavily dependent upon food imports."[8] As the central and southern grain areas succumb to permanent desiccation, the total income from plains agriculture will decline, as will the ability of Americans to buy foreign-made goods and to pay for them with the products of our land. Other countries, such as Canada, may gain a comparative advantage over U.S. producers and shippers in the international grain trade. The people who once looked to us for a reliable source of nutrition may have to tighten their belts or seek other suppliers.[9]

I can hear the optimists object to these predictions: "Say it won't be so. Say the scientists don't know enough to be sure. Say that we common folk really know better. Say, above all, that we can handle anything; even if these predictions come true, we can adapt our crops and tools and survive."

They have words of encouragement from a few other scientific and economic experts. A panel appointed by the National Academy of Sciences has declared that, as the world gets warmer, Americans should be able to adjust their lives and agricultural production at a reasonable cost. Although some panelists expressed sharp dissent from that conclusion, the majority were optimistic about the future survivability of agriculture and industry. They pointed out that humans have long managed to live in both polar and tropical climates, proving that we are an adaptable species. They held out hope that farmers could easily diversify their crops, find new water supplies for them, and air-condition their houses. "The capacity of humans to adapt," said the panel's chairman, Paul Waggoner of the Connecticut Agricultural Experiment Station, "is evident in the rapid technological, economic and political changes of the past 90 years." He recommended a greater willingness of people to move from one region to another, taking their farm machinery with them, along with a government program to give disaster aid to those most affected by the desiccation and a more flexible, open market economy in which rising or falling prices will enforce adaptation. Take all those precautionary measures, he suggested, and we should weather the future nicely.[10]

Those are precisely the kind of encouraging words people want to hear, and if they also encourage Americans to begin taking rational steps to cut down on fossil fuel consumption along with investing in new crop research, they are sensible. But they are also naive in their assumptions about the easy adaptability of culture and institutions to climate change and about the resourcefulness of the human mind. Will people dwelling in rural Colorado readily move to Canada? Will adequate disaster aid always come, and come promptly, from taxpayers living across the nation? Will more freedom in the market economy make the plains farmer better off, or put him or her at a competitive disadvantage with farmers in other regions—and out of business? Can we really depend on plant breeders to discover a miracle wheat that can grow without rain? Will consumers learn to eat some newfangled, high-protein syrup made from prickly pears in a lab? And what can we really learn from the example of Eskimos' adapting to cold and snow that will help us adapt our farming to a desiccated plains? The Eskimos took thousands of years to adapt their technology and thinking to their limited resources and difficult climate. We, on the other hand, have only a few decades to make our adjustments—to discover what kind of population density, standard of living, and agronomy the plains can support. A lot has to be invented quickly. The model of the Eskimo is, therefore, more inspirational than immediately practical. We won't need igloos in the days ahead.

Moreover, the happy notion put forward by Waggoner's panel that the last ninety years of experience with rapid change proves that everything will turn out all right is wildly misinformed about that history. In the first place, those years have not been all that glorious or successful in adaptation. They have witnessed the rise of the United States to global economic supremacy, all right, but also the subsequent loss of some of that leadership to the Japanese and Germans and, in recent decades, a dispiriting decline in living standards for most Americans. During those tumultuous years, crime has gone up, drug use up, divorce up, and military armaments up, all of which might be seen as manifestations of failure to adapt to modern life. To be sure, the technologies that have appeared so rapidly, so bewilderingly, over that span of time have been more or less absorbed into our lives, but at what cost to our psyches and the natural environment? All along the way the adaptations to change have been far more difficult than the panel remembers. And the evidence of abandoned farm houses and dwindling small towns on the plains and growing slums and social decay in the cities suggests that many people have not been able to make those adaptations at all. The road from 1900 to the present has required massive demographic dislocations and great human and ecological costs and has produced a sharp-toothed anxiety gnawing at our national self-confidence. Panels looking toward a future of global warming ought to remember the full truth of what we have been through already.

Will the next ninety years present an easier transition than the last, or will they see even more dislocations, a more rapid downward spiral of bust and decline, and many more human failures? If some people will be winners in the global-warming era, who will be the losers? The plains are our most likely candidate in the United States. To what extent can they survive the coming trends and endure as living space? The past gives no comforting answers to those questions.

As A HISTORIAN, I recommend that the Connecticut agronomist and his fellow scientists avoid making shallow analogies with other peoples and other times and seek the aid of my profession to better understand the full, complex reality of how well we have adapted to atmospheric change in the past. They might, for example, read more of the history of the Great Plains and ask what lessons it holds for the future. To be sure, a warming trend that exceeds anything that has happened over the past ten thousand years stands outside history as we have known it. Nonetheless, we do have considerable experience with severe droughts to draw on—in the 1890s, the 1930s, and so forth. We can learn from that history what people went through in such periods, what they learned or didn't learn about nature, and how well they adapted and where they failed, and we can draw from that experience some insights for the years ahead.

So what are some lessons from the history of white settlement in this central region of North America? How successful have we been in adapting our institutions, technology, and thinking to the not so simple Great Plains?

The first lesson we should learn from plains history is that it is hard to adapt to a climate that you do not fully understand or do not fully want to accept. This is a lesson that applies to experts and common people alike. Both groups have a long record of talking about the plains climate with more confidence than is warranted. Indeed, the history of climate thinking in the region ought to give even global climate modelers pause as they make their predictions.

Throughout the nineteenth century, the dominant image of the plains bounced between two oversimplified abstractions, each naive, rigid, and epic in its implications. The plains are a "desert," said some of the earliest white travelers into the region; not at all, insisted others, they are a "garden," present or in the making. The geographer Martyn Bowden has argued that the first of those abstractions was one favored by eastern elite opinion, expressed in gazetteers and topographical reports, but that the westward-moving "folk" in all their practical wisdom ignored the implied warning in the desert notion, the pessimism about the land's potential, and went out to settle the country, anyway. The experts had wanted to counteract the anarchic, unsettling impulses of American society by inventing a formidable desert on the western frontier. Their negative images of the environment, which dominated national thinking until 1860, thus had an ideological content: they exaggerated the harsh natural conditions in order to serve a conservative social end.[11]

Undoubtedly, Bowden is right about the elite's manipulation of data, but the common people were driven by ideology as much as were those councilors of restraint. What the people moving west in covered wagons wanted was a land of unlimited economic opportunity—an abundance of free soil ready for free labor by free men. They needed to find or create a climate that would support that dream, and so they chose to talk about the plains as the "garden of the West." If the land was not yet garden-like, able to support all their traditional crops, they would make it so with their plows and enterprise. Rain would follow the moldboard plow. Climate would be reformed by human technology working with the aid of providence. As Wright Morris puts it, "God and man, working in close collaboration, first settled and then improved God's country."[12] From the time the first pioneers arrived on the plains, during the late 1870s and 1880s, that folk ideology of environmental improvement has been by far the dominant one in public discourse. Though it comes from deep impulses in the nation's popular culture, which rejects the idea of any natural limits on economic aspirations, it has managed to find support from many experts. By the 1880s, in fact, the experts were almost all converts to climate reformism. Hardly anyone dared to classify the region as a veritable desert anymore, while millions had come to settle, believing that, given enough faith and technology, it could become a garden producing an overflowing cornucopia of food and wealth.

As the dust began to blow in 1932, that victorious image of the plains as a garden in the making was profoundly shaken, though not fatally so. Climate suddenly seemed to be more treacherous than many settlers had supposed. They felt betrayed by providence, nature, the government, railroad companies, agronomists, anyone who had encouraged them in their efforts to make the plains sprout wheat instead of buffalo grass. The rest of the nation shared their sense of disappointment and has never quite recovered from it. Henceforth the climate of the plains became a leading symbol of a nature that was fickle, dark, threatening, dangerous, and undependable. The old confidence that nature was reliably on the side of America, eager to help in its own subjection and amelioration, disappeared, and ever since then, the climate of the plains has loomed as a wicked witch of the West, smiling beguilingly on occasions, then suddenly shrieking out curses and visiting terror on an innocent country.

In the aftermath of the dirty thirties we have talked about and studied the North American climate intensely, but we have not yet achieved a full understanding of its complexity, despite spending millions of dollars on scientific research. Climate, it has been said, is only average weather, but what that average is we still want to know. Even if we could discover an average, what would that tell us about the specific weather we can expect tomorrow?

Until very recently, the science of plains climatology has doggedly proceeded in the confidence that it could discover what nineteenth-century gazetteers and homesteaders ardently wished for: a predictability in the climate that could be neatly categorized, wet or dry, so that a stable civilization could be built on it. Nature must have a clear steady pattern, it has been assumed; nature must offer us a clear, single norm, even if we want to change it. That expectation has its roots in ancient ideas, going back to the Greeks, of a designed universe, a world that is rational and orderly in terms we humans can understand. Scientist after scientist has hoped that, because the earth turns regularly on its axis, runs smoothly in its orbit, behaves itself circumspectly as all planetary bodies must, the earth's weather must also ultimately be a predictable phenomenon. If we could only gather enough facts, we could locate those regularities in the climate, just as Newton precisely described in mathematical language the motions of the heavens.

Among serious students of the plains climate was the Nebraska ecologist Frederick Clements, who invented the idea of a "climax" plant community to characterize the native vegetation growing on the plains before the white man's invasion. The climax was supposed to be the end point of plant succession, when, after the pioneering species had entered and broken ground, the plant community matured and settled down to an equilibrium condition. Unfortunately for the theory, the climax stage was determined by climate, and climate, Clements eventually had to acknowledge, did not manifest any final equilibrium stage. Droughts came and droughts went. Was there any order to be found in those droughts, Clements wondered, an order that the plant community had evolved to handle? In other words, was drought regular? If it was, farmers might still follow the model of the native plants and anticipate the coming of drought, adapting their behavior as the plants did.

Clements thought he had found that key to order in the cycles of sunspots, dark blotches on the surface of the sun that are associated with magnetic fields and intensified solar activity. Regularly every eleven years (scientists now calculate one sunspot cycle of eleven years and another of twenty-two years) the sun bursts out with spots. Gathering rainfall records from all over the western states, Clements thought he saw in them a pattern of increased sunspot activity correlating with drought and one of minimum sunspot activity correlating with increased rainfall. For the rest of his life he collected such data in the hope of predicting exactly the natural rhythms of drought. If he could make those predictions firm, he could help farmers devise "a scientific system of expansion and contraction."[13]

Since Clements's day, other scientists have spent a lot of time and money chasing the same will-of-the-wisp. The atmosphere around us, they are sure, must have the same orderliness as the distant heavens we admire. So far, however, no one has been able to discover what that order is. As it turns out, the sunspots do not exactly correlate with drought, or their absence with precipitation, at least over the full historical record of plains climate. The strongest evidence for a twenty-two-year periodicity in drought exists for the short period from the 1890s to the 1970s, but before that the evidence of regularity is weak or nonexistent. A severe drought occurred in the 1750s, another in the 1820s, and then another in the 1860s, all of them more severe than that of the Dust Bowl years; in fact, the 1930s do not rank in the top five droughts known to have occurred since the early eighteenth century. Scientists at the Laboratory of Tree-Ring Research in Arizona have concluded that "drought appears to recur at ill-defined intervals of from 15 to 25 years," and they add that "whatever effect solar variability may have on drought, it is overwhelmed by other factors at particular locations."[14] Those "other factors" remain unexplained, but undoubtedly they are far more complex and numerous than scientists have been able to identify. We may know a lot more about the mechanics of drought than we did a few decades ago, but because of the complexity of factors involved we can still predict almost nothing from week to week.

Climate, we are now beginning to acknowledge, is so complicated a series of events that we may never be able to make predictions that a farmer can rely on. The sky above us is in perpetual, random motion—

it is non-Newtonian, after all. In the view of contemporary meteorologists, the climate is innately "chaotic," which is not to say that it has no structure or pattern at all but that its patterns are nonlinear, stochastic, and dependent on too many variables to locate simple order in them— the simple order that the old science of meteorology expected.[15] So argues one of the most influential students of the subject, Edward Lorenz of the Massachusetts Institute of Technology. "There is little question but that the real atmosphere is an irregular system," he has pointed out, and "the most obvious influence of irregularity is its limitation on the extent to which the weather may be predicted."[16] Despite a considerable increase in climate and weather research funding from the 1960s to the present, scientists are no more ready than before to venture solid forecasting beyond about a week into the future, let alone anticipate drought cycles over a period of decades or centuries.

Even the predictions of global warming I have mentioned, though worked out by powerful Cray computers and many brilliant human minds, must be taken with a few grains of salt. Though scientists can calculate how much carbon we are putting into the atmosphere and what its effects would be in a simplified computer model, they cannot say where all the carbon really goes—how much of it gets absorbed by the oceans, for example, or how much gets taken up by plants, or how much stays in the atmosphere, creating a kind of greenhouse roof over our heads.[17]

The history of thinking about the climate on the plains, therefore, leaves one with a humble sense of the mind's inadequacy before nature. We fail again and again when we try to transcend our mental limits and locate some enduring truths, some overall picture of order, in nature. Our perceptions of climate have always been clouded by ideology, by a willingness to believe in popular and elite notions alike, and even by the scientist's faith in a comprehensible, rational order in the universe. If a hundred years of settler experience and systematic scientific investigation have still not given us any great ability to predict droughts, how can we expect to prepare readily for that long desiccation that may lie ahead? The answer is, we cannot. Instead, we must develop at all levels of society a more adequate awareness of the complexity of the causes of drought, both natural and anthropogenic. And we must try to respect what we do not fully understand, which is most of the world around us. We cannot expect to make a smooth, easy adaptation of settlement, agriculture, or economy to a climate that will always be a turbulent chaos of cloud, heat, and gas.

A SECOND lesson derivable from the past is that trying to control nature through technology is never a fully adequate or long-term approach to successful adaptation. We should not rely exclusively on our mechanical ingenuity to get us through drought or permanent desiccation or to establish harmony with the natural world, for if we do, the outcome may be even more catastrophic. Adaptation to the environment, if it is to be lasting, must be cultural and social as well as technological.

One of the most important documents ever published about the western region appeared in 1936, *The Future of the Great Plains*, written by a committee appointed by President Franklin Roosevelt and headed by Lewis C. Gray of the Bureau of Agricultural Economics. No other study collected as much social and economic information about rural communities or understood so clearly and fully the root causes of the Dust Bowl disaster. The causes, according to the report, lay primarily in attitudes of mind brought by settlers to the region. The report argued that avoiding more disasters in the future required more than simply enduring until better times arrived but also changing those attitudes, no matter how deep-seated they were. Number one on their list of attitudes needing reform was the notion that "man conquers Nature." "It is an inherent characteristic of pioneering settlement to assume that Nature is something of which to take advantage and to exploit; that Nature can be shaped at will to man's convenience." We now know, they concluded, that "it is our ways, not Nature's, which can be changed." [18]

Since the 1930s, people on the plains have often ignored that admonition and instead looked for quicker, easier solutions—for a simple technological panacea that would not require any searching self-examination or moral reassessment. So far, in fact, no fundamental reform of attitudes has taken place. The conquest of nature through technology is still the dominant way of thinking. As global warming commences, that same old faith in technology's ability to manage nature completely may continue and may lead to foolish investments in one expensive, short-lived panacea after another, wasting time and capital in a vain effort to postpone the ultimate day of reckoning.

The most important technology adopted since the 1930s has been deep-well irrigation. More than any other innovation, it has allowed the plains to overcome the threat of severe drought and thereby encouraged the confidence that water not only can be conserved but also can be "invented." If there is no water in sight, it can be found somewhere else. That way of thinking was, of course, implicit in the old slogan "rain follows the plow," but by the thirties people were beginning to lose faith

in the plow as a water maker and were looking for new mechanical sav-
iors, including a steel pipe to bring water from someplace else—from a
river or reservoir. The Great Plains Committee warned that "the current
popular emphasis on new supplies of water . . . by which irrigation farming may widely replace dry farming, rests on hopes inevitably doomed
to disappointment . . . Sound water-mindedness will recognize the basic
facts of nature which man is powerless to alter."[19] The committee knew
the checkered history of surface irrigation on the plains, illustrated, for
example, in the failure of the nineteenth-century irrigation boom on the
Arkansas River, which flows through western Kansas; it knew the extrav-
agant hopes that had once gathered around places like the town of Garden City; and it was sure that there would never be enough dams and
reservoirs to satisfy the water demand or any cheap or easy imports available from other regions.[20]

In retrospect, the committee clearly underestimated the resourcefulness of the region's people in finding water, as it failed to anticipate the
extraordinary changes that irrigation would introduce in the next half-
century of plains agriculture. Over the short run, technology would indeed make a greater abundance possible where there had been extreme
scarcity. Nonetheless, in the long run the committee was absolutely
right: irrigation farming would be doomed to disappointment. After
many decades of development, water remains a severely scarce resource
on the plains, and it will be more scarce in the future than today. Ironically, part of that future scarcity will be due to the atmospheric pollution
caused by the fossil-fuel–consuming technology that gave a passing
victory.

The water that was "invented" as a response to the Dust Bowl, a supply that was not well understood or widely appreciated in the 1930s, was
the High Plains aquifer, which underlies the region from Texas to South
Dakota and includes as its largest unit the Ogallala Formation. During
the Miocene epoch (5 to 24 million years ago), deep beds of sand and
gravel were deposited on the plains by streams flowing down from the
Rockies. Today, those beds lie buried underground and are saturated
with the rainfall of millions of years, accumulating at rates from less than
one-tenth of an inch in Texas to six inches a year in south-central Kansas.
Like the rate of recharge, the thickness of the saturated beds varies considerably; Nebraska and Wyoming have by far the largest share of the
water.[21]

Pumping the High Plains aquifer first began on an extensive scale in
the Texas Panhandle right after World War Two. By 1957 there were
over forty thousand pumps there, pouring water on the fields. Pumps

were working noisily night and day over in New Mexico, too, and all the way north to the Sand Hills of Nebraska, pumping water in wet and dry years alike. Some of the pumps burned gasoline, and others, diesel fuel or natural gas. Though originally touted as an emergency relief mechanism to be turned on in bad times only, the irrigation pumps soon became a permanent, everyday tool of production, and on them a new way of rural life, very different from that of the old sodbuster and dryland farmer, came to depend. They stabilized the agricultural economy and population. By the late 1950s deep-well irrigation regularly contributed more than 20 percent to the region's income—several hundred million dollars a year that meant for many farmers and towns the margin between prosperity and bankruptcy. No wonder, with that miraculous achievement all around them, the people of the plains put their faith in technological adaptation more than ever. If nature had laid down those underground deposits of water and fuel, it was humans who had discovered them and brought them to the surface, turning them into wealth. The mayor of Lubbock, Texas, expressed the region's confidence in technological salvation as well as anyone when he declared, "The history of this country is that as the need arose for anything, somebody was there with the right tool to take care of it. This is the way this country was built."[22] The unacknowledged problem in that confidence was that this country was not built to last. Irrigated agriculture was a mode of production that could not be sustained.

In 1978 there were about 170,000 wells punched down into the aquifer, and they were annually withdrawing 23 million acre-feet (enough water to cover 23 million acres one foot deep). The total irrigated acreage amounted that year to 13 million. Though only one in 10 acres was irrigated, almost half of the region's crop value came from those acres. The extraordinary crop production allowed by that irrigation supported a vast livestock industry, worth $10 billion a year.[23] In fact, the raising and slaughtering of American beef had been revolutionized by plains irrigation, as the stockyards of Chicago closed down and the leading packing houses moved west to be nearer the supply of grain and as cattle moved off the range and onto feedlots where they were mass-fed and mass-fattened. All this took a heavy toll on the aquifer. By 1980 Texans had pumped up 114 million acre-feet of water. Kansans had pumped up 29 million acre-feet—more water than two Colorado Rivers could furnish in a year. The pumping intensified. By the mid-1980s, farmers in parts of those states had so depleted the underground supply that they had to go down three hundred feet to find the water table. Many pumps had to shut down, and new ones could not be set up.

With a multimillion-dollar grant from the federal government, six of the plains states undertook to study the economic impact of a future decline of their water and energy resources.[24] The study for my own state, published in 1982 by the Kansas Water Office, opened by accepting what had been for so long an unacceptable idea: that nature, even the deep underground aquifer of the plains, is limited and exhaustible. By the year 2020, the study indicated, 75 percent of the irrigated acreage in western Kansas will be lost, or 1.6 million acres. Long before the water is gone, the natural gas to pump it will be depleted. Remarkably, despite those grim facts, two economists at Kansas State University concluded that depletion will have no ill effect on plains communities; population will increase by 20 percent and total personal income by 130 percent, they predicted, despite the fact that farming on those 1.6 million acres must revert to dryland methods, which involve fallowing half of a farmer's acreage every year (allowing that land to lie idle to accumulate soil moisture) and planting wheat and grain sorghum rather than corn. This amazingly rosy conclusion was based on a U.S. Department of Agriculture (USDA) computer program that projected commodity price increases of 20 to 50 percent for the major dryland crops. In the real world of international markets, however, American farmers have watched their prices *fall*, not increase, as they have experienced increasingly intense competition from other nations. When and if global warming becomes evident, that competitive situation may get even worse for the plains. But then the economists ignored completely the possibility of global warming, as they did the lessons of the past, of changing international relations, and of foreign agricultural development. Despite its enormous expenditure of money, the study failed to take into account almost all of the critical uncertainties we face and thus failed to confront fully the stark, unhappy fact that the future will see a return to the nonirrigated, high-risk farming that prevailed prior to the 1930s.[25]

A similar story, with varying terminal dates, could be told for every section of the Great Plains, so that by the time global warming begins to show up in higher summer temperatures and lower rainfall, the irrigation empire will already have nearly collapsed. We will have to face the desiccation without the aid of the Ogallala, without affordable fuel, without the miracle of pipe-and-pump technology. That short-term technological miracle, in other words, by stimulating so much development and investment so fast, may have made the future calamity worse than it might have been.[26]

Never mind for the moment all the environmental costs that intensive irrigation has entailed, including an altered soil structure, fertilizer and herbicide pollution of groundwater, and heavy nitrate runoff from cattle feedlots—environmental costs that future generations must pay without enjoying the benefits of the water. Never mind the social costs that this expensive mode of production has entailed, which can be seen in increasing farm sizes, declining rural population, and shrinking small towns. Concentrate only on the simple fact, agreed on by almost everybody, that we are beginning to see the end of the irrigation era on the plains. That technological triumph over nature has about run its course, leaving us vulnerable once more to blowing dust, poverty, and out-migration.[27]

Every panacea, it should be clear, has hidden within its promises the potential disappointment of unforeseen costs and the possibility of unanticipated catastrophe. This is not to say that technology is always evil or dangerous or that new technological solutions should always be rejected, but only that adapting to a volatile environment through technology is a far more unreliable strategy than Americans often realize and one more filled with ambiguity and defeat. Technology appears to offer an easy way out of our difficulties with nature, but easy ways are usually hard ways to sustain and more hazardous than we anticipate.

A THIRD lesson we can draw from history is that the best hope for avoiding another Dust Bowl lies in restoring more of the plains to their natural, preagricultural condition. We have a habit, rooted in our economic institutions, of pushing agricultural development too far, repeatedly going beyond what the environment can bear, and now we must overcome that habit and lower our demands on this fragile land, returning much of the country to a more natural state. Probably the only practical way that can be done is through federal purchase of conservation easements, rather than outright acquisition of title, on those portions of the Great Plains that will be the most ecologically vulnerable in the next century, taking them permanently out of production, out of the marketplace as much as possible, and out of the reach of short-term self-interest.

If early predictions that the entire Great Plains would be forever off limits to agriculture proved false, the opposite view that every single acre might be made to produce food and profit has been equally wrong. Severe wind erosion has continued to plague the region from time to time, following cycles of plow-up and abandonment. What's more, the too-optimistic view has been very expensive. It has cost American taxpayers billions of dollars in bailout money whenever drought has appeared, and in accepting that money again and again the region has become vitally dependent on the federal government for its survival. This has perhaps been the most important outcome of the 1930s. A greater panacea than even irrigation, which brings in only water and then only for a minority of farmers, the government has brought in cash—pumping it through an elaborate system of pipelines called "farm programs" and spreading that cash over virtually every farm and ranch in order to keep agriculture alive on the plains. A common conclusion drawn from the Dust Bowl experience was to "seek federal assistance," and that conclusion has now become entrenched policy. No technological innovation, no local spirit of rugged determination, has produced nearly as much income security as Washington, D.C., has done. But relying on government subsidies is a strategy that has serious flaws, too, which will become manifest as long-term desiccation sets in.

The federal government has agreed to provide massive amounts of assistance because Americans have been a generous people toward those who take risks and fail. Risk is, after all, widely proclaimed as a national virtue. In our schools, competitive sports, market economy, government policies, and entertainment, we have constantly encouraged such behavior, believing as we do that by encouraging individual risk we all prosper. Such was the assumption directing the land policies of the nation in the nineteenth century, and the assumption has not altered much in the last decade of the twentieth century. Western mythology still celebrates those men and women who risk all to push the frontier forward, putting family and community behind them if necessary, itching to get out of the older settlements and their confining ways, venturing onto the plains to acquire property, in many cases grabbing it illegally and with impunity, then pushing against its ecological limits to determine what it can yield in the way of wealth. Unfortunately, a land policy that encourages such risk taking also produces a lot of failures. The Great Plains have always had, by world standards, a high number of arrogant failures— men and women colorful, brash, domineering, self-reliant in their rhetoric, resentful of any criticism—and again and again they have gone beyond what the land can bear. Our persistent failure to discourage such behavior and put communal and environmental stability ahead of individual risk taking and acquisitiveness as a social good has been by far the biggest obstacle to successful adaptation on the plains.

Many plains people, to be sure, have not been such foolhardy risk takers or self-seekers. And some of those who have taken large risks have been unfortunate, powerless people forced by the American lottery of land and opportunity to try to extract a living out here on the fringes of the good earth. Since settlement first began on the plains a hundred years ago, such people have come and gone with the volatile cycles of climate. There are fewer of them now than before on the land, but they are still here, trying to scrape an income out of sand dunes or other marginal ground.

One of the major achievements of the 1930s was to extend the protective hand of the federal government indiscriminately and liberally to both the arrogant fools and the desperately poor, and indeed to the whole spectrum of farm operators whenever they suffer from the chaotic play of drought and winds. I have mentioned the WPA handing out help in the Depression, but there have been many other, more permanent, forms of agricultural assistance, which either began in the thirties or have appeared since then, including not only all the well-known programs to support farm commodity prices but also what has become a regularized, permanent system of disaster assistance.

Each new drought since the 1930s has seen the level of relief, and the disaster bureaucracy that administers it, increase exponentially.[28] During the most recent dry spell of 1988, severe but short-lived and centered mainly in the Dakotas, the government sent out $4 billion in relief money. All of that sum was supposed to go to alleviate drought conditions, but Associated Press reporters uncovered massive fraud around the country; some farmers collected cash payments of hundreds of thousands of dollars for damages from wind, rain, hail, frost, fungus, insects, "anything related to nature," as one county administrator explained.[29]

Farming, we are frequently told in farm journals and USDA publications, must be regarded as a business like any other, but that is not quite the way things have developed. Instead, farming is a business that passes many of its risks onto the taxpayers. Great Plains farmers are unique only in the extent of their dependency: they regularly pursue a business that relies on people in other regions of the country to step in and save them from bankruptcy whenever nature, or their own miscalculation, threatens. They have managed to shift a critical part of their burden to millions of total strangers who have consented to cushion them from a hard world.

What has been the ecological effect of so much kindness? Has it lessened the pressure to adapt agriculture to local conditions? Has it shielded farmers from the disciplining hand of natural forces, even encouraged them to try to push their land to the breaking point? Removing risk from the marketplace, we are told by many economists, results in inefficiency and waste, though for many good reasons the government does so, anyway. What happens when we remove risk from the physical environment? Does it lead to an overextension of agriculture that cannot be sustained? I think it does.

This conclusion is not one that some scholars of natural hazards would draw. Like floods and other calamities, they argue, droughts are evils visited on innocent humankind by nature, and the best remedy is to develop a complex, risk-spreading social system that can alleviate the pain and suffering they cause. According to this view, precisely such an order has emerged since the 1930s in the form of a responsive federal government. We now live, it is said, in a "mature" society with a centralized authority that is well organized to give assistance in time of need. Tracking the history of distress indices on the plains, these scholars note a lessening of bankruptcy, out-migration, and social instability over time. In the 1890s crisis, when laissez-faire was the predominant national philosophy, millions had to abandon the region simply to avoid starvation. More recently, however, hardly anyone has had to leave because of adverse ecological conditions. The next severe drought, according to this view, will have even less of an impact than the last one, due to the continuing integration of the American economy and the growth of organized disaster management. This organization of risk, they say, is how modern, complex societies learn to adapt to nature: they concentrate enough power and wealth at the center in order to overcome most natural vicissitudes. They learn how to create stability out of chaos by sending out money regularly to compensate for local loss.[30]

That has indeed been one of the most important strategies followed over the past half-century, but it may be a poor one to rely on as an era of global warming approaches. Desiccation is a long-term process. It may last for a century or more. Calling on other areas of the nation to share their income with the plains over that period of time may be unrealistic; as those regions encounter their own climate-related difficulties, it may be like calling on other western watersheds to release some of their precious water. They may have little money to release, or they simply may resent the idea of their taxes going to support an agriculture where it seems marginal, uneconomic, and irrational. Helping the plains farmers through an occasional hazardous period is one thing, but supporting them decade after decade in an effort to defy the laws of nature is another. Instead of continuing to function as a compensatory, risk-spreading force, the government, reflecting its own straitened circumstances, may begin to insist that the plains people work out for themselves how to adapt and survive.

The notion that a complex, highly integrated and centralized society could, or would, reliably alleviate the problems of a future Dust Bowl is itself open to question. If one examines the long historical record of social adaptation to nature, complex societies seem to be more vulnerable than simple ones. Despite their well-developed administrative structure and their capacity to call on vast pools of labor and resources, such societies may actually be quite vulnerable to long-term environmental adversity and may collapse or withdraw instead of adapting. They commonly lack a detailed, intimate understanding of the many local environments on which they depend or which they exploit. They substitute money for that knowledge. They find it hard to make fundamental changes of direction. They grow discouraged and fall apart. This conclusion emerged from discussions among a number of anthropologists at a Canadian conference on how modern civilization might cope with rapid climate change. The report of that conference argued:

> The idea that complex societies—modern technological societies included—are *more* vulnerable to climate change than less complex societies will undoubtedly come as an unwelcome surprise to many people. The rapid pace of technological development in the industrialized nations since the second world war has created the comfortable illusion of increasing human *invulnerability* to the vagaries of nature. This illusion is not only wrong but dangerous. There's every reason to believe the complex nature of modern societies renders them more susceptible to internal collapse than hunter-gatherer or agriculture societies of the past, or even those groups which today preserve hunter-gatherer or subsistence agricultural roots—for example, native populations of the Canadian North and traditional agricultural groups such as the Amish or the Pueblos.[31]

However unreliable it may prove in the future, the federal government will for the foreseeable future be expected to help with the process of adapting agriculture to the future plains. What then should its aims be? What should be left to the market? How can the government help achieve the best possible adaptation for the longest time and under the worst possible adversity? Should the government continue to underwrite an overextended farm economy at the cost of ecological degradation, or should it seek to reduce that economy to fit the best-estimate, long-term limits of the land? I believe the clear answer is the latter.

Altogether, the Great Plains states of the United States contain about

500 million acres. Almost none of that expanse has been permanently put off limits to row-crop agriculture. A few parks and wildlife refuges and a few national grasslands, altogether not exceeding 1 percent of the total—that is all we have designated by law for perpetual protection. The rest is in private hands, and we still leave it up to the marketplace to determine how much of that 99-plus percent of the land should be used. If the world price of wheat goes up, the plow gets to work turning grassland into wheat land. If the price goes down, the land reverts to some form of cover, weeds or grass, and may for a while support a herd of cattle. Whatever the decision made, it is still the private landowner who makes it, with government inducements and pressures, and he or she decides according to the principle of self-interest, rational or not so rational, in response to a world market economy.

In recent years, the government has tried to become a more active counterforce to that market economy, not only by supporting prices artificially but also by renting private land for nonproductive purposes. The Soil Bank program of the 1950s, for example, idled over 21 million acres across the nation, though its purpose was to reduce crop surpluses rather than to protect vulnerable soils. During the 1970s international grain markets boomed, thanks in part to Soviet purchases, and for the first time in decades the government encouraged farmers to increase production, even if it meant plowing from "fence row to fence row." Following that advice, they plowed up millions of fragile rangeland and other highly erodible acreage; nationally, the great plow-up of the 1970s added 56 million acres to production.[32] Later, even free market ideologues began to question the wisdom of this unrestrained expansion, and during the Reagan presidency a movement gathered support from both conservatives and liberals to set up controls over "sodbusting" and "swamp busting" (draining wetlands to plant crops). The Conservation Reserve Program (CRP) was the outcome, established under the Food Security Act of 1985 to help adjust agriculture to its environmental conditions. It does this by renting some of the most marginal lands—up to 25 percent of the area in any single county—and putting it back into grass, keeping it that way for a period of ten years. The goal of the CRP is to retire 45 million acres from production, about half of that on the Great Plains. The program is only temporary, however, and all those acres could revert to cropping when the program expires in 1995. Renewing the program is a possibility, but the CRP is a very expensive remedy. Should we go on paying such rents to farmers through the next century in order to help them adjust to the coming desiccation trends? That could be the most expensive program the Department of Agriculture has ever embarked on, and even then it would always be a stopgap measure. As Earle J. Bedenbaugh, deputy administrator of the USDA's Agricultural Stabilization and Conservation Service, has admitted, we have been making "short-term solutions to long-term problems," and "at some point in time we are going to have to take a basic sound philosophy concerning agricultural legislation, put it into effect, and see it to its ultimate end."[33]

What should that permanent philosophy be? Looking at the long history of plains settlement, the historian concludes that only a program of permanent grassland restoration has any economic or ecological rationality to it. We need to put a larger part of the burden of adaptation on nature, which has been at that work for hundreds of millions of years. Such a program could be carried out only by collective means through the state or federal *purchase of land title* or the purchase of conservation easements in perpetuity, controlling all future development of that land.

In the midst of every drought there have been knowledgeable observers who have called for changing the pattern of land ownership to some substantial degree. They have estimated that several million acres—the estimates vary considerably, though a tenth of the whole region, or some 50 million acres, would be a rough consensus—ought to be retired permanently from all crop use and restored to something like their presettlement state of vegetation. Those observers have recommended that the restored acres be kept under close public supervision as the best guarantee against a future Dust Bowl of drought, denudation, and wind erosion. Nature, they have suggested, can teach us something important about how to adapt more perfectly to the plains, particularly in those areas most vulnerable to climate change. That recommendation, however, has never really been heeded. To do so would require changing our reliance on self-interest and private ownership as the primary basis of land policy.

Private landowners cannot in most cases undertake such a program of market withdrawal and ecological restoration. They will never find a market that will make such a program profitable or attractive. Only some disinterested group of individuals, acting in the public interest, or some federal, state, or local government agency could ever undertake it. They would do so not because they were more intelligent than the private owner, but because they would have more freedom to ignore the abrupt swings of market pressures.

In my view, the most compelling lesson we can learn from the history of the Great Plains is that the best adaptation to a volatile climate can never be achieved merely by a system of private property and market-place economics. Nor can it be achieved by supplementing that system with expensive, endless government relief or subsidies. To insist otherwise would be to turn the ideals of private ownership and free enterprise into a religious-like fundamentalism that ignores all historical experience and rejects critical thinking.

In drawing this conclusion from the past, I have not really addressed the ultimate question of how the threat of global warming might be prevented or lessened at its source: the tailpipes and smokestacks of the global industrial economy. Obviously, any comprehensive solution to the plains situation, and that of other regions, must lie in curtailing the pollution of the earth's atmosphere with carbon, a solution far beyond the power of local people to carry out alone. What they can do, however, in cooperation with public officials, is examine their history critically and ask what it suggests as their best strategy for mitigating the effects of global change.

The challenge of cultural adaptation to environment, contrary to the opinion of many technological and scientific experts, has never been easy to meet anywhere in the world. It has been harder on the plains than most places. For more than a hundred years now, men and women have been working hard at the task of farming the plains, and still they have not really figured out how to meet the basic challenge of the land for very long. We have settled the landscape with farms and towns but then seen many of those settlements dwindle away, and more such decline lies ahead. We have made a great deal of money out of the land, but in recent decades more and more farm income has come from government sources. And for every generation there has been the threat of blowing dust and ravaged crops. If we have not solved the riddle of the country so far, how can we expect to solve it easily in the future, without a significant change of thinking, in those remorseless, unclouded summers that may lie ahead?

NOTES

1. Intergovernmental Panel on Climate Change, *Climate Change: The IPCC Scientific Assessment*, ed. J. T. Houghton, G. J. Jenkins, and J. J. Ephraums (Cambridge: Cambridge University Press, 1990). See also Stephen H. Schneider, "The Greenhouse Effect: Science and Policy," *Science* 243 (10 February 1989): 771–781.

2. Intergovernmental Panel, *Climate Change*, xi-xii. Also see Geoffrey Wall, ed., "Symposium on the Impacts of Climactic Change and Variability on the Great Plains," Occasional Paper No. 12, Department of Geography (University of Waterloo, Waterloo, Ontario, 1991); Environmental Protection Agency, *The Potential Ef-*

fects of Global Climate Change on the United States (New York: Hemisphere, 1990), Chapter 7.

3. William W. Kellogg and Robert Schware, *Climate Change and Society: Consequences of Increasing Atmospheric Carbon Dioxide* (Boulder, CO: Westview Press, 1981), 65, 157–161. Areas that seem likely to become wetter than now, improving their agricultural prospects, include central Mexico, the Sahara and East Africa, Saudi Arabia, India, and western Australia. Thus, the agricultural impact of global warming may be far more adverse for the United States than for a large part of the developing world.

4. Thomas R. Karl and Richard R. Heim Jr., "The Greenhouse Effect in Central North America: If Not Now, When?" in *"Symposium on the Impacts of Climatic Change and Variability on the Great Plains,"* ed. Geoffrey Wall, Occasional Paper No. 12, Department of Geography (University of Waterloo, Waterloo, Ontario, 1991), 19–29.

5. I owe these figures to Marty Bender of the Land Institute, Salina, Kansas. See also David Pimental, ed. *Handbook of Energy Utilization in Agriculture* (Boca Raton, FL: CRC Press, 1980). It should be added that only 2 percent of all energy consumed in the United States is consumed on the farm; clearly, agriculture is going to be more victim than perpetrator of global warming. On the other hand, the heavily rural Great Plains states are among the most intensive users of energy in the world.

6. Donald Worster, *Dust Bowl: The Southern Plains in the 1930s* (New York: Oxford University Press, 1979), 35.

7. According to Karl W. Butzer, "Dust Bowl conditions would become commonplace on the Great Plains, with an increased incidence of drought in the Midwest." See Butzer, "Adaptation to Global Environmental Change," *Professional Geographer* 32 (August 1980): 271.

8. Butzer, "Adaptation," 273.

9. Two agricultural economists calculate that "a 3°C increase in regional temperatures and a small reduction in summer precipitation tends [sic] to shift the Corn Belt into droughty soils, hence to reduce the region suited for such production by about 30 percent." With a 2°C rise in temperatures and a 10 percent decrease in precipitation, Great Plains wheat yields fall 8 to 15 percent and 6 to 8 percent, respectively. "The soil resource base of the granary, especially on the Great Plains, may [also] be at risk with a major warming. Wind erosion of soils . . . may intensify under drier and droughtier conditions, leading to increased erosion, intensification of current desertification trends, and in the worst case, to dust bowl conditions. . . . Insect populations may increase as much milder winters and frost-free seasons increase the number of insect generations per summer. Grasshoppers, cutworms, corn borers, and aphids may be important as may viruses vectored by insect populations." As for the prospects of overcoming these conditions through plant breeding, they add,

> It has not yet been proven possible to incorporate significantly enhanced heat or drought tolerance into present cultivars without lower average yields . . . Adaptive capability in this regard has been limited both by technological and income constraints, and given current trends in agricultural income, these appear unlikely to lessen in the future. Traditional plant breeding techniques might be able to adapt present cultivars somewhat at the margin, perhaps to several tenths of a degree Celsius per decade. However, adaptations to more rapid changes, for instance 0.3° to 0.8°C per decade, seem beyond the power

of these techniques. Revolutionary alterations in the genetic makeup of American crops, one frequently cited potential adaptation, might alter this situation. However, such alterations have yet to be demonstrated. The adaptive capability of the agricultural system, in this case, probably ought to be considered rather limited from the perspective of major alterations in crop physiology until the prospects of genetic engineering have been substantiated.

Peter Ciborowski and Dean Abrahamson, "The Granary and the Greenhouse Problem," in *The Future of the North American Granary: Politics, Economics, and Resource Constraints in North American Agriculture*, ed. C. Ford Runge (Ames: Iowa State University Press, 1986),70–73.

10. *New York Times*, 7 September 1991. More recently, the economists Robert Mendelsohn, William Nordhaus, and Daigee Shaw argue that the United States as a whole will adapt easily to the agricultural effects of global warming, though they do not deny that the corn and wheat belts may be hit severely. The rest of the nation, consequently, may not see any crisis for them and be unwilling to take the steps necessary to reduce carbon pollution. See the *New York Times*, 8 September 1994.

11. Martyn J. Bowden, "The Great American Desert and the American Frontier, 1800–1882: A Problem in Historical Geosophy," in *Anonymous Americans: Explanations in Nineteenth-Century Social History*, ed. Tamara Hareven (Englewood Cliffs, NJ: Prentice-Hall, 1971), 48–79. See also Bowden, "The Great American Desert in the American Mind: The Historiography of a Geographical Notion," in *Geographies of the Mind: Essays in Historical Geosophy*, ed. David Lowenthal and Martyn J. Bowden (New York: Oxford University Press, 1976), 119–148; and the various essays in *Images of the Plains: The Role of Human Nature in Settlement*, ed. Brian W. Blouet and Merlin P. Lawson, (Lincoln: University of Nebraska Press, 1975).

12. Wright Morris, *God's Country and My People* (New York: Harper & Row, 1968), n. p. The standard account of this overwrought optimism is Henry Nash Smith, "Rain Follows the Plow: The Notion of Increased Rainfall for the Great Plains, 1844–1880," *Huntington Library Quarterly* 10 (February 1947): 169–193. The most influential scientist supporting this climate meliorism was Samuel Aughey of the University of Nebraska, who believed that, through planting trees, humans could lend "a helping hand to the processes of nature for the development and utilization of the material wealth" of the region (Smith, "Rain Follows," 184–185). See also Clark C. Spence, *The Rainmakers: American "Pluviculture" to World War II* (Lincoln: University of Nebraska Press, 1980).

13. Frederick E. Clements, "Drought Periods and Climatic Cycles," *Ecology* 2 (July 1921): 181–188. For a more recent study of the same question, see J. M. Mitchell, C. W. Stockton, and D. M. Meko, "Evidence of a 22-Year Rhythm of Drought in the Western United States Related to the Hale Solar Cycle Since the 17th Century," in *Solar-Terrestrial Influences on Weather and Climate*, ed. B. M. McCormac and T. A. Seliga (Dordrecht, Holland: D. Reidel, 1979), 125–143. For an overview of the history of weather forecasting, see Alan D. Hecht, "Drought in the Great Plains: History of Societal Response," *Journal of Climate and Applied Meteorology* 22 (January 1983): 51–56.

14. Charles W. Stockton and David M. Meko, "Drought Recurrence in the Great

Plains as Reconstructed from Long-Term Tree-Ring Records," *Journal of Climate and Applied Meteorology* 22 (January 1983): 17–29.

15. This argument is made in James Gleick, *Chaos: The Making of a New Science* (New York: Viking, 1987), especially at pages 15–20, which discuss the revolutionary work of meteorologist Edward Lorenz.

16. Edward N. Lorenz, "Irregularity: A Fundamental Property of the Atmosphere," *Tellus* 36A (1984): 108.

17. This conclusion is admitted by a leading climatologist, F. Kenneth Hare, "Drought and Desiccation: Twin Hazards of a Variable Climate," in *Planning for Drought: Toward a Reduction of Societal Vulnerability*, ed. Donald A. Wilhite and William E. Easterly, with Deborah A. Wood (Boulder, CO: Westview Press, 1987). Hare writes: "Why does rainfall sometimes fail? Amazingly there is no clear answer. . . . drought is still largely unpredictable, and its causes are obscure" (5). See also Schneider, "The Greenhouse Effect," 775–776, where he admits that the actual warming that has occurred over the past 100 years is less than the present models would have predicted.

18. House, *The Future of the Great Plains*, 75th Cong. 1st sess. 1936, H. Doc. 144, 63–64.

19. House, *The Future of the Great Plains*, 76.

20. James Earl Shcrow, *Watering the Valley: Development among the High Plains Arkansas River, 1870–1950* (Lawrence: University Press of Kansas, 1990), 79–100; Anne M. Marvin, "The Fertile Domain: Irrigation as Adaptation in the Garden City, Kansas Area, 1880–1910" (Ph.D. diss., University of Kansas, 1985).

21. Edwin D. Gutentag et al., *Geohydrology of the High Plains Aquifer*, U.S. Geological Survey Professional Paper 1400-B (Washington, DC: Government Printing Office, 1984).

22. Cited in Donald E. Green, *Land of the Underground Rain: Irrigation on the Texas High Plains, 1919–1970* (Austin: University of Texas Press, 1973), 220. The expansion of irrigation after World War Two is discussed on pages 145–164.

23. Jonathan Taylor, Mary W. Downton, and Thomas R. Stewart, "Adapting to Environmental Change: Perceptions and Farming Practices in the Ogallala Aquifer Region," in *Arid Lands Today and Tomorrow*, ed. Emily E. Whitehead (Boulder, CO: Westview Press, 1988), 666. By the 1970s cattle feedlots in the Ogallala Aquifer area supplied 40 percent of the total production of grain-fed beef in the United States.

24. High Plains Associates, *Six-State High Plains-Ogallala Aquifer Regional Resources Study: Summary* (Austin, TX: Camp Dreser and McKee, 1982); High Plains Study Council, *A Summary of Results of the Ogallala Aquifer Regional Study, with Recommendations to the Secretary of Commerce and Congress* (1982); U.S. Army Corps of Engineers, *Six-State High Plains Ogallala Aquifer Regional Resources Study: Summary Report* (Albuquerque, NM: U.S. Army Corps of Engineers, Southwestern Division, 1982).

25. Michael H. Glantz and Jesse H. Ausubel, "The Ogallala Aquifer and Carbon Dioxide: Comparison and Convergence," *Environmental Conservation* 11 (Summer 1984): 123–131.

26. A task force on responses to drought, headed by Harold E. Dregne, concluded that irrigation technology may have increased vulnerability in some cases. They pointed in particular to center-pivot sprinkler systems, which have permitted the breaking of soils too sandy for irrigation and highly susceptible to wind erosion,

systems that have required the removal of shelterbelts planted in the 1930s because they interfere with the circular sweep of the sprinkler devices. They also have created a false sense of security—a belief that agriculture has achieved freedom from the vicissitudes of climate. Consequently, the task force concluded, "A protracted and widespread drought in western North America is certain to have a far greater impact now than would have been the case twenty years ago." See Norman J. Rosenberg, ed., *Drought in the Great Plains: Research on Impact and Strategies* (Littleton, CO: Water Resources Publications, 1979), 20–21.

27. One recent study found considerable disillusionment with irrigation technology among Great Plains farmers. Thirty percent of those questioned agreed that "Irrigation, in this country, is a short-term blessing but a long-term curse." The authors of the study concluded that irrigation "is certainly not perceived to be a universal panacea. It involves significantly greater expense, more work, and more time commitment. In the north, where irrigation came in later, farmers could be locked into an excessive debt load. With rising costs, especially for energy, several farmers described irrigation as 'just chasing dollars,' where every dollar increase in output requires at least another dollar input." Taylor, Downton, and Stewart, "Adapting to Environmental Change," 681–682.

28. The percentage of farm income that comes from government payments has never exceeded 20 percent across the "grassland" states as a whole, including Iowa, Illinois, and Missouri, writes John R. Borchert in "The Dust Bowl in the 1970s," *Annals of the Association of American Geographers* 61 (March 1971): 14. In the mid-1930s government payments came close to 20 percent; they fell to a mere 1 to 2 percent in the 1950s drought, and then rose to about 10 percent in 1968. By far the largest source of farm income—over 60 percent—has derived from livestock. However, Borchert's figures do not reveal the true picture on the western plains, the Dust Bowl area proper. There, government payments constitute a much higher proportion of farm income, and particularly so in droughty years, when it may exceed 50 percent in some counties.

29. Scott McCartney and Fred Bayles, "Drought Relief Stretches Far Beyond the Need," *Lawrence (Kansas) Journal-World*, 17 December 1989. A useful review of state and federal assistance is given by Donald A. Wilhite, "Government Response to Drought in the United States: With Particular Reference to the Great Plains," *Journal of Climate and Applied Meteorology* 22 (January 1988): 40–50.

30. I am summarizing here an argument that has been made by such scholars as Robert Kates, Martyn Bowden, Richard Warrick, and William Riebsame, all associated with the Climate and Research Group at the Center for Technology, Environment, and Development, Clark University. See, for example, Warrick's "Drought in the Great Plains: A Case Study of Research on Climate and Society in the USA," in *Climatic Constraints and Human Activities*, ed. Jesse Ausubel and Asit K. Biswas (Oxford: Pergamon Press, 1980), 93–123. The scholars mentioned above refer to this argument as the "lessening hypothesis" and have accumulated some valuable data to support it; like most hypotheses in historical research, however, it may describe the patterns of the past better than it predicts the future. See also William E. Riebsame, "Managing Drought Impacts on Agriculture: The Great Plains Experience," in *Beyond the Urban Fringe: Land Use Issues of Nonmetropolitan America*, ed. Rutherford H. Platt and George Macinko (Minneapolis: University of Minnesota Press, 1983), 257–270; Hecht, "Drought in the Great Plains."

31. Lydia Dotto, *Thinking the Unthinkable* (Calgary, Alberta: Wilfrid Laurier University Press/Calgary Institute for the Humanities, 1987), 37–38.

32. From 1978 to 1983, nearly 600,000 acres of fragile grassland (classes IVe, VI, and VII) in eastern Colorado and over 4.5 million acres in the northern and central plains went under the plow. The most important reason for that plow-up was, according to a Colorado State University resource economist, "simple economic profit maximization": raising a wheat crop brought more money than grazing cattle. Paul C. Huszar, "Nature and Causes of the Plowout Problem in Colorado," in *Arid Lands Today and Tomorrow*, ed. Emily E. Whitehead (Boulder, CO: Westview Press, 1988), 663.

33. Earle J. Bedenbaugh, "History of Cropland Set Aside Programs in the Great Plains," in *Impact of the Conservation Reserve Program in the Great Plains*, ed. John E. Mitchell, USDA, Forest Service, Rocky Mountain Forest and Range Experiment Station, General Technical Report RM-18 (Washington, DC: Government Printing Office, 1988), 14.

STEPHEN J. PYNE

Consumed by Either Fire or Fire

A Prolegomenon to Anthropogenic Fire

> We only live, only suspire
> Consumed by either fire or fire.
> —T. S. Eliot, *Four Quartets*

Ends and Beginnings

THE CAPTURE of fire by the genus *Homo* changed forever the natural and cultural histories of the Earth. Nothing else so empowered hominids, and no other human technology has influenced the planet for so long and so pervasively. A grand dialectic emerged between the fire-proneness of the biota and the fire capacity of humans such that they coevolved, welded by fire to a common destiny. If humans assimilated nature's fire into their biological heritage as a species, it is no less true that virtually all biotas have come to accept anthropogenic fire, and not a few demand it. Deny anthropogenic fire and you deny humanity and many of its ancient allies anyplace on the planet.

Keeper of the Flame

In effect, a pact was struck, the first of humanity's Faustian bargains and the origin of an environmental ethos. Humans got fire and, through fire, access to the world's biota; that biota, in turn, got a new regimen of fire, one disciplined by passage through human society. Of course, humans did not control all fires and had to suffer wildfire from their own or natural sources. Nor did humans always use fire prudently or with the regularity their allied environments expected. Nevertheless, once begun, the synergism proved irresistible. A uniquely fire creature became bonded to a uniquely fire planet.

Everywhere that humans went—and they went everywhere—they carried fire. The hominid flame propagated across the continents like an expanding ring of fire, remaking everything it touched. Within that ring lived humans; outside, the wild still reigned. Humans occupied preferentially those sites susceptible to fire and shunned, or tried to restructure, those that were less amenable. They sought out what needed to be burned, and burned it. Through its human agents, the biosphere in effect wrested control over ignition from lightning and other inorganic phenomena, and the structure of fire regimes responded to new patterns, its rhythms yielding new cadences.

If humans became dependent on fire (and in some cases intoxicated, even addicted to it), it is no less true that many biotas came to depend on anthropogenic fire practices for their own survival. Organisms adapted not to fire in the abstract or to individual fires but to fire *regimes*. A sudden change in a fire regime—its load, its frequency, its seasonal timing and intensities—could propagate catastrophically throughout the system. This of course is exactly the revolution that human firebrands made possible. But if fire granted them new power, it also conveyed new responsibilities. It was vital that the flame neither fail nor run wild.

The firestick was different from other implements in the hominid toolkit. It took on the attributes not only of *Homo* but of the larger biota. An ax head has the properties of the rock from which it comes; an arrowhead, of its source flint or obsidian; an awl, of bone. Fire is more protean than these. Confined as a torch or oven, it can substitute for ax or drill or desert sun. Broadcast as free-burning fire, however, it has the properties of the weather and winds that drive it, the terrain across which it propagates, and the fuels—living, dead, and dormant—that sustain it. Through these fuels, fused by a kind of biotic weld, fire shares the properties of life. Virus-like, fire becomes as complex, dynamic, subtle, and varied as the biotas that sustain it. Like life, it can propagate, amplify, reproduce. Its reach is pervasive, penetrating into every crevice of an ecosystem. Its effects are plural, and often indeterminate. Anthropogenic fire resembles less a tool in the conventional sense than a colossal symbiosis, exclusive to humans, the source of their unique power over terrestrial biotas.

In effect, humans domesticated fire. Once tamed, it had to be fed, housed, cared for, and bred. It could no longer range for its own food, or be left to fend for itself in the elements. It had to be tended. Its reproduction, too, could be guided into select "breeds"—fire for cooking, fire for ceramics, fire for heating, lighting, hardening; fire for hunting, farming, herding, clearing the land. Wherever humans ranged, domesticated fire began to replace wild fire. Much as humans killed wolves and propagated dogs, so they drove back the domain of wildfire and substituted a regime based on anthropogenic burning. Once begun, the process could become self-reinforcing; the very act of burning helped reshape ecosystems, redesigning fuel complexes in ways that shielded humans from unwanted wildfire and further promoted domesticated fire. Domestic fire could serve humans only if humans served it. The domain of one determined the domain of the other.

Fire as Power

The quest for fire is a quest for power. Virtually all fire-origin myths confirm this fact. In words that in one form or another find echoes everywhere, Aeschylus has Prometheus declare that, by giving fire, he founded "all the arts of men." To possess fire is to become human, but fire is almost always denied, only rarely granted by a fire-hoarding potentate, more typically stolen by some culture hero through force or guile. Once acquired, the balance of biotic power began to shift in favor of the otherwise meagerly endowed genus *Homo*.

For all the manifold feebleness of this species, fire compensated, and more. It made palatable many foodstuffs otherwise inedible or toxic; with smoke or heat, it made possible the preservation of foods that would soon spoil; it promoted a cultivation of indigenous forbs, grasses, tubers, and nut-bearing trees; it stimulated hunting; it hardened wooden tools, made malleable shafts to be rendered into arrows or spears, and prepared certain stones for splitting; it kept at bay the night terrors, promoted and defined the solidarity of the group, and made available the evening for storytelling and ceremony. It even allowed humans to reshape whole landscapes as, in effect, humans slowly began to cook the Earth. Everywhere humans went, fire went also as guide, laborer, camp follower, and chronicler.

Fire's danger matched its power. If untended, once-domesticated fire could go wild. The extraordinary pervasiveness that made fire universally useful also threatened humanity's ability to control it. The relationship was truly symbiotic; if humans controlled fire, so also fire controlled humans, forcing the species to live in certain ways, either to seize fire's power or to avoid its wild outbreaks. In domesticating fire, humanity also domesticated itself; fire's power could come only by assuming responsibility for fire's care. The danger of extinction was ever present. The feral fire lurked always in the shadows. Even today the control of fire is far from satisfactory.

Imagine, for example, the reception that would greet someone who announced that he had discovered a process fundamental to the chemistry and biology of the planet, a phenomenon that could grant to humans the power to intervene massively in the biosphere and atmosphere, that could, in effect, allow humans to reform the living world. Some critics would object on the grounds that humans were congenitally incapable of using such power wisely and would only harm themselves and the world. Even those who would cheer such an announcement might be quieted by the qualifying fine print, which stipulated that control would be inevitably incomplete; that even in the built environment lives and property would be lost; and that the hazards attendant with the use of this process would be so great that emergency crews and expensive equipment would need to be stationed every few city blocks, buildings would need emergency alarms and exits illuminated by separate power sources, every residence should be outfitted with instruments to detect it, special insurance schemes would be necessary—the list goes on and on. Anyone advocating the universal adoption of such a discovery would be denounced as a lunatic or locked up as a menace to society. The process itself would never make it through federal regulatory agencies. Yet of course this is exactly our relationship with fire. Its power was too great to refuse, and its nature too protean to control completely.

As domesticated fire became indispensable, moreover, its flame had to be kept inextinguishable. The communal fire joined the communal well as the earliest of public utilities. Much as the hearth became the symbol of the family, so the ever-burning vestal fire became the symbol of tribe, city, or state. Perpetual fire-keeping, or the tending of eternal flames, expressed not only the continuity of human society but its differentness. If fire was universal to humans, it was also exclusive to them. Almost certainly we will never allow any other creature to possess it.

Fire as Morality

With power came choice, and with choice anthropogenic fire entered a moral universe. Humans were genetically equipped to handle fire, but they did not come programmed knowing how to use it. Their capacity for colossal power lacked an equivalent capacity for control. Environmental conditions imposed some limitations on the land's ability to accept fire, and human societies established still other parameters by which their cultures could absorb fire. Still, the range of options remained huge; individual choices, neither obvious nor singular. Fire practices would reflect values, perceptions, beliefs, economics, institutions, politics, all the things that guide humanity through a contingent universe about which it has incomplete knowledge. The capture of fire became a paradigm for all of humanity's interaction with nature.

For fire there was no revealed wisdom, only an existential Earth that could accommodate many practices and a silent creator who issued no decalogue to guide proper use. After all, fire had not come to humans accompanied by stone-engraved commandments or an operating manual; it had been stolen. Humans were on their own. The fire they seized they had to maintain. They could nurture their special power into a vestal fire for the Earth or use their torches like the spray cans of environmental vandals. They did both.

Fire became a pyric projection of themselves. It could be used wisely or stupidly, and it illuminated and assayed other acts as sound or reckless, but, until very recently, it was not itself condemned. There were good fires and bad fires but no escape from the imperative to use fire. The belief that free-burning anthropogenic fire was *ipso facto* an expression of environmental abuse by humans, that the suppression of fire was humanity's primary duty, is a very recent invention, an outcome of new combustion sources and ideologies that consider humans inevitably "unnatural." Instead, most peoples have exploited fire to define their relationship to their environment. From medieval Icelanders to the twentieth-century Kwakiutl, people have carried fire around their lands to announce their claim to them. Like Australia's Gidjingali, aboriginal peoples explain fire use as a means of "cleaning up the country," of housekeeping, of exercising their ecological stewardship; as Robert Logan Jack expressed it, burning is "the alpha and omega of their simple notion of 'doing their duty by the land'" (Jack 1881, p. 3). Not to burn is as irresponsible as improper burning. Early fire codes in colonial America, for example, required routine burning as a social obligation. Good citizens used fire well; bad ones, poorly or not at all.

In this way, fire became not merely a projection of human will but a test of human character. Thus fire history, like fire itself, is a maddening amalgamation of human and ecological history; it belongs with the humanities as much as with the sciences. Track fire, and you track human history. However, it is the special promise of fire history that one can do more, that it is possible to use anthropogenic fire to extract information out of the historic record that might otherwise be inaccessible or overlooked, much as burning often flushes infertile biotas with nutrients and cooking renders palatable many otherwise inedible foodstuffs. Fire can remake raw materials into humanly usable history. It can drive out of archival scrub the vital character of humanity. Around it—around that informing fire—humans tell the stories that make up their history, that say who they are.

Universal Fire

In a universe informed by fire, fire becomes a universal tool. To the fire the planet contained, humans have added, subtracted, redistributed, and rearranged. Human societies have inserted fire into every conceivable place for every conceivable purpose, and they have done so for so long and so pervasively that it is impossible to disentangle fire from either human life or the biosphere. The alliance between hominids and fire is ancient, apparently dating from the time of *Homo erectus,* a part of our biological inheritance.

Fire and Society

Fire restructured the relationship between humans and their world. It furnished light and heat. It made possible a social life after dark; redefined social roles; warmed against the cold; demanded shelter and sustenance; served as communication media; supplemented ax, knife, and drill; and allowed cooking, which revolutionized diets and food gathering. Fire assisted almost all branches of technology. The tools for recreating fire more or less at will—drills, saws, and flints—obviously derived from or coevolved with the technologies for striking, scraping, and drilling in stone and bone. Cooking inspired new technologies such as ceramics and metallurgy. The preservation of fire was an intensely practical as well as symbolic act.

Of course, fire also entered into cultural life, reshaping the cognitive world of humans as fully as their physical landscape. Fire worship and divination constituted a primitive religion; trial by fire served as a primitive legal system, and fire-based philosophies and myths formed a primitive science and literature. As a source of heat and light, fire inevitably accompanied ceremonies, and in time became indelibly associated with them as a part of their symbolic milieu. Burned offerings carried sacrifices to the heavens. The trying fire segregated dross from essence.

No dimension of human existence was untouched by fire, directly or indirectly, and the more limited was a people's technology, the more pervasive and apparent was their fire dependence. Remove fire from a society, even today, and both its technology and its social order will lie in ruins. Strip fire away from language, and you reduce many of its vital metaphors to ash. It should come as no surprise that from the Aztecs to the Stoics, from the Christian Apocalypse to the Nordic *Ragnarok,* the myth of a world-beginning or world-consuming fire is nearly universal. Humans' dominion began with fire, and it may well end with it.

Fire and Land

Fire's influence on the environment extended beyond its valued service as an aide-de-camp to wandering hominids. It was also applied directly to the landscape, and it was this capacity that defined humanity's special ecological niche, that made fire something more than a surrogate for talons, fangs, fleetness, or massive muscles. Anthropogenic fire endowed whole ecosystems, not merely a species.

Of course, anthropogenic fire built upon fire adaptations already resident in the biota, humans' fire practices mimicked nature's own fire drives and slash-and-burn cycling by windstorm and lightning, and human-kindled fire rarely produced results through its own, isolated actions. Instead, anthropogenic fire reshaped the structure and composition of landscapes, recalibrated their dynamics, and reset their timings of growth and decay; fire accelerated, catalyzed, animated, and leveraged; it combined with other processes to multiply their compounded effects beyond the sum of their individual impacts, choreographing new rhythms and steps for the partners of this biotic ballet. Fire was a remarkably intricate and pervasive enabling device, without which other technologies or practices were often incompetent; for example, there could be no slashing without the capacity for burning, and no hunting without the means by which to maintain the habitat. Similarly, humans' ability to manipulate fuels redesigned the environment within which fires (either theirs or nature's) had to operate.

Thus fire interacted with the new flora and fauna that migrating humans introduced, with livestock in search of browse, with ax and sword as weapons of conquest, with plow and seed, and with humans' understanding of how they should behave. It passed into the natural environment through the cognitive and moral world of humans. What people knew about fire affected how they used it. Knowledge (or error) acquired in one place could be transferred to another. Fire ecology had to incorporate the pathways of human institutions and knowledge as fully as biogeochemical cycles of carbon and sulfur.

Aboriginal Fire

Anthropogenic fire made the world more habitable for those who held the torch. Aboriginal fires kept open corridors of travel; they assisted hunting, both by drives and by controlling the extent and timing of browse; they helped cultivate, after a fashion, many indigenous plants and simplified their harvest; not least, anthropogenic burning shielded human societies from wildfire by laying down controlled fields of fire around habitations in regions prone to natural ignitions. Fire was both cause and effect for the fact that humans preferentially lived where burning was possible and shunned unburned regions as uninhabitable. The nomadism of hunting and gathering societies was intimately interdependent with the cycle of growth and regrowth that was itself contingent on a cycle of burning.

Fire made the land inhabitable, as cooking helped make environment into food and the forge reworked rock into metals. When John Smith asked a Manahoac Indian what lay beyond the mountains, he was told "the Sunne"; his informant could say nothing more because "the woods were not burnt," and without fire, both travel and knowledge were impossible (Barbour, 1936, II, p. 176). John DeBrahm noted also the purgative effect of burning, the purifying passage of humanity's ring of fire: "The fire of the burning old Grass, Leaves, and Underwoods consumes a Number of Serpents, Lizards, Scorpions, Spiders and their Eggs, as also Bucks, Ticks, Petiles, and Muskotoes, and other Vermins, and Insects in General very offensive, and some very poisonous, whose Increase would, without this Expedient, cover the Land, and make America disinhabitable" (DeBrahm 1971, p. 80). To a remarkable extent, humans were able through fire to shape wholesale the environments in which they lived, to render that land accessible.

To this end, fire was both subtle and pervasive. It promoted favored grasses, forbs, tubers, and fruits such as wild rice, sunflowers, camas, bracken, cassava, and blueberries; it helped harden the sticks that dug them up; it stimulated the reeds that, woven into baskets, carried the harvest; it cooked the gatherings, leached them of toxins, or boiled them down into oils or sap; it yielded the light by which the crop was discussed and celebrated. Fire helped humans gather chestnuts in the Appalachians, acorns in California, and mesquite beans in the Southwest. It drove off insects—in certain seasons, humans lived in smoke, as reluctant to leave its sheltering cloud as to walk away from a campfire into a moonless night.

Perhaps the most spectacular practices involved hunting. It may be said that any creature that *could* be hunted by fire *was* hunted by fire. Torches assisted evening hunts and made fishing at night notoriously productive. Smoke flushed bears from dens, sables from hollowed firs, possums from termite-cored eucalypts. Fire drives were practiced for springboks in southern Africa, elephants in Sudan, turtles in Venezuela, rheas in Patagonia, kangaroos and maalas in Australia, boars in Transbaikalia. In North America, fire hunting targeted bison, deer, and antelope; aboriginal Alaskans used it against moose and muskrat, Yuman Indians for wood rats, Californians for rabbits, Great Basin tribes for grasshoppers, and Texans for lizards.

Apart from the hunt itself, fire controlled browse, recalibrating the calendar of renewal and the rich flush of nutrients that the springtime brought. Applied correctly, fire could inaugurate that first growth or stimulate a second flush in the autumn. Regardless, the pattern of burning dictated the pattern of feeding, that is, the annual migration of grazers and browsers and of course the hunters that followed them. Fire hunting and fire herding involved enticing as well as driving. Select sites could even be baited with the smoke that meant relief from flies or with the fresh grasses recharged by burning. Apaches set such traps for mule deer; Samme, for reindeer; contemporary African poachers, for rhinos.

This fire-mediated relationship was profoundly reciprocal. If humans used flame to promote their food stocks, wildlife often depended on anthropogenic burning to fertilize and ready the landscape and stimulate the fodder that they, the indigenous fauna, also required in order to thrive. The land reflected this symbiosis, reshaped into a fire-sculpted hunting grounds both large and small. Where fire could burn freely—where winds, terrain, and biota favored routine, expansive fires—great steppes, *campos*, *llanos*, prairies, savannas, *cerrados*, and grassy veld could result.

Agricultural Fire

The Neolithic revolution adumbrated these practices with others. Of course, fire alone could not create agriculture, but it is no accident that agriculture originated in fire-susceptible environments and reworked fire-adapted grasses into cereals at its biotic forges. Outside of flood-recharged riverbanks, early agriculture was impossible without burning. At a minimum, fire was mandatory for the expansion of farming and herding beyond their special environments of origin. Agriculturists needed fire to convert and catalyze, to impose an alien flora and fauna, to forge a stubborn biome into new shapes. The metamorphosis was sometimes easy, for small adjustments could reconfigure fire practices suitable for hunting into those to serve herding or allow the shift from the harvest of a fire-cultivated native flora to the harvest of a fire-catalyzed alien one. More often, the transformation was messy and complex.

The really revolutionary changes occurred not from fire *per se* but from fire in conjunction with other practices. The violence of farming was expressed with torch and ax, the fire and sword of biological imperialism. Slash-and-burn agriculture was carried into the most remote landscapes by wandering farmers: Maoris in New Zealand; Melanesians in Fiji; American indigenes where maize or cassava was cultivated; Bantus across tropical and subtropical Africa; [jhum]-practicing tribes throughout the Indian subcontinent and southeast Asia; and, of course, Europeans, who practiced it across their varied frontiers, from the Finns in Karelia to the Norse in Iceland and Greenland, from Russians in the *taiga* to overseas colonists in the Americas, Africa, and Australia. In mixed economies fields once cleared for crops might, through subsequent burning, be kept in grass for pasture.

Meanwhile, flocks of domesticated livestock advanced like shock troops. They reclaimed range previously fired for the hunting of wildlife; they frequently forced herders to keep open fields, first cleared for farming, as pasture, again through routine burning; they forced the creation of new grazing lands or degraded the old ones. In humid environments, from Australia's northern tropics to Siberia's *taiga*, the introduction of domesticated animals typically inspired more fire, which was used as a flaming axe to slash back the encroaching woods. In more arid grasslands, from Morocco to the American West, the introduction of livestock commonly brought a reduction in burning as animals cropped fuels that would otherwise feed flames.

Eventually fire was itself domesticated no less than land, flora, and fauna. Wildland fire became agricultural fire. Field burning obeyed a new calendar, operated at reduced intensities, and altered the frequency of broadcast fire. For the most part, humans dictated these parameters. Where fire had once reflected the subtlety and complexity of the natural world, it increasingly assumed the regimen and personality of human society, responsive to the dynamics of the human mind or what a flawed human will could impose of itself on nature.

What Survey-General T. L. Mitchell said of 1840s Australia can stand as an epigraph for all the environments humans have fashioned through flame:

> Fire, grass, and kangaroos, and human inhabitants, seem all dependent on each other for existence in Australia; for any one of these being wanting, the others could no longer continue. Fire is necessary to burn the grass, and form those open forests, in which we find the large forest-kangaroo; the native applies that fire to the grass at certain seasons, in order that a young green crop may subsequently spring up, and so attract and enable him to kill or take the

kangaroo with nets. In summer, the burning of long grass also discloses vermin, birds' nests, etc., on which the females and children, who chiefly burn
the grass [,] feed. But for this simple process, the Australia woods had probably
contained as thick a jungle as those of New Zealand or America. (Mitchell
1969, p. 306)

When those fires were removed, the jungle reclaimed what the new inhabitants did not remake. Reclaiming the land meant redefining its fire
regime.

Imperial Fire

Within the last half millennium, two events have rewritten the history
and geography of fire. The expansion of Europe, begun with the Renaissance voyages of discovery, set in motion a colossal mixing of the world's
flora, fauna, and peoples. Old orders disintegrated, and new ones continue to emerge that have yielded a global economy, a global ecology,
and a global scholarship. Even while these events were proceeding, imperial Europe became industrial Europe, and the pressures for global
change accelerated. Anthropogenic fire was, as always, catalyst, cause,
and consequence of these processes.

Imperial Fire

Access to a global market exposed large hinterlands to revolutionary reforms. Inevitably, economic exchange brought ecological change. By
both accident and deliberation, the world's flora and fauna became hope-
lessly intermingled and selectively exterminated, usually with the help of
or at least in the presence of fire. Livestock from the Old World went to
the New; cultigens from the New World went to the Old; plants and
animals from both invaded Australia; Africans were forcibly shipped to
the Americas, British convicts to Australia, and Russian exiles to Siberia,
and an unstable alloy of hope and despair inspired a century of emigra-
tion that reduced Europe's population in roughly the same proportion
as the Black Death; weeds, vermin, diseases, and insects, no less than
cereals, cattle, citrus, and medicines, crossed over long-separated biotas
with results that were sometimes productive, usually unexpected, and occasionally catastrophic.

The new geography of global fire followed the evolving geography
of European expansion. In some places, fires increased to epidemic proportions as immigrant fires mingled with indigenous ones, all now cut
loose from traditional moorings. In others, the fires flared, like a Bunsen

burner speeding a critical reaction, and then ceased, the experiment completed. In still others, they vanished into field and ceremony, little more than vestigial symbols. The permutations in how fire, flora, and fauna could interact were infinite, and no single formula can encompass all the outcomes. The general effect, however, has been to reduce the amount of open burning. The Earth has far less free-burning fire now than when Columbus sailed.

This observation is counterintuitive, or, more accurately, it runs against the grain of Western colonial mythology, populated by noble savages and virgin forests, which requires that a discovered world of innocence be ravished by a decadent and cynical Europe. Thus an influential figure has explained the fire policy of America's National Park Service as a program to restore the state of nature that existed prior to the advent of "technological man," an ambiguous state that includes the native peoples who burned sequoia groves annually but not the German immigrants who swatted those fires out with pine boughs.

Certainly, European contact led to environmental change. In locales where land clearing became extensive, horrific fires often resulted. The transitional period—when fire practices and fire regimes mixed—was metastable and prone to violent reactions. However, this scenario fails to recognize the geographic extent and longevity of anthropogenic fire. It sees the violent flash of European gunpowder but not the already smoldering land behind it. In fact, when the smoke cleared—as it soon did—there was typically less fire after settlement or colonial rule than before. Sediment cores from off-shore Central America, for example, demonstrate that burning collapsed after the Spanish conquest and has not yet returned to pre-Columbian levels. With local exceptions, this scenario could probably be generalized throughout the Americas.

As often as not, the expulsion of fire was itself frequently a cause of environmental degradation and social disintegration. Fire's abolition subverted the habitat of fire-reliant humans as it did that of such fire-adapted organisms as longleaf pine, eucalypts, fynbos, kangaroos, and rhinos. The more universal fire was a tool—that is, the fewer the alternative technologies that were available to a people—the more devastating was its removal. With the environment changed, at least partly through a new regimen of fire, recovery was impossible. Bison could not return to tallgrass prairies that had reverted to woods. Springbok fled lands overrun with scrub. The change in fire regimen meant a change in habitat.

The reasons for Europe's pyrophobia are several. For one, temperate

Europe—a climatic oddity—did not experience a routine fire season. Temperature, not precipitation, defined its seasonality. Fire existed in Europe because people put it there. It thrived because agriculture required it. It appeared to exist as a useful tool, not as an ecological necessity. To intellectuals and officials fire stigmatized land practices as primitive; they knew wildfire as an index of social disorder; they experienced it in cities, not the countryside. They saw the burning of fallow as simple waste, as indulgent, slothful farming. If they could extinguish fire from the garden, they would. In time Europe came itself to resemble a kind of fire, burned out in the center, flaming only along its perimeter. But that smoldering core held the great powers of Enlightenment Europe, its new wave of colonizers, its scientific centers, its industrial inventors. Its ideal became the eccentric norm for the planet.

For another, Enlightenment Europe had created a species of engineering, forestry, among whose charges was the management of fire. Silviculture was a graft onto the great rootstock of European agriculture, and it correspondingly condemned burning of every sort. According to European agronomists, fire not only killed trees, it destroyed the humus upon which all life depended. Without that spongy soil, erosion was inevitable, and environmental decay irreversible. Far better to exploit biological agents such as sheep and cattle, fertilize with compost and dung rather than with ash, and sponsor labor-intensive surrogates such as weeding—anything but fire.

Besides, fire threatened fixed property, or social relationships of rigidly ordered societies. Broadcast fire encouraged nomadism, the seasonal cycling of pastoralists, the long-fallow hegiras of swidden farmers, a mobility of population that made political control and taxation difficult. The ideal of the garden that prevailed in central Europe demanded that every person, like every plant, have his or her assigned place. Finnish and Russian swiddeners fled king and tsar by burning into the boreal forest. Greek, Corsican, Sardinian, and other Mediterranean pastoralists ranged as freely as their flocks, defiant of political authority and confident in the power of fire to intimidate settled communities and harass agents of the state. Control over fire meant control over how people lived.

Forestry absorbed these doctrines, and in its shock encounter with other, oft-fired biotas, it hammered these precepts into a catechism of fire exclusion. Some wildfires would inevitably occur from the striking of native flint and European steel and from lightning, arson, and accident, as, for example, railroad-powered logging sent its iron tentacles everywhere. In addition, some broadcast burning might be unavoidable in the early years of institution building, an expedient compromise until surer control was possible. However, the ultimate ambition was a forest without fire, an orchard of saw wood and pulp. As soon as it was politically and technically feasible, foresters instigated fire control measures. As often as not, fire suppression was one of the most powerful means of controlling indigenes.

This enterprise, like other imperial exercises, proceeded most smoothly where the indigenous populations disappeared before the swarm of invading colonizers and their biological allies and camp followers. Where native peoples and biotas persisted, so did native fire. Traditional burning persisted in defiance of European desires; often, in such circumstances, colonists adopted native fire practices or hybridized with them in mutual defiance of the edicts of colonial administrators and the theories of European intellectuals. Fire policy became an expression of colonial rule, and firefighters a species of frontiersman, especially where land was reserved from folk access. In response, natives burned illicitly, at once a protest and an attempt to restore traditional lands to traditional purposes. Regardless of whether they retained political access to reserved lands like forests, unless they had fire they lacked biological access to those potential resources. The character of the land changed, often irrevocably.

All this did not pass unnoticed. Typically a debate ensued, often formal and even published, that pitted European standards against local practices, fire control against fire use. The most dramatic confrontations flared in British and French colonies, where foresters nurtured in Franco-German traditions were most aggressive at imposing policies that aspired to fire exclusion. In North America, fire control triumphed; in India, Australia, and South Africa, awkward compromises resulted that first denounced controlled burning, only to recant and ultimately absorb them into official doctrine. In time, controlled burning returned to North America as well, although in a language ("prescribed burning") that denied the legitimacy of its earlier incarnation ("light burning").

Industrial Fire

Not least of all, industrial Europe began to sublimate the power of fire into machines. Controlled combustion began to replace controlled burning, and the fossil fallow of coal and oil, the living fallow of traditional agriculture. Technology invented new devices to illuminate rooms, warm houses, bake bread, harden ceramics, shape metals, and the myriad other tasks fire had once performed. New sources of energy led to new sources of mechanical power and transportation, the pathways by which nutrients would be cycled in this revolutionary model of nature's economy. Guano could be shipped from Pacific islands to fertilize European fields; clover could replace indigenous grasses unpalatable to Eurasian livestock; chemical herbicides and tractors could substitute for free-burning fallow to control weeds.

The ecology of fire was no longer confined to burned sites, but cycled and ramified throughout human institutions as well. Through foresters in the service of the British Empire, the impact of teak burning in Burma could be transmitted to the fynbos of Cape Colony. Through technology transfer—the European education of American students—the fire history of the Schwarzwald could influence fire policy in the chaparral of California. Through French colonialism and scientific publications, fires in the Riviera could be felt in Madagascar, Chad, and New South Wales. Postfire succession in Nevada and Natal could be influenced by exotic grasses from Central Asia and acacias from Australia. Fire's evolutionary future had to incorporate grazers from Europe and fertilizers from petrochemicals.

Industrialization profoundly reworked the geography of fire through its engineering of fire itself. Increasingly combustion depended on fossil fuels, long abstracted from the rhythms of free-burning fire, its emissions outside the mechanics of ecological scavenging. Always in the past, fire had depended on life, on the interaction with the oxygen and fuels that life generated, on the ecological fugue between fire and fuel forged over eons of evolutionary trials. While this symbiosis had made fire possible—even necessary—it had also restricted its dominion. In contrast, industrial combustion burned without regard to the living environment. It literally stood outside the ancient ecology of fire—and outside the traditional social mores that had guided fire use.

Industrialization's impact extended still further; its reach exceeded its grasp. It redefined what in nature was a resource and what was not, what land uses were appropriate and what were not, what regions were accessible and what weren't. It compelled a full-blown redefinition of nature, rank with new values and new perceptions, many derived from the rowdy efflorescence of a global scholarship, modern science, that accompanied and sought to explain European expansion. It set into motion a counter-reclamation that redefined the basis of European land use away from traditional agriculture. The revolution demanded not only new combustion technologies for furnace and forge but new fire practices for field and forest. Europe's intellectuals had achieved their agronomic goal. They had apparently transformed fire from an ecological process into a human tool.

Like planets orbiting binary suns, ecosystems now had to revolve around the gravitational field of industrial combustion as well as biomass burning. The interplay between the two became increasingly dense, inextricable, unpredictable, even contradictory. Industrialization could strip forests through logging and, equally, restore woodlands by abolishing the need for fuelwood. Industrial societies could subject the most remote site to exploitation, yet simultaneously propose special categories of wildland that were, by law, to remain untrammeled by human consumption. For these new landscapes new fire practices had to be invented, sanctioned by neither nature nor the precedent of human history. The upshot has been the most fundamental restructuring of anthropogenic fire and global fire regimes since the Neolithic.

Vestal Fire

One outcome of this extraordinary expansion was that Europe established itself as a standard and censor of planetary fire. The fire practices of its industrial and agricultural heartland—Germany, northern France, the Low Countries, England—were accepted either as the norm or as the goal to which developing nations should aspire. Not incidentally, these nations were themselves among the important colonial powers, and they, unlike Spain and Portugal, distrusted free-burning fire and promulgated modern science. To a remarkable extent, fire practices and norms have flowed one way, from Europe outward. That trend continues, qualified by the emergence of North America as a fire power, although until recent decades North Americans also accepted the anomalous European fire scene as legitimate. It is perhaps no accident that current alarms over fire as an instrument of global change—with the implied condemnation of biomass burning—have emanated from Europe.

Global Fire

The contemporary geography of Earth's fire is far from stable. Perhaps it never has been. Even where the rhythms of returning fires show some regularity, two idiographic tendencies have worked to upset any putative cycles. One is climate, and the other humanity. It is among the alarming trends of contemporary times that, through fire, climate and humanity have begun to interact in new ways.

Under the impress of industrial combustion, the ancient dialectic between hearth and holocaust has skewed. In the past decade, large wildfires have ravaged landscapes as diverse as Borneo and Canada, Australia and Manchuria. Fire has been implicated in world-ending scenarios, the fast burn of nuclear winter, and the slow burn of a greenhouse summer, the epoch-ending extinctions of the Cretaceous/Tertiary boundary fires. The public imagery of fire has become both vivid and confused, with conflagrations applauded at Yellowstone National Park and denounced in Amazonia. Nearly everywhere, free-burning fire is identified with global environmental havoc, a pilot flame to apocalypse. Again, it will be both cause and consequence; uncontrolled combustion may not only provoke global warming but result from it, in the form of wildfire, as boreal and temperate forests become colossal tinderboxes.

The transition from flame to furnace has demanded new indices by which to measure the new world order of fire. In particular, air has replaced humus as the ultimate yardstick of fire effects. The role soil served in indicating environmental health for agricultural societies the atmosphere has assumed for industrial states. Combustion will be distinguished as good or bad according to whether it aggravates or ameliorates the world's airsheds and, through them, planetary weather and climate. Fires for foraging and hunting must compare not only with fires for farming and herding but with those that power turbines and automobiles; the by-products of field and forest burning, with those from the burning of coal and oil. Even so, the Earth remains a fire planet, and humans remain fire creatures. Neither can forsake fire and be what they are.

Contemporary Pyrogeography

It is clear that the dominion of fire is changing with more than usual speed, and that it is pursuing pathways not sanctioned by evolutionary trial. This new pyrogeography takes many forms and is rewriting the registry of fire excesses and deficits. Where change is sudden, where new land use or misplaced fire practices introduce ignitions, upset fuels, or scramble the wild with the rural or urban, wildfires rage. While the particulars vary, the outcome has been a global surge in uncontrolled burning—in Siberia; Indonesia; Amazonia; the Mediterranean littoral; and Oakland, California. Elsewhere, anthropogenic fire, once rare, has more or less permanently established itself like a naturalized weed. Still other landscapes—grasslands, brushlands, and forests—suffer from a deficiency of burning, a fire famine. After all, biodiversity can be lost as surely through fire exclusion as through fire excess. In the United States the best minds and most aggressive programs over the past two decades have sought to *restore* fire to wildlands.

But industrialization has gone further. Much as early hominids sought to replace the flame with the torch, and early agriculturalists sought to substitute domesticated burning for wild fire, so modernized societies have striven to replace wildland combustion with industrial combustion. The furnace supersedes the hearth; the power of fire engines, the power of torches. The critical landscapes are mechanical environments, literally within machines, and those portions of the atmosphere and biosphere that directly exchange gases with them. Combustion is no longer necessarily even associated with flame.

For fire history, modernization is associated with the liberation of fossil hydrocarbons. Other aspects of industrialization, whatever their social dimension, were secondary, because they still relied for fuel and fallow on existing biomass and were thus enmeshed in the ecological cycles of growth and decay. Their output was limited by their input. European agriculture, for example, was forever trying to close the circle between fertilizer and harvest (Mediterranean agriculture seemingly tried to square it). However much damage might be done, there was a sense in which it was self-limiting.

The exhumation of fossil hydrocarbons out of the geologic past transcended that closing cycle. Combustion (and the pyrotechnologies it supported) was no longer limited by biomass; landscape features that had previously existed to grow fuel—the woods and fallow—could now be put to other purposes. Here was biochemical bullion that could accelerate nature's economy. That it has also induced biotic inflation is also true. Its excess emissions could no longer be absorbed. Agricultural fallow was lost, and its fabled biodiversity with it, except where industrial created an alternative in nature reserves.

This substitution of fossil fuels and fossil fallow for biomass, however, has been incomplete and unsynchronized, and will likely remain so. Industrial combustion burns with savage indifference to fire ecology, without regard to time of day, season, climate, or biota. Because it depends on fossil fuels, there is no ecological feedback between fuel and fire, no biological linkage between combustion, nutrient cycles, or pathways of succession. The "fires" burn outside the parameters of natural ecosystems, often beyond the capabilities of organic scavenging and biochemical recapture. Instead of liberating nutrients, their by-products may overload or poison the environment. For industrial combustion, unlike wildland varieties, it is not sufficient that ecosystems be reshaped; they must be wholly reconstituted, complete with new pathways of energy; new cycles for biogeochemical compounds; new creatures; and, of course, new fire practices.

Those practices are far from established. Worse, they are challenging more traditional expressions—reforming the domain of anthropogenic fire, subjecting the Earth, particularly its atmosphere, to a combustion load in excess of what it can likely absorb—all this while operating outside the folkways that have traditionally guided anthropogenic fire. The new practices are rewriting the history of fire on Earth and redefining the special relationship between humans and fire. Humans are less keepers of the flame than custodians of the combustion chamber. This change, if it continues, has enormous ramifications for the natural world. The transition from torch to furnace has unsettled the fire geography of the planet, and this, in turn, has upset the moral geography of anthropogenic fire.

Over the millennia, anthropogenic fire mediated between human society and its natural environment. If it broke one regime, it fused another. Through a kind of ecological pyrolysis, it dissolved certain relationships among species, yet through an equally compelling process it had welded new ones. If fire segregated humanity from the natural world, it also bonded humans to the living world through the fuels they shared and the fire-powered dynamics it made possible. What it granted in power, anthropogenic fire demanded in responsibility. The new world order is breaking this legacy in fundamental ways. Industrial fire threatens to strangle agricultural fire, much as agricultural fire practices once weeded out aboriginal fire.

What may be lost is anthropogenic fire altogether. Contemporary primitivism typically denies anthropogenic fire a role in the management of parks, wilderness, and natural reserves. As a projection of human agency, such fires, so it is argued, represent an unwarranted and unbalancing intervention into the natural order that these sites seek to preserve. Equally, the magnitude of industrial combustion is forcing biomass burning to compete with fossil-fuel burning for limited airsheds. Taken together, these processes suggest that fire is unnecessary and undesirable. But while one pyrotechnology can substitute for another, it is more difficult to substitute one pyro-ecology for another. The ecology of fire within the built environment is not identical with that of the natural environment. Replacing a wood-burning stove with an electronic oven works within the home. Using bulldozers and herbicides instead of open flame does not work so cleanly in woods or prairies. Without an appropriate fire regime, the biodiversity of many sites may fail. Without controlled burning, many biotas will not store carbon but only stoke wildfires. In one form or another, combustion will persist, and humans will remain its keepers.

Trial by Fire

The history of anthropogenic fire poses two fundamental questions for any assessment of global climate change: What is the quantitative history of fire on the planet? And what is the natural fire load? It is not obvious that there is more fire on Earth now than in the past. In fact, a good case can be made that there is not enough fire, or rather that there is a maldistribution of fire—too much of the wrong kind of combustion in the wrong kind of places, too little of the right kind of burning in the right places or at the right times. In particular, the competition between furnace and flame has thrown into confusion the combustion calculus of the Earth. Put differently, it is difficult to establish a baseline for burning.

How much combustion can the Earth absorb? What is the relationship between fire and combustion? How much industrial combustion can be added and how much wildland burning withdrawn without ecological damage? What, in brief, is an appropriate standard for anthropogenic fire, and can guidelines be found in Earth's fire history, or must humanity confront an existential Earth, silent as a sphinx? The longevity and pervasiveness of anthropogenic fire not only complicate the quest for a natural norm but may condemn it as metaphysically meaningless.

Nevertheless, norms of some kind are needed. Which fires are good and which are bad? What is the right proportion of fires? How much fire should be tolerated, or even promoted? Increasingly intellectuals are looking to the atmosphere to supply these standards, and to the threat of anthropogenic climate change as a means to furnish the moral and political arguments for reform. Air has replaced humus as the European measure of environmental health. It is not just that fire may be changing world climate, but that the climate of world opinion is compelling a change in our ancient relationship to fire.

In this emerging ecological credo, burning is bad because it releases greenhouse and other noxious gases either through outright combustion or through the destruction of vegetation, particularly forests, which are viewed as carbon vaults. Fire enthusiasts might argue that fire control, like gun control, misplaces the blame, that fires don't kill forests, people kill forests. But fire's medium is also its message. The means do make possible the ends. Extinguish anthropogenic fire, and the human impact on the Earth would be rapidly snuffed out.

Measured by critical emissions, biomass burning constitutes a significant fraction of global combustion. Some 8,680 million tons of dry matter burn yearly, which in turn release 3,500 million tons of carbon yearly, about 40 percent of the world's annual production of carbon dioxide. In the past, the domain of anthropogenic fire was almost certainly much larger. How was it that the Earth accommodated so much fire? How did it breathe when its Amazonian lungs ten thousand years ago were *cerrado*, a Gondwanic savanna congested with shrubs and grasses? How did it stockpile carbon when ice sheets blotted out the boreal forests? How can the atmosphere—itself given to extraordinary variations—become a standard of reference? Just which of the many climates experienced even during the past millennium is the norm? Surely excessive emissions are a kind of environmental debt, but it is not obvious how exactly overconsumption works in this model of nature's economy. It could be argued, plausibly, that anthropogenic fire has retarded—or may be necessary to retard—the advent of a new ice age, which is surely closer to the climatic norm in which *Homo sapiens* has had to live. The protest against global warming is a cry against changing the status quo.

Regardless, Europe's air pollution has become so serious that it can project that local deterioration across the globe, just as in centuries past it projected the reckless destruction of its soils. There are good causes for alarm over atmospheric abuse, but it would be tragic if global powers once again misapplied their ignorance and mistreated fire. European fire is an anomaly, not a norm. However humans try to manipulate the Earth's climate, fire management, not fire extinguishment, will be vital.

The attempted suppression of fire by Europe's colonial powers was an ecological disaster, though one often camouflaged because it accompanied other, equally damaging and more visible practices. The extinction of indigenous burning was often a critical prelude or catalyst to other extinctions. Today European-dominated policy stands to repeat that error if, once again, it categorically seeks to expunge fire from the landscape. Carbon cannot be sequestered like bullion; biologic preserves are not a kind of Fort Knox for carbon. Living systems store that carbon, and those terrestrial biotas demand a fire tithe. That tithe can be given voluntarily, or it will be extracted by force. There can be net changes in the Earth's fire load, but to speak of eliminating burning is not only quixotic but dangerous. Eliminate fire and you can build up, for a while, carbon stocks, but at probable damage to the ecosystem upon whose health the future regulation of carbon in the biosphere depends. Stockpile biomass carbon and you also stockpile fuel, the combustion equivalent of burying toxic waste. Cease controlled burning and, paradoxically, you may stoke ever larger conflagrations. Refuse to tend the domestic fire and the feral fire will return.

For millennia—in much of the Earth, for hundreds of millennia—humans have directed those obligatory fires. They have paid the fire tithe. Now the situation has blurred; the liberation of fossil fuels has changed the identity of *Homo sapiens* as a fire creature; new fire technologies have broken down the old order without yet installing a replacement. The issues involve more than just atmospheric pollutants and human-governed combustion; fire binds humans to the biosphere. There is a legitimate place for anthropogenic fire, and it must be defended against both those who seek to replace the flame with the furnace and those who wish to abolish the torch for the lightning bolt.

It remains a valuable intermediary with regard to humanity's relationship to both the atmosphere and the biosphere. The escalation of greenhouse gases is the outcome of industrial combustion burning fossil fuels. It burns in a different ecological context, for which biomass combustion can only partially compensate; air, not fuels, links them. The suppression of biomass burning cannot, except briefly, substitute for fossil-fuel burning. On the contrary, fire yields elemental carbon as one of its products, which is then stored in soils and constitutes almost the only long-term carbon sequestration available, the only means by which to return to mineral state something akin to the exhumation of geologic carbon. It could be argued, in fact, that more rather than less fire is necessary, at least in grasslands where the fuel cycle is annual rather than decadal.

The living world also argues for the preservation of fire. After all, fire is not exclusive to humans, and in fact it binds humans to a complicated biosphere for which hominids, as unique fire-keepers, enjoy a special niche and obligation. Much of the natural world that preservationists seek to protect coevolved with anthropogenic fire. To remove that fire regime can be catastrophic: replacing anthropogenic fire with lightning fire alone does not restore a former, prelapsarian era but more likely fashions an ecosystem that has never before existed.

These are special environments of course, a new order of sacred groves. Their practical dimensions may be small. But their symbolism is important, a kind of vestal fire for virgin lands. They testify that humans were intended to be the keepers of the flame, not its extinguishers; they insist on an active role for humans. To preserve and use fire is the oldest of humanity's ecological duties, its most distinctive trait as a biological organism, the first of its quests for power and knowledge, the genesis of its environmental ethics.

Fire is imprecise, and anthropogenic fire imperfect. In Plato's Allegory of the Cave he describes the human condition by imagining that humanity lives in a dark cave, divorced from the true light of the sun; we shackled humans can see and know only what we see in the fire or in the corrupt shadows cast by the fire. Yet without fire humans would be helpless and hopeless. It is through fire that we still know the world, and it is through fire that we can, and ought to, relate to that Great Other, Nature.

REFERENCES

Barbour, Philip L., ed. 1986. *The Complete Works of Captain John Smith (1580–1631)*. 3 vols. Chapel Hill: University of North Carolina Press.

Bartlett, Harley H. 1955–61. *Fire in Relation to Primitive Agriculture and Grazing in the Tropics: Annotated Bibliography*. 3 vols. Ann Arbor: University of Michigan Botanical Gardens.

———. 1956. Fire, Primitive Agriculture, and Grazing in the Tropics. In *Man's Role in Changing the Face of the Earth*, edited by William L. Thomas Jr., Vol. 2. Chicago: University of Chicago Press.

Batchelder, Robert, and Howard Hirt. 1966. *Fire in Tropical Forests and Grasslands*. *Technical Report 67–41-ES*. Natick, Massachusetts. U.S. Army Natick Laboratories.

Clouser, Roger A. 1978. *Man's Intervention in the Post-Wisconsin Vegetational Succession of the Great Plains*. Occasional Paper No. 4, Department of Geography-Meteorology, University of Kansas, Lawrence.

Crosby, Alfred. 1987. *Ecological Imperialism. The Biological Expansion of Europe from 900 A.D.* Cambridge: Cambridge University Press.

DeBrahm, John Gerar William. 1971. *Report of the General Survey in the Southern District of North America*, ed. Louis De Vorsey Jr. Columbia: University of South Carolina Press.

diCastri, Francesco, and Harold Mooney, eds. 1973. *Mediterranean-type Ecosystems: Origin and Structure*. Heidelberg: Springer-Verlag.

diCastri, Francesco, David W. Goodall, Raymond Specht, eds. 1981. *Mediterranean-type Shrublands*. Heidelberg: Springer-Verlag.

Hall, Martin. 1984. Man's Historical and Traditional Use of Fire in Southern Africa. In *Ecological Effects of Fire in South African Ecosystems*, edited by P. de V. Booysen and N. M. Tainton. Heidelberg: Springer-Verlag.

Hallam, Sylvia. 1979. *Fire and Hearth: A Study of Aboriginal Usage and European Usurpation in South-Western Australia*. Canberra: Australian Institute of Aboriginal-Studies.

————. 1985. The History of Aboriginal Firing. In *Fire Ecology and Management in Western Australian Ecosystems*, edited by Julian Ford. Perth: Western Australian Institute of Technology.

Harris, David R., ed. 1980. *Human Ecology in Savanna Environments*. New York: Academic Press.

Jack, R. L. 1881. *Report on Explorations on Cape York Peninsula, 1879–80*. Brisbane: Queensland Parliamentary Paper.

Jones, Rhys. 1969. Fire-Stick Farmers. *Australian Natural History* 16:224–228.

————. 1975. The Neolithic, Palaeolithic and the Hunting Gardeners: Man and Land in the Antipodes. In *Quaternary Studies*, edited by R. P. Suggate and M. M. Cresswell. Wellington: Royal Society of New Zealand.

Komarek, E. V. 1967. Fire—and the Ecology of Man. In *Proceedings, Tall Timbers Fire Ecology Conference*. Vol. 6. Tallahassee, FL: Tall Timbers Research Station.

Lewis, Henry T. 1985. Burning the "Top End": Kangaroos and Cattle. In *Fire Ecology and Management in Western Australian Ecosystems*, edited by Julian Ford. Perth: Western Australian Institute of Technology.

Lewis, Henry T., and Theresa Ferguson. 1988. Yards, Corridors, and Mosaics: How to Burn a Boreal Forest. *Human Ecology* 16(1):57–77.

Mitchell, T. L. 1969. *Journal of an Expedition into the Interior of Tropical Australia*. New York: Greenwood Press. Reprint of 1848 edition.

Nichols, Phillis H. 1981. Fire and the Australian Aborigine—an Enigma. In *Fire and the Australian Biota*, edited by A. M. Gill, R. H. Groves, I. R. Noble. Canberra: Australian Academy of Science.

Pyne, Stephen. 1982, 1988. *Fire in America: A Cultural History of Wildland and Rural Fire*. Princeton, NJ: Princeton University Press.

————. 1990. Fire Conservancy: The Origins of Wildland Fire Protection in British India, America, and Australia. In *Fire in the Tropical Biota*, edited by J. G. Goldammer. Berlin: Springer-Verlag.

————. 1990. Firestick History. *Journal of American History* 76:1132–1141.

————. 1991. *Burning Bush*. New York: Henry Holt.

————. 1995. *World Fire. The Culture of Fire on Earth*. New York: Henry Holt.

Sauer, Carl. 1963. Fire and Early Man. In *Land and Life*, edited by John Leighly. Berkeley: University of California Press.

Schule, W. 1990. Landscape and Climate in Prehistory: Interactions of Wildlife, Man, and Fire. In *Fire in the Tropical Biota*, edited by J. G. Goldammer. Berlin: Springer-Verlag.

Stewart, Omer C. 1956. Fire as the First Great Force Employed by Man. In *Man's Role in Changing the Face of the Earth*, edited by William L. Thomas Jr. Vol. 1. Chicago: University of Chicago Press.

————. 1963. Barriers to Understanding the Influence of Use of Fire by Aborigines on Vegetation. In *Proceedings, Tall Timbers Fire Ecology Conference*. Vol. 2. Tallahassee, FL: Tall Timbers Research Station.

Tall Timbers Research Station. 1962–76. *Proceedings, Tall Timbers Fire Ecology Conference*. 15 vols. Tallahassee, FL: Tall Timbers Research Station.

JOHN F. RICHARDS

Only a World Perspective Is Significant

Settlement Frontiers and Property
Rights in Early Modern World History

IN THE FIVE centuries since Columbus's first voyage to the New World, the global landscape has undergone an unprecedented transformation as a direct result of human action (anthropogenic change). With accelerating velocity, human agency has drastically altered the world's lands, flora, and fauna in order to improve their usefulness or productivity. Carried by ever more mobile humans, plant and animal invaders have breached the barriers that formerly separated the world's major regions in a new mixing of species. Humans have eliminated herds of wild herbivores and their predators on grasslands and savannas to make room for domesticated animals. Forest clearing for settlement and timber has swept remorselessly around the world at an accelerating pace. Plow cultivation and annual cropping have displaced woodlands, jungles, grasslands, and even arid deserts. New single-crop permanent horticulture has indelibly changed the landscape of the tropical world. Wetland drainage and perennial irrigation have altered the configuration and flows of fresh water resources.[1]

Now, in the late twentieth century, human control is imposed over every hectare of land on the earth's surface. Animal and plant species live or die as the result of continuing management decisions made by various human groups. Wilderness and wild animals exist only in our collective imagination. Today, however, the dimensions of this vast transformation are becoming clearer and the consequences more problematic. Countless management decisions and hours of human and animal labor have been invested in a massive upgrading of the productive capacity of the world's lands.[2] From another viewpoint, that of global ecology, the world's lands have been badly degraded. In 1992, the total biomass and total numbers of species of both plants and animals were substantially reduced from those present in 1492. Annual net primary production of world biomass is less now than it was in the fifteenth century. Increased human domination over the landscape has tended to simplify, and possibly render more sensitive and less resilient, the resulting ecosystems around the world.

This lengthy process of land transformation has had significant consequences for biogeochemical processes and cycles. The cumulative effect of human action on the land contributes to the possibility of atmospheric change in the near term. Early modern and modern forest clearing, soil disturbance, and biomass burning have released billions of metric tons of carbon into the atmosphere on a scale far exceeding anything postulated from human activity in the past.[3] In the past two centuries, fossil-fuel burning has become the principal source of atmospheric carbon, but forest clearing and other land use changes continue to contribute substantially to the rising concentration of carbon dioxide in the atmosphere. Expanded areas devoted to wet-rice cultivation or more intensive raising of bovines are two of the principal sources of methane, another important atmospheric trace gas.[4]

Reliable, long-term, global land use data are of the utmost importance for scientific climate modeling of the past, current, and future effects of increased trace gases in the atmosphere. Unlike data for fossil-fuel consumption, which are relatively easy to compile and check, changes in world forests, agriculture, and other forms of land use are beset with uncertainties and complexities. How many billion metric tons of carbon have been released into the atmosphere by land use changes (the terrestrial source) as opposed to fossil-fuel burning since 1500? What have been the relative rates of land clearing in the tropics as opposed to the temperate zones? What has been the regional trajectory of global land transformation? Questions such as these must be better resolved before more reliable global climate change modeling is possible.

It is important to recognize that atmospheric change is a global problem. That is, a ton of carbon or methane released anywhere on earth circulates throughout the earth's atmosphere for a prolonged residence time (possibly as much as two hundred years for carbon). Therefore, although regional and local studies are important, in the end only a world perspective is significant. Among ecologists exists a growing recognition that integrative, world-scale approaches are a necessity.[5]

At this juncture, ongoing scientific research into global climate change intersects with the concerns of social scientists, historians, and humanists in two ways. First, scientists must have reliable quantitative data on past world land use to base their estimates. These data are best retrieved and refined in cooperation with historians and other social scientists.[6] Second, understanding the forces in human society and history that have driven these transformations is primarily a task for students of human behavior, not physical and biological scientists. And most assuredly, successful national, regional, or international policies for reducing the release of atmospheric trace gases or for adapting to future climate change rest on our understanding of human motivation and behavior in the past and present.

Since George Perkins Marsh's remarkable synthetic work *Man and Nature; Or, The Earth as Modified by Human Action* first appeared in 1864, geographers have recognized and chronicled the global alteration in the natural world caused by human intervention. In the mid-1950s Carl Sauer, Lewis Mumford, and Marston Bates organized a weeklong symposium at Princeton University to update Marsh's assessments. Published in 1956 under the title *Man's Role in Changing the Face of the Earth*, the collected papers in this volume made a significant impact on post–World War II geography.[7] Subsequently, as recognition of environmental problems grew in the 1960s, a spate of conferences and publications addressed these concerns. Recently, the 1987 conference "The Earth as Transformed by Human Action" brought geographers, historians, social scientists, ecologists and other physical scientists together in an attempt to emulate and move beyond the Princeton symposium. The 1990 published volume, bearing the conference title, compiles systematic assessments of various changes in the world environment since 1700.[8] Unfortunately, most historians have been largely immune to the questions or the data emerging from this approach to world geography.

Today, however, questions of atmospheric change nudge us inexorably toward a unified history of humankind, a world history. Concern over global climate change meshes with rising scholarly interest in a shared world history, rather than a sharply bounded national or regional history. Understanding the rise of an industrial, capitalist, world economy has been the formative question for historians and social scientists. Karl Polanyi's *The Great Transformation*, first published in 1944, was one of the most influential works to analyze the movement toward a world market economy. In the past two decades, Wallerstein's "world-systems" approach and the economic/structural analyses of Braudel, Jones, and Wolf, among others, have stimulated a vigorous debate about the primary forces that have shaped modern human history.[9] Moving beyond analysis of European expansion and the effects of colonialism, the salient issues for this group are socioeconomic, centered on the growth and integration of a world capitalist economy in the early modern and modern periods.

As yet the newly emerging field of environmental history has had little influence on this discourse. Albert Crosby's well-known study, *Ecological Imperialism*, drew attention to the ecological consequences of European world domination. Volumes of essays originating in conferences on the fate of world forests in the nineteenth and twentieth centuries have underscored common global trends. But in general the entire question of land and resource use in the context of an evolving world economy has been either ignored or minimized. Access to land and resources is not viewed as an essential aspect of the growth of world capitalism. Commercialization of land, the creation of land markets, has been much studied and analyzed since the classical economists Ricardo, Malthus, and Smith. But larger worldwide processes and patterns of land use transformation have not been treated in a powerful, synthetic fashion by historians or even historical geographers. Especially noteworthy is the failure to integrate copious national and regional historical literatures devoted to settlement frontiers and changing property rights in land.

Around the world, episodes of frontier expansion brought large tracts of additional land into more intensive cultivation and settlement. During these transient periods, new peripheral areas were created or extended. "Wild" forests, woodlands, and wetlands became aesthetically pleasing and fertile landscapes of settlement and cultivation. In the first stage of frontier expansion, pioneers often borrowed heavily from indigenous techniques for less intensive forms of horticulture, like shifting cultivation. As the frontier began to close, accepted, familiar models for agriculture, livestock raising, and horticulture were imposed on the reshaped landscape—sometimes at very great cost if the model was entirely exotic. Commercialized extractive frontiers for fur, timber, mining, and other commodities preceded and accompanied the march of pioneer settlers on these new lands. Populations of indigenous "tribal" peoples, whether shifting cultivators, hunter-gatherers, or nomadic pastoralists, were either assimilated, displaced, or exterminated.

Every expansive settler frontier depended on several essential conditions: (1) the power of a centralizing state to seize and declare lands available for occupation; (2) capital resources sufficient to provide the material means for pioneer settlers to proceed; (3) one or more successful cultural models of intensive agriculture; (4) the attractive power of state-nurtured and -encouraged, usually monetized, commodity markets to capture frontier production; and (5) the availability of surplus populations eager for new lands. In region after region these conditions were met and frontiers of settlement opened and closed with regularity.

Frontier expansion relied heavily on state power to survey, register, and guarantee property rights in lands newly claimed and occupied by settlers. Frontiers were zones where rival sets of property rights collided. After a more fluid, transitional period, state-legitimized tenures swept aside the complex layered rights of access to land characteristic of tribal societies throughout the world. At the frontier new forms of tenure conceded the nearly absolute right to dispose of all resources—animals, soil, water, forests, minerals—on land. Generally to secure and maintain more intensive use and greater productivity, the state has carefully demarcated land in a new expression of territoriality, that is, "a powerful geographic strategy to control people and things by controlling area."[10] Officials of the state have named, surveyed, mapped, bounded, and recorded location and ownership of parcels of land. Some form of taxation on newly claimed land, unless specifically exempted, has usually accompanied demarcation and registration.

Land rights obtained by pioneers acquired new monetary value in a market for land. "[C]ommercialization of the soil" has been an essential step in meeting the ever-growing needs of an urbanizing and industrializing world economy.[11] As frontiers moved on, tenures ratified by the state rose in value. Responding to buoyant long-term regional and international commodity demands, land managers, often those who have replaced the original pioneer settlers, have engaged in unimpeded exploitation of their holdings.

World land clearing and settlement can be divided conveniently into three broad periods. First, the early modern world from 1500 to the end of the eighteenth century saw numerous, often dramatic frontier episodes. Second, explosive worldwide land transformations urged on by colonialism and the new world economy characterized the period from 1800 to the Great Depression of the 1930s. Finally, diminishing available land and dwindling frontiers in all save the humid tropics mark the post–Depression/World War II era. For much of the twentieth century, for most of the world, expanding production has necessitated investment in production on existing land, rather than frontier expansion and extension of cultivation.

Land Clearing and Settlement in the Early Modern World: 1500–1800

During the early modern centuries, prior to 1800, the mainland areas of Eurasia and Africa displayed a common frontier pattern. Ming China, the Ottoman Empire, Mughal India, Russia, and Sub-Saharan Africa experienced noticeable rises in population. The material needs of growing populations were largely met by frontier expansions that brought more land into cultivation. Rising urban demand for food, water, energy, and other commodities drove land clearing and resource extraction in the urban hinterlands in each region. Intensive garden cultivation adjacent to cities pushed more extensive grain farming and livestock raising outward as the population grew. Beyond these belts new pressures were put on forests and other ecosystems for timber, game, minerals, fish and other extractive resources. Economic growth generally kept pace with growing populations in these regions.

The agricultural/settlement model or blueprint followed by pioneer settlers tended to be a variant of the peasant smallholder employing mixed livestock and grain farming. To take a well-known example, to the south and east of Muscovy, Slavic peasant pioneers and their landed backers relied on the manorial, three-field grain farming model of northern Europe. Between 1550 and 1700, Russian pioneer settlers moved slowly into the forest and then forest-steppe and eventually encroached on the steppe grasslands south of Muscovy. East of the Dnieper River and south of the Oka River lay a belt of broad-leaved oak forests interspersed with pine woodlands on sandy river terraces. Below this lay the mixed oak woodlands and grasslands of the forest-steppe shading off into the semi-arid grasslands of the true steppe leading to the Caspian and Black Seas. Soils throughout this region were predominantly deep, black, humus-rich chernozem with high natural fertility. By the first half of the sixteenth century, these zones were sparsely inhabited by two groups of horse nomads. Central Asian Muslim Tartars, part of the Nogay Horde in the Crimea, pressed hard against Muscovy's defenses in the first half of the sixteenth century. To the northeast on the lower Don and Donets Rivers, the Cossacks had adopted the equestrian culture of the Tartars. Slav fugitives from Muscovy and escaped slaves had created autonomous hordes, or "Hosts." Both groups grazed their herds north into the forest-steppes and raided deep into the forests. The Tartars even burst through and sacked Muscovy in 1571.

The rapidly centralizing Muscovy state offered land to state servitors who could organize a military defense against Tartar raids. Between the spring thaw and first snowfalls of the winter, militia patrols and guard posts manned defensive lines that were several hundred kilometers long and fortified with felled trees and earthen ramparts. By 1650, aggressive state action had pushed the border to the Belgorod line, with its new frontier forts. Behind this barrier Ukrainian Cossacks and other frontier settlers were encouraged to infill and clear land for cultivation. Similar expansion took place against the Tartars in the Volga region as Slavic landowners settled serfs on their newly claimed manorial lands. As the Russian state maintained its cohesion and increased its power vis à vis the Tartar nomads, the frontiers to the south continued to advance to the steppe borders in the late seventeenth century. Pastoral herdsmen gave way to Slavic manorial lords and peasants. Around the larger settlements, steady forest clearing put up to 80 percent of the land under pasture or plow cultivation.[12]

In Asia, a pervasive model was that of the peasant smallholder growing paddy rice as his principal crop. This was balanced by dry-rice cultivation, often a variant of shifting cultivation, in steeper, hilly zones. Peter Perdue's detailed study of frontier expansion in early modern Hunan Province under both the Ming and Ch'ing dynasties reveals the extent to which the state and increasing numbers of migrants brought new lands under more intensive cultivation prior to the nineteenth century.[13] Adoption of maize and other new world food crops helped to make dryland farming more productive. Han Chinese settlers steadily pressed against the indigenous Miao tribespeople, who, despite occasional violent resistance, lost large chunks of their ancestral lands. On the South Asian subcontinent, in Mughal India during the seventeenth and early eighteenth centuries, rice-growing Bengali peasant-pioneers pushed steadily into the upper delta of the Ganges and Brahmaputra Rivers. Led by Muslim saints or Sufis, newly converted Bengali peasants cleared the dense wetland forests of eastern Bengal and built water control systems to permit wet-rice cultivation. Scattered indigenous tribes of non-Muslim fishermen relinquished their property rights to these pioneers backed by the power of the Mughal state.[14]

Western Europe—especially the domestic territories of England, France, the Low Countries, Italy, and Spain—benefited from new settlement frontiers, but with a significant difference. To a much greater proportion than their counterparts in the Far East, South Asia, Africa, or even the Middle East, western European cities extracted resources by sea from dependent regions. In far-flung regions around the world, settler frontiers resulted from the stimulus of early modern European maritime trade, colonization, and state intervention. Pioneer settlers seized and held new lands along the eastern North American seaboard and even in such unlikely places as the Cape of Good Hope at the southern tip of Africa.

In some areas, rather than peasant smallholding, large estates employing coerced labor became the dominant cultural model for pioneers. Imitating the methods used in estate-grown sugar in the Canary Islands, sixteenth-century Portuguese planters and Dutch financiers in northeastern Brazil introduced Mediterranean cane and deployed black slaves to produce sugar for European markets.[15] Plantation colonization began in the West Indies when parties of English colonists occupied the two small islands of St. Kitts and Barbados in the 1620s. By midcentury, sugar growing supplanted tobacco in Barbados to become the favored crop of the island's planters. By the 1660s virtually all the forest cover on the island had been cleared and the land apportioned into sugar estates. Barbados became an off-shore factory producing a single crop for voracious western European and North American consumers. Indentured white servants, black African slaves, draft animals, and foodstuffs were imported to meet the industrial needs of the new factory complex. Subsequently, in the seventeenth and eighteenth centuries, the sugar frontier extended the plantation system to every island in the West Indies with the exception of Puerto Rico. Sugar and its by-products, including rum and molasses, dramatically changed the landscapes and ecosystems of one after another of the West Indies islands.[16]

The flourishing sea trade that moved large quantities of grain from eastern Baltic ports to Amsterdam and other western European markets is well known. Increasing market demand drove land clearing and grain farming in Poland and the Baltic states. Less recognized is Amsterdam's impact on frontier expansion in the western Baltic in forested Scandinavia. Between 1500 and 1700, Finnish pioneers settled the unpromising boreal forest reaches of the Finnish peninsula, then ruled by the Swedish monarchy. Recently two American geographers have argued that Finnish pioneer forest culture provided the model for North American backwoods frontier techniques and culture. Jordan and Kaups have traced this influence to the Finnish immigrants who settled New Sweden along the west bank of the Delaware River in the mid-seventeenth century.[17] Early modern Finland neatly illustrates the structural features of frontier expansion and is well worth a more detailed review.

Although blessed with vast coniferous forests, Finland lacked a riverine system suited for flotation of logs to the coastal ports. Rather than timbering, Finnish peasants resorted to processing tar, pitch, and resin. By 1600 the fir and spruce forests of Finland and Sweden had virtually supplanted the fast-depleting woodlands of Poland as a source of the indispensable caulking and waterproofing agents needed by the fleets of early modern Europe. Steadily rising consumption of naval stores moved prices upward for these commodities.

Viborg, the port at the head of the Gulf of Finland, became the leading export center for tar and pitch in the seventeenth-century Baltic.[18] The quantities involved in the trade became more impressive with each passing decade. In 1600 shipments of tar, pitch, and potash passing through the Sound tolls reached five thousand lasts; by 1618, the total had tripled to fifteen thousand lasts. Between 1618 and 1632 a normal year saw about ten thousand lasts, or twenty thousand tons, of these commodities shipped from Baltic ports.[19] Toward the end of the century Finnish and, to a lesser extent, Swedish peasants relinquished 100,000 to 120,000 barrels of tar to the Swedish royal chartered company that in 1648 was given a monopoly of sales.

Tar, pitch, and resin consumed considerable quantities of wood each year. Peasant producers employed large brick or stone hearths in a crude distilling process to produce tar. As the fire burned coniferous fuel, tar ran into the container beneath the hearth. The charcoal remaining was used by smiths who preferred it to other types. To refine pitch, the tar maker simply boiled tar until its residue was the required consistency. Resin could be produced by boiling small, knotty slivers of coniferous wood in water. The resulting turpentine thickened and hardened on cooling.[20]

The standard formula assumed fifteen good-sized conifers to produce a barrel of tar. Exporting 100,000 to 120,000 tons of tar meant that 1.5 to 1.8 million trees fell to the peasant's axe each year in late-seventeenth-century Finland.[21] The overall effect of the tar export was to slowly, but steadily, reduce the extent and density of forest cover in Finland.

The boreal forest gave the peasantry access to an important auxiliary source of income, in cash. Peasant tar producers carried their filled barrels to the nearest collection point in the coastal towns—Viborg, if feasible. There German merchants paid their supplier's living expenses, bought his product, and advanced him money against the next season. The merchants provided salt—a necessity for man and beast—at the fixed government monopoly price. The German traders in turn acted as agents for the Swedish tar monopoly and shipped their accumulated barrels of tar and pitch to Stockholm. Thus enriched, the peasants could proceed with cultivation of barley and rye in the course of the next agricultural season.[22] The latter were the means by which they paid their state land taxes.

It is clear that both primary processor—the peasant—and the monopolizing corporation—the Swedish crown—directly benefited from the export trade in naval stores. From another perspective, both market and state directly encouraged the frontier expansion of the Finnish pioneer settler.

Commercial demands for forest products added impetus to large-scale pioneer settler movements already pushing into the Finnish boreal forest. During the fourteenth and fifteenth centuries Finnish-speaking peasants populated Tvastland, Satakunta, and Nyland Provinces the warmer, more fertile provinces of the southwest coast and interior.[23] By 1500, some scattered penetration and settlement of Savo and Karjala (Karelia) had occurred, especially around the shores of Lake Ladoga, but little more. These lands, populated only by dispersed groups of Lapp hunters and pastoralists, invited the frontiersman. In the course of the sixteenth century, Finnish pioneers moved energetically into the forests of the eastern provinces and into the northernmost regions of the four provinces stretching from the Gulf of Bothnia to Lake Ladoga. They readily brushed aside the property claims of the peaceable Lapp inhabitants of these forests. The brisk pace of settlement and dispersal continued steadily into the 1600s and beyond. Between 1500 and 1700, the sparse population tripled from 100,000 to between 300,000 and 400,000 persons.[24]

Finnish pioneering success rested in large part on state encouragement and the opportunities afforded by cutting and refining Baltic fir and spruce for tar and pitch. Other state policies encouraged expansion of agriculture. Between 1560 and 1660, the Swedish monarchy alienated huge sections of its central domain lands to the aristocracy, largely in payment of old debts or unpaid salaries. Peasants who had been royal tenants now became virtually serfs of the nobility. The latter busied themselves in creating large landed estates. In Finland the process of alienation of lands under sedentary cultivation—especially in the southwest—was more sweeping than in Sweden proper. Thus tenant or even freeholding Finnish peasants found that their dues or land taxes were now assigned to be collected by aristocrats rather than by the crown. In effect, they had lost their status as independent peasant proprietors.[25]

For a century (1560 to 1660) Swedish landed aristocrats consolidated their hold on manors in southern Finland. The new landlords were themselves a force for expansion and intensification of cultivation within and adjacent to their estates. Simultaneously, however, the prospect of pioneering forest settlement attracted those peasants who had been caught in this system. Finnish peasants could move boldly into the forest to claim new lands. If they persisted through the first generation of extensive shifting cultivation they could establish more permanent farmsteads. State recognition of bounded farm holdings, often named for the original patriarch who settled them, added another incentive for the pioneer. The state did recognize individual ownership and inheritance rights of the forest colonist.[26] When in the latter decades of the seventeenth century the Swedish monarchy resumed many alienated estates and reaffirmed the tenure rights of peasant proprietors, the Finnish peasantry found its status generally improved.[27]

Forest settlement in Finland relied more directly on startlingly productive, specialized forms of shifting cultivation adapted to the high-latitude boreal forest. Unfortunately, the fifty- to hundredfold yields of seed to grain common in the first two years of "burn-beating" could not be sustained, and the pioneer farmer had to shift to another tract. Frontier settlement in Finland, like settler movements in other regions, underwent two distinct phases: a fast-moving diaspora of small numbers of settlers practicing extensive shifting cultivation, followed by a slower process of stabilization as a growing population of pioneer farmers moved gradually toward more intensive forms of cultivation. In the latter period, profits from felling and refining conifers for the royal Swedish tar monopoly offered tangible incentives to those peasants anxious to build more stable, populous settlements in the forests.

The thinly scattered first-generation pioneers used a specialized type of shifting agriculture referred to as "burn-beating" in this part of the world. *Huuhta*, one form of burn-beating, evolved to grow rye in virgin coniferous forests—predominantly spruce (*Picea excelsa*) mixed with pine, birch, and alder. In the interior, spruce stands were ubiquitous on the wetter, more fertile moraine soils in Finland and also in the peatlands so characteristic of the Finnish landscape. Cultivators using this method first ringed the bark on the larger spruce trees within the area selected; then in the spring of the year they felled the smaller to midsized conifers with care so that the trunks lay facing in the same direction. They allowed the largest notched trees to stand. Thereafter the cultivators left

the plot undisturbed for two full years. In the third summer they returned and fired the entire tract. Immediately after burning, they sowed
rye seed broadside, without any tillage. The next summer they returned
to harvest a crop that could range from a 1:20 to a 1:100 yield of seed
to grain.[28]

This was a technique best suited to pioneers moving into unclaimed
and undisturbed spruce-dominated boreal forests. In addition to its occupation of the more fertile soils, spruce burned well and yielded a great
deal of ash. Spruce stands in general are unable to tap the full range of
nutrients in acidic soils. Spruce humus and litter are generally acidic and
contribute to the buildup of nitrogen-rich nutrient reserves in the forest
soil. Burning reduced soil acidity, mineralized organic nitrogen compounds, and increased the value of phosphoric nutrients. Thus, for one
crop, at least, the cultivator released the stored-up potential of the forest
soil. *Huuhta* depended also on a specific indigenous variety of rye known
as *korpiruis*. This variety grew up to thirty shoots on a single stem and
was resistant to brown rust. Such heavy tillering helps to explain the
extremely large yields.[29]

The *huuhta* cultivator had to have access to great expanses of unclaimed forest. He was engaged in a life of constant movement into the
woodlands. Since this method permitted only one crop in the same tract,
to obtain continuous annual rye crops the frontier farmer had to work
with four tracts in a four-year cycle of felling, burning, sowing, and harvesting. Each year he abandoned the burnt-over harvested woodland
and notched the largest trees before burning a new tract of coniferous
forest. This form of shifting cultivation permitted new settlers to obtain
nearly unbelievable fifty- to hundredfold yields of rye in the first years
of settlement. Thus assured of initial subsistence, the pioneers could
move with confidence into the northern forests.

In the second phase, Finnish pioneers continued to burn, but with
a different technique (*kaski*) aimed at growing barley in mixed forests
dominated by broad-leaved deciduous trees. Second-growth alder and
birch succeeded spruce stands within thirty to fifty years after the pioneer *huuhta* farmers had moved on. In the *kaski* burn-beating method,
second-generation farmers looked for stands of deciduous trees of
roughly uniform age from fifteen to thirty years. They felled the trees
in midsummer to permit faster drying. They lopped branches and shrubs
to leave an even layer of vegetation on the ground. The next summer,
the farmers fired their plots. To promote even burning they often rolled

flaming tree trunks over the ground. Sowing of rye or barley took place immediately after burning. The *kaski* cultivator obtained yields of 1:8 to 1:15 for two to as many as seven years before the soil depleted.[30]

During this phase the settlers cleared and plowed some plots for full-blown sedentary cultivation in a variant of the general European three-course system. These tended to be scattered arable strips with attached commonly-held woodlands and forest lands. *Kaski* cultivation seems to have flourished side by side with more permanent forms of cereal growing. Burn-beating had the added advantage of supplying better forest grasses for stock grazing in the summer and fodder for cutting and storage for winter. Climate and soils together encouraged relatively extensive farming practices in early modern Finland. Burn-beating in one form or another persisted. Despite this necessary compromise, the goal of Finnish farmers has been access to demarcated, stable fields of cleared and plowed land.[31] In general, the Finnish pioneer settler saw the forest as an enemy to be cut and burnt back.

The result of two centuries of pioneer assaults on the Finnish forest can be seen in the thousands of hectares put to two- and three-course cultivation and grazing. More sweeping was the far greater expanse of forest land subjected to felling, burning, and cropping first on *huuhta* and later on sequences of *kaski* cycles. The frontiersman's techniques proved themselves. A threefold greater population was supported by forest agriculture. Even the great famine of 1697, when one hundred thousand persons died of starvation, did not slow expansion for long.[32] Mixed deciduous cover succeeded conifers; tree cover of lesser biomass succeeded that of greater; nutrients stored in the soil released in bursts of cultivation exhausted the cropped soils. If, as sometimes occurred, the burn-beater saw his fire leap to other areas of the dry summer woodlands, a forest fire could result and thousands of hectares could be burnt. If, however, peasant pressures on the forests did not intensify, the system seemingly was sustainable over extended periods of time.

Scrutiny of other frontier episodes in other parts of the world would reveal structural features similar to the Finnish case as well as obvious cultural and ecological variation. Like the Finns, the Bengali, Han Chinese, and other pioneers the world over have responded to the promise of land and a new lifestyle with enthusiasm. Techniques evolved for frontier survival in one ecosystem could be transferred to another similar situation. In every instance, we find a highly-productive, essentially transient frontier culture at the periphery. Nevertheless, the driving forces of early modern and modern frontier expansion—state, market, and population increases—are always discernible.

Conclusion

Settlement frontiers in the nineteenth century far surpassed in number, scale, and impact those of the early modern world. In just over a century, landscapes around the world were changed beyond recognition. Steam-driven transportation by rail and ship, metropolitan finance, colonial state power, and migrant populations combined to create a global economy and world commodity markets. North American forest clearing and expansion of cultivation from the eastern seaboard to the Pacific coast shaped the destiny of the United States and Canada in these years.[33] At the antipodes, New Zealand and Australia were settled and fertile lands cleared and cultivated. In Latin America, the Argentine grasslands were opened to more intensive ranching and wheat farming. In Brazil, coffee planting ate into the mahogany rain forests. Throughout the Indian subcontinent under British colonial rule, land clearing and perennial canal irrigation proceeded along multiple settlement frontiers. In South and Southeast Asia, the richly fertile river deltas of Bengal, Burma, Thailand, and Vietnam were cleared and brought into wet-rice cultivation by pioneer settlers.[34] In Sub-Saharan Africa, well before colonial rule was imposed, pioneer settlers were successfully battling a hostile, disease-ridden environment.[35]

The world depression of the 1930s marked the closure of many frontiers around the world. The pace and scale of land transformation slowed. After World War II the locus of land transformation shifted to tropical and subtropical forests. Dense forests in Central America, Amazonia, Southeast Asia, and Central Africa were opened to frontier settlement. A recent shift in values signified by a growing world environmental ideology has altered our perception of settlement frontiers. No longer are they seen as clear signs of human progress. Instead, the destruction of tropical rain forests today is viewed as a tragedy.

Present-day environmentalism has moved beyond older conservation movements with their concern for the wise use of natural resources and preservationist movements that seek to set aside habitats as wilderness. Today's coalescing environmental movements, found in virtually every country, have a global vision that places humankind within and dependent on the natural environment, and not its master. The new environmentalists fear for human survival in light of the newest technical prowess of our species and the amazing scale of human interventions into natural processes. Much of this vision rests on the insights of ecological science that stress the interconnectedness of all living things and our inability to predict the consequences of seemingly innocuous actions. Therefore the drive toward more intensive land use, with reduction in species and biomass, found in frontiers of settlement is often deplored by environmentalists. They see this as a process of degrading complex ecosystems, not an upgrading of the land's productivity to meet human needs. The settler frontier assault on the forest is no longer an unquestioned symbol of civilized advance.[36]

Older values still persist, however. The frontier is a shared, recent memory that retains a profound influence on human values and perceptions. Before the frontiers began closing in the early years of the twentieth century, nearly all human societies had access to abundant, underused land. Nearly all societies vigorously claimed that asset by investing combined collective and individual energy in frontier expansion. Frontier life presents endless possibilities for land, wealth, and prosperity. New reserves of human energy and spirits are marshalled. Access to low-cost resources—wood, water, soil, wild game, and fish— reinforce buoyant optimism. Frontier societies take on a life and culture of their own that alter the myths of the larger, cosmopolitan society. When frontiers close, these trends are halted, even reversed. The modern state exerts its control over new territory. By the end of this millennium, we will have reached the end of our land frontiers and must adjust to a more constrained world.

NOTES

1. B. L. Turner, *The Earth as Transformed by Human Action: Global and Regional Changes in the Biosphere over the Past 300 Years.*(Cambridge; New York: Cambridge University Press with Clark University, 1990).

2. Piers Blaikie and Harold Brookfield, *Land Degradation and Society* (New York: Methuen, 1987).

3. R. A. Houghton and David Skole, "Carbon," in Turner *The Earth as Transformed by Human Action*, 396–97.

4. T. E. Graedel and P. J. Crutzen, "Atmospheric Trace Constituents," in *Ibid.*, 300–301.

5. H. A. Mooney, "Emergence of the Study of Global Ecology: Is Terrestrial Ecology an Impediment to Progress?" *Ecological Applications* 1 (1991):2–5.

6. For an example of this type of data recovery, see John F. Richards and Eliza-

beth P. Flint, "A Century of Land-Use Change in South and Southeast Asia," and Elizabeth P. Flint and John F. Richards, "Trends in Carbon Content of Vegetation in South and Southeast Asia Associated with Changes in Land Use," in *Effects of Land Use Change in Atmospheric Concentrations: Southeast Asia as a Case Study*, ed. Virginia H. Dale (New York: Springer-Verlag, 1994). Also see Elizabeth P. Flint and John F. Richards, "Historical Analysis of Changes in Land Use and Carbon Stock of Vegetation in South and Southeast Asia," *Canadian Journal of Forest Research* 21 (1991): 91–110.

7. W. L. Thomas Jr., ed., *Man's Role in Changing the Face of the Earth* (Chicago: University of Chicago Press, 1956).

8. B. L. Turner II. *The Earth as Transformed by Human Action*. See Table 1-1, "Global Studies of Human-Induced Environmental Transformation, 1957–1987" (4).

9. Immanuel Wallerstein, *The Modern World System* (New York: Academic Press, 1974) and subsequent volumes; Fernand Braudel, *Civilization and Capitalism 15th–18th Century*, 3 vols. Translated by Sian Reynolds (New York: Harper and Row, 1981); E. L. Jones, *The European Miracle*, (Cambridge: Cambridge University Press, 1981), and *Growth Recurring: Economic Change in World History* (Cambridge: Cambridge University Press, 1988); Eric Wolf, *Europe and the People without History* (Berkeley: University of California Press, 1982).

10. Robert Sack, *Human Territoriality: Its Theory and History* (Cambridge: Cambridge University Press, 1986), 5.

11. Karl Polanyi, *The Great Transformation* (1944; reprint, New York: Farrar & Reinhart, Inc.) 178–179.

12. D. J. B. Shaw, "Southern Frontiers of Muscovy, 1550–1700," and R. A. French, "Russians and the Forest," in *Studies in Russian Historical Geography*, ed. J. H. Bater and R. A. French (New York: Academic Press, 1983), vol. 1.

13. Peter C. Perdue, *Exhausting the Earth: State and Peasant in Hunan, 1500–1850* (Cambridge, Mass: Harvard University Press, 1987).

14. Richard Eaton, "Human Settlement and Colonization in the Sundarbans, 1200–1750," in *Development Pressures and Ecological Constraints: The Deltaic Forests of India and Bangladesh*, ed. Ronald J. Herring, *Agriculture and Human Values* 7 (Spring 1990): 6–16.

15. Wolf, *Europe and the People without History*, 149–151.

16. David Watts, *The West Indies: Patterns of Development, Culture and Environmental Change Since 1492* (Cambridge: Cambridge University Press, 1987).

17. Terry G. Jordan and Matti Kaups, *Backwoods Frontier: An Ethnic and Ecological Interpretation* (Baltimore: Johns Hopkins Press, 1989).

18. See Artur Attman, *Russian and Polish Markets in International Trade, 1500–1650* (Göteborg, Sweden: Institute of Economic History, Gothenburg University, 1973), 17–18. Attman quoting Sven Erik Astrom, *From Cloth to Iron*, I:43ff., argues for the displacement of pitch from Danzig and Konigsberg by about 1600 and the emergence of Swedish-Finnish dominance in this market. In the closing decades of the sixteenth century, Viborg began to export sizeable amounts of tar for the first time. Thus, in 1584 the port sent out 2,023 barrels of tar in a modest beginning to a lucrative export trade.

19. Pierre Jeannin, "Les comptes du Sund comme source pour la construction

d'indices généraux de l'activité économique en Europe (XVIe–XVIIIe siecle)." *Revue Historique* 88 (1964) 1:55–102, 2:307–340.

20. Charles Joseph Singer, and Trevor Illtyd Williams. *A History of Technology* (Oxford: Clarendon Press, 1954).

21. Pierre Jeannin, *L'Europe du Nord-ouest et du Nord aux XVIIe et XVIIIe Siecles,* (Paris: Presses Universitaires de France, 1969), 93. Jeannin cites E. F. Heckscher, *Sveriges Ekonomiska Historia Fran Gustav Vasa* (Stockholm, 1935–49), I:431–434.

22. Braudel, *Civilization and Capitalism,* 3:253–254, relying on an unpublished paper of Sven Erik Astrom.

23. F. Skrubbeltrang, "The History of the Finnish Peasant," *Scandinavian Economic History Review* 12 (1964): 166–167. This essay is a synopsis of the findings of the leading Finnish agrarian historian, Eino Jutikkala, from his *Bonden i Finland Genom Tiderna* (Helsingfors, 1963).

24. Colin McEvedy, and Richard Jones. *Atlas of World Population History* (London: A. Lane, 1978).

25. Eli F. Hecksher, *An Economic History of Sweden* (Cambridge, MA: Harvard University Press, 1954) 126–128.

26. Skrubbeltrang, "Finnish Peasant," in *Farming in Finland,* W. R. Mead, (London: University of London, Athlone Press, 1953) 167–170.

27. Hecksher, *Economic History of Sweden,* 127–128.

28. Soininen, "Technical Basis of Colonization," 153–154. A variant provided for two burnings: one in the summer of the second year and one in the summer of the third year. Apparently this was somewhat more productive.

29. *Ibid.,* 164.

30. *Ibid.,* 163–164.

31. For the layout of the Finnish village of Broby prior to the enclosure or Great Partition of 1755, see Mead, *Farming in Finland,* fig. 41. According to Mead, "Historically, the forest has been the enemy and the agricultural *legende* which fires the imagination of Finnish farmers has been based on cleared land" (*Ibid.,* 84).

32. See Eino Jutikkala, "The Great Finnish Famine in 1696–97," *Scandinavian Economic History Review* 3 (1955): 48–63. The cause seems to have been exceptionally cold and prolonged winters that prevented the crops from ripening properly for three years in a row.

33. Michael Williams, *Americans and Their Forests: A Historical Geography* (Cambridge: Cambridge University Press, 1989).

34. John F. Richards, "Agricultural Impacts in Tropical Wetlands: Rice Paddies for Mangroves in South and Southeast Asia," in *Wetlands: A Threatened Landscape,* ed. Michael Williams, (London: Basil Blackwell, 1990).

35. Helge Kjekshus, *Ecology Control and Economic Development in East African History* (Athens: Ohio University Press, 1995).

36. For a lucid statement of this change, see John McCormick, *Reclaiming Paradise* (Bloomington and Indianapolis: Indiana University Press, 1989).

II. SOCIAL STUDIES

INTRODUCTION

THE ESSAYS in this section remind us of a point taken for
granted in discussions of the environment: environmental
degradation is in large measure caused by, and afflicts, real
human beings, who live with other human beings in groups
and communities, and who share common cultures with their
fellows. As we noted in the Introduction to this volume, the
prevailing discourses of environmental studies are too often
"distancing discourses"—that is, ways of talking about environment that obscure living, breathing people behind the
technical vocabularies of the biophysical sciences or the
equally specialized languages of the social sciences. The majority of studies of environmental problems at our own institution, the Massachusetts Institute of Technology, naturally
enough stress the biophysical causes, manifestations, and
possible technological solutions to environmental degradation. Even studies of what are sometimes awkwardly called
"human factors" or "anthropogenic environmental problems" are often phrased in terms that obscure the people
who breathe exhaust fumes, the women who walk four hours
a day for firewood, the children who live downwind from
a polluting industrial site, the Brazilian people who drink
poisoned water from an upstream gold mine. But these essays remind us, for example, that the wives, mothers, and
daughters who constitute the backbone of the "toxics movement" described by Barbara Epstein feel menaced in their
own persons by the potentially lethal environments in which
they live. As Terence Turner makes clear with regard to the

Brazilian Kayapo, decisions to pollute, permissions to log and mine, are not the result of scientific laws, divine edict, impersonal forces, or technological necessity, but rather of individuals acting in their own self-interest or in what they regard as the best interests of their communities.

The historical studies in the preceding section examine some of the ways that the biophysical world changed and was understood and dealt with in the past. The studies that follow are more contemporary: they reveal how men, women, and institutions create and respond to environmental problems nowadays. Contemporary social scientists are interested in two kinds of relations between people and the natural environment. They study how the environment (and especially environmental damage) affects and is affected by individuals and their communities. They also examine how people protect themselves from environmental degradation and thereby try to improve the condition of their biophysical world. The essays in this section exemplify some of the ways humanistic social scientists look at the relations between environment and people, and especially how people respond to environmental damage.

The "humanistic social sciences" (or "social studies") constitute only one part of the larger enterprise of understanding human, social, and cultural behavior, an effort that involves psychologists, sociologists, anthropologists, political scientists, and economists of many stripes. In each of these fields there also is another tradition not represented by these essays—a scientific effort to quantify, build abstract models, formulate precise theories, and deduce hypotheses to be tested by exact empirical and statistical investigations. We have no doubt about the value of these techniques. They typify the work of those who emulate the scientific rather than the humanistic side of social studies. Here, however, we feature what might be called "humanistic studies" by social scientists. They focus on people in a broad "ethnographic" or "clinical" sense: that is, they are intensive studies of particular times, places, or developments, explored in depth by a single researcher using the insights of subjectivity, personal understanding, and ethical commitment as well as disciplinary training, self-aware reflection, and, at times scholarly distance.

All the essays in this section touch on what are called "new social movements." Traditionally, social movements were exemplified by communism, fascism, and liberalism, that is, by all-embracing worldviews organized into mass political parties with a hierarchical structure, designed to provide their militant adherents with a way of life and an agenda of social and political change. A "new" social movement, in contrast, tends to focus on a specific issue or problem; to recruit its membership selectively, usually from young, middle-class participants; and is said to be organized "democratically" rather than hierarchically. These movements tend to disappear when the urgency of their concerns has been dissipated, their vitality undermined by opposing groups, or their objectives achieved. Despite the efforts of a few deep ecologists to elevate environmentalism into the status of an all-inclusive philosophy, most current environmental movements focus on particular wrongs, and they frequently disappear along with the evils they combat.

The authors of these essays are committed participants in the causes they record, study, and analyze. Bina Agarwal, who has studied the position of women in the village economies of India, also is actively involved in efforts to improve their status. Oleg Yanitsky, a leading scholar of the sociology of Soviet and Russian environmental movements, manifestly is a supporter of some of these movements as well as a critic of those he deems wrongheaded. Terence Turner, one of the foremost students of the Brazilian Kayapo, also is an advocate who has helped them make their case to the world. Both Barbara Epstein and Richard White make clear their ethical and political positions in evaluating environmental movements and tendencies. All of these scholars are committed to some aspects of the environmental cause.

To critics of this approach, the term "committed scholarship" is an oxymoron, for to them "scholarship" is incompatible with advocacy or partisanship. This ideal of the completely dispassionate scholar, who analyzes the world without betraying any commitment, judgment, or "personal bias," is repudiated by the writers represented here.

But they do not believe that intellectual insight, critical perspective, and personal commitment necessarily are incompatible. On the contrary, each scholar combines personal beliefs with a critical spirit that includes her or his own intellectual commitment. Thus although White is sympathetic with many aims of contemporary environmentalism, he nonetheless deplores the romanticized, historically inaccurate image of the "environmental Indian"; he would have us pay more attention to the actual environment—the broken wine bottles; siting of toxic waste dumps; and, in general, the environmental degradation—of today's actual Native American communities. Epstein, writing of the "toxics movement" with which she clearly sympathizes, nonetheless criticizes it for its lack of a broad perspective, just as she criticizes its academic allies (like herself) for their "rather precious and self-referential version of radicalism." Turner, a supporter of the Kayapo's struggle to preserve some portion of their inherited cultural and ecological world, is nonetheless critical of the environmental policies of a whole generation of Kayapo leaders. In short, these writers show that they can be committed and yet self-aware and self-critical and that it is possible to start from one set of assumptions and learn from experience the need to modify or reject them.

These essays pose two key questions: How does environmental degradation affect people, communities, and values? How do people, groups, and cultures respond to environmental threat? The essays show that the answers vary from place to place, from time to time, and from one level of analysis to another. The cases of the creation of the "environmental Indian" and the Kayapo illustrate the complexity of human and cultural responses to environmental menace. As White argues, the popular contemporary image of the "environmental Indian" is not in any sense to be taken as a reliable portrait of the actual pre-Columbian native living harmoniously in a state of nature. Rather, it is a complex creation, largely constructed in the last two centuries, that melds elements of pre-Columbian Native American cultures with Victorian romanticism and, more recently, the intellectual and rhetorical preferences of conservationists and environmentalists. Turner's account of stages in the evolution of environmental conservation adopted by the Brazilian Kayapo similarly demonstrates that any simple generalizations about "indigenous cultures" vis-à-vis the environment must be locally and temporally qualified: the reckless exploitation of Kayapo resources encouraged by one generation of Kayapo leaders was more recently ended by a popular revolt that united the elders of the tribe with the younger generation. So, too, Agarwal's account of Indian women's collective action to protect local environmental resources, in particular the communal lands on which the poorest villagers depend, shows that it is a struggle whose outcome remains unpredictable. And Yanitsky's pessimistic conclusions about many contemporary Soviet/Russian environmental outlooks underscores the fact that environmental movements may appear successful at one moment and yet weaken or evaporate in response to new political conditions or to the disruptions resulting from their internal contradictions.

Each of these essays makes a critical point, obvious and yet often neglected in writings on the environment: namely, that every environmental outlook or action occurs in—and partly reflects—a particular human, social, biophysical, and cultural situation. It therefore may change as that situation changes. At the most obvious level, living downwind from a polluting smokestack in Russia, or on top of a leaking hazardous waste site in Los Angeles, is a context that provides powerful incentives for mobilization. But as Yanitsky argues, membership in the old Soviet nomenclatura disposes one group of Russians to a bureaucratic, antiactivist response to Russian environmental problems—a response consistent with their professional and ideological self-interests. Turner shows how one generation of Kayapo leaders saw the granting of rights for mining and lumbering as an opportunity to escape the confines of Kayapo society, to enrich themselves, and to enjoy some of the privileges of the non-indigenous Brazilian elite. Agarwal's analysis is particularly subtle. She shows that the social status of poor women in Indian villages assigns them such duties as cooking and therefore finding firewood for the stoves. Since gathering firewood is socially defined as an exclusively female responsibility, it is women who pay the price of the destruction of communal forests by having to travel farther and farther to find fuel. Thus, Agarwal shows, the concern of Indian women for the preservation of communal forests does not derive from women's so-called essential attachment to nature, but rather from the particular conditions of their lives—specifically, their social role as cooks and firewood gatherers. Environmental outlooks and actions often have less to do with abstract reflection on the problems of the world than with people's interests and positions in particular social and cultural contexts.

These essays cast doubt on the assumption that contemporary environmental problems are attributable to impersonal, uncontrollable forces such as "history," "capitalism," "modernity," or "technology." Rather, they make clear that environmental degradation is, above all, the consequence of the specific actions, decisions, will, neglect, or errors of specific people. Similarly, if solutions to today's most pressing environmental problems are to be found, it will be finally be by people who act to change their outlooks, behavior, and institutions.

KK

RICHARD WHITE

Environmentalism and Indian Peoples

I

THE GREAT appeal of Indian peoples to the modern environmental movement is their ecological otherness. For environmentalists, Indian peoples represent a dramatic alternative to modern relations with the natural world. Bill Devall and George Sessions include in their book *Deep Ecology* a chapter titled "Primal Peoples and Deep Ecology." They seek "inspiration from primal traditions."[1] When Barry Lopez in *Crossing Open Ground* wants to praise the environmental attitudes of people he meets, he writes that like "the Athabascans in this country before them, they have a manifestly spiritual relation with the landscape."[2]

Through Indians and their traditions, environmentalists acquire, as it were, a new set of eyes; nature as a category separate and distinct from the category human largely disappears.[3] Many Indian communities have conceived of themselves as one group of conscious persons living in the midst of other groups of persons, most of whom are not human. Conceiving of other species, and indeed of some natural objects such as rivers, waterfalls, or rocks, as persons with whom personal relations can be established eliminates purely instrumentalist attitudes toward other species that characterize much European thought and behavior.[4]

There is also a second sense in which Indians emerge as an environmental other. In *The Tewa World* and other writings, Alfonso Ortiz has stressed how the Tewa Pueblos conceive of their bounded worlds in terms of sacred landscapes. Various aspects of the homeland, the known places, become elements of a story told through the land. The landscape itself becomes a gateway to the supernatural and a holy story that calls up a sacred past. This sense of a sacred landscape is, of course, not foreign to other cultural traditions, but in North America it is largely peculiar to Indians.[5] Modern environmentalists have seized on this sacred landscape

---

This paper amounts to a long synopsis of a short book that I will never write. Given the format of the seminars, I have decided that it is better to give a broad account of why and how a very small segment of the American population occupies such a large place in American environmental consciousness than to present a more confined and focused study on particular aspects of Indian environmental practices or attitudes.

as a tool for protecting lands from development.[6] In their enthusiasm they sometimes verge on making the entire landscape holy.[7] They have made "Mother Earth" synonymous with Indian environmental attitudes.

The creation and representation of this Indian environmental otherness are historically quite complicated. Modern environmentalists have not simply studied Indian cultural traditions and then borrowed from them. Indeed, few contemporary environmentalists have acquired more than a superficial acquaintance with Indian practices or beliefs. Nor is the division between Indian and non-Indian as clear as the image of otherness would indicate. Although the environmental Indian exerts appeal because the image exists as an alternative to the European vision of nature, the two are historically linked. The environmental attitudes of modern Indian peoples, as much as Euro-American attitudes, have been formed in part within the cauldron of capitalist expansion and western European imperialism. Environmentalists at least implicitly recognize such connections, for they very rarely point to modern Indians as environmental models. Like Raymond Williams's pastoral English countryside, environmentalist Indians usually exist in the past.[8] Individual Indians, to be sure, act as modern environmental spokespersons, but models of environmental behavior taken from contemporary Indian communities are rare.

Apparent exceptions to such a broad statement—the Mistassini Cree in Adrian Tanner's *Bringing Home Animals* and the Koyukons in Richard Nelson's *Make Prayers to the Raven*—are, in fact, quite compromised. Tanner's work is still the best study of the cognitive world of Indian hunters, but the Mistassini, as Tanner shows, occupy a far more complicated world than that of environmentalist Indians. The cultural world of the Mistassini does not consist of an aboriginal survival, but has instead been formed over the centuries within the capitalist relations of the fur trade. And it does not form the whole world of the Mistassini. The modern Mistassini Cree inhabit a bifurcated social world of settlement/bush, and they are available as models of environmental Indians only in the bush.[9] Nelson's Koyukon have become probably the leading current ethnographic example of environmental Indians, but Nelson himself purposefully decided "against discussing the negative elements and the malefactors, which of course exist in every culture." Going still further—fruitlessly, as it turns out—Nelson urges readers "against using this account to represent the relationship between all native Americans and their natural environments."[10]

Thus even when modern examples of desired environmental otherness seem to occur, they actually gesture back toward purer and less compromised worlds. Historians have tried to reconstruct such worlds. These investigations have looked at both environmental attitudes and practices in this presumably purer past.[11] But, again as with Williams's pastoral, pristine aboriginal relations with nature and the world they produce always seems to exist just prior to the period under reconstruction. Historians find different cognitive worlds to be sure, but they also find declarations of declension. In 1807, for example, the Ottawa visionary The Trout conveyed a message from the Great Spirit to the Indians living around Michilimackinac.

> You complain that the animals of the Forest are few and scattered. How should it be otherwise? You destroy them yourselves for their Skins only and leave their bodies to rot or give the best pieces to the Whites. I am displeased when I see this, and taken them back to the Earth that they may not come to you again. You must kill no more animals than necessary to feed and cloathe you.[12]

In 1807 such denunciations were nearly a century-and-a-half old in the Great Lakes region. One of the justifications that the Iroquois offered the French in 1681 for their attacks on the Miamis was that they slaughtered all the beaver and left no breeding stock.[13]

Because the whole presumption of ecological otherness is that other cognitive worlds will not yield the environmental damage that Western civilization does, declarations of damage mean a world in decline. Ecological otherness presumes that overhunting to point of extinction or near-extinction is not reconcilable with a belief in the existence of animals as persons with whom personal relations exist. We must, therefore, be dealing with Indians after their ecological fall. Environmentalists, after all, do not just desire an example of a different cognitive world; they want an example of a different cognitive world that does not permit what Kirkpatrick Sale has called the European warring against species.[14] Indian overhunting must be a mark of declension or else the environmentalist Indians do not fulfill their purpose.

This whole environmentalist position of asserting that different beliefs will yield purer actions is questionable. Indeed, if we are to enter sympathetically into the cognitive worlds of historic Indian peoples, we must abandon contemporary beliefs about why species prosper and disappear and adopt temporarily other logics of cause and effect. It is possible that hunting can be regarded as a holy task; that hunters can supplicate and honor their prey; that respect can be accorded animal persons; and that, despite all this, species can still be hunted to extinction or near-extinction.

Studies of contemporary hunting peoples as well as historical accounts make it clear that many Indian cultures do mandate that humans must respect their prey and that they owe ritual obligations to animals who die that humans might live. If hunters fail in these ritual obligations, either the animals will not surrender themselves to the hunter in the future or else a keeper or master of the game or some other deity will withdraw the animals from the earth or visit humans with disease. Very often embedded in such codes are strictures against taking any more than a person needs.[15]

Such codes, however, do not necessarily yield the ends that modern environmentalists desire; within them overhunting—killing to depletion or extinction—can be rationalized. For example, precisely because animals are persons capable of mutual obligation with humans, actions of animals can invite retaliation by humans. I am not arguing for the fur trade as a war against beaver, as Calvin Martin has, but there are instances where Indian hunters have taken revenge on animals.[16] In 1677, Father Allouez learned that a young Potawatomi near Green Bay had been killed by a bear in a particularly gruesome manner. The bear had "torn off his scalp, disemboweled him, and dismembered his entire body." The bear had, in effect, treated the young man as a warrior treated the body of an enemy. Afterwards, "by way of avenging . . . this death," the relatives and friends of the dead man declared war on the bears. They killed more than five hundred of them, giving the Jesuits a share of the meat and skins, because, they said, "God delivered the bears into their hands as satisfaction for the death of the Young man who had been so cruelly treated by one of their nation."[17]

Even greater difficulties arise for environmentalists looking for alternative models to modern instrumentalism when this logic of personal relations, mutual obligation, and ritual respect merges with the economic requirements of commercial hunting. Indian hunters working in the fur trade could kill animals for their skins, leave the meat to rot, and still consider themselves blameless for the decline in game populations if they accorded the animals the proper ritual respect and if they "needed" the skins to provide for their own and their communities' wants. In the early nineteenth century, even as overhunting in the deerskin trade was rampant, Shawnee hunters continued to believe that each deer, when killed in proper ritual manner, had four lives. It would immediately reincarnate itself and return. They blamed the whites for the disappearance of deer not because whites overhunted, but because they did not observe the proper rituals.[18] When the Delawares, whom other Indians considered among the earliest and most ruthless market hunters, were unable to kill deer despite their use of powerful *besoins*, or medicines, they blamed their failure on the presence of white missionaries who defeated the hunting medicine, and not on their own depletion of game.[19]

Indian peoples on the Great Plains applied a similar logic to the disappearance of the buffalo. The account of the rapid decline in the buffalo herds has recently undergone an interesting and convincing reassessment by Dan Flores. He argues that the largest drop in bison population preceded the orgy of market hunting that followed the Civil War. In the early nineteenth century, bison found themselves under severe environmental pressure that had little to do with overhunting. The adoption of horses by the Indians and the movement of cattle into key areas caused competition in key riverine areas where bison had once been the only grazers. Even more significantly, cattle and horses introduced diseases such as brucellosis and tuberculosis which ravaged the herds. The bison were reeling and in decline when the railroads and the use of their hides for leather for machine belts and other uses caused white buffalo hunters to assault the herds. Indian peoples faced with the rapid decline of the animals sought an explanation.[20]

The Lakotas, correctly enough, placed the blame for the final near-extinction on whites, but they did not blame overhunting per se. Instead, they reasoned that because the whites did not pay proper respect to the animals, the bison had withdrawn underground and would not return until the whites had vanished. When the Ghost Dance prophets, therefore, promised the return of the buffalo and the disappearance of whites, their message fit well with existing Lakota understandings. Respect and proper ritual retreatment were the critical issues. Once ritual balance was restored, the bison would return.[21]

In all of these examples, a radically different way of understanding relations between humans and other species can coexist with actions we might regard as harmful to the well-being of these species. The critical duties are ritual, and such ritual duties may, or may not, contribute to the biological well-being of the animals.

Ritual duties are not always irrelevant. In the Oregon country, ritual obligations to salmon did mandate that fish traps be left open and fishing cease at certain periods, allowing spawning salmon to escape upstream.[22]

However, it would be a misguided and outdated functionalism to think that these elaborate rituals were an elaborate mask whose real purpose was only to allow an escapement of fish that could spawn.

The point here is not to blame, or exonerate, Indians from particular episodes of environmental "decline," but only to point out that the different cognitive worlds they created did not necessarily prevent what we now regard as environmental damage either prior to or after the spread of capitalist markets. In some instances, beliefs and rituals might indeed have preserved a necessary breeding population and the conditions necessary for their survival, but this was not a necessary outcome, and decline could be reconciled with existing beliefs and practices.

In any case, as Calvin Martin has argued, even if Indian practices did yield results that modern environmentalists find admirable, they did so in a way that is impossible for modern Western society to imitate.[23] The belief that we might literally borrow the worldview of selected Indian peoples seems but a deformed child of functionalism. Where functionalists treat culture as an act of self-deception designed to obtain a smoothly functioning order and desired material ends, imitation of Indians calls for conscious self-deception in order to obtain desired ends.

If simply borrowing from Indian cultures seems naive, more complicated cultural conversations involving Indians and whites can yield new environmental formulations that pose as beliefs borrowed from Indians. The current popularity of sacred landscapes and appeals to "Mother Earth" serve as an example. Twentieth-century environmentalists may, as some historians have argued, create what amount to sacred landscapes in national parks or wilderness areas, but here I am more interested in the pervasive use of a supposedly Indian-derived "Mother Earth" in environmental writings. Sam Gill, in arguing that the appearance of the figure of Mother Earth in so many Indian religions is a relatively recent phenomenon, may have overstated his case in some instances, but his overall position is probably sound. Gill dissects numerous versions of "Mother Earth," and while he finds metaphorical references, he rarely finds Earth conceived of as a concrete mother figure or goddess. "Mother Earth" rarely played a prominent part in particular Indian cosomologies until the twentieth century.[24]

If Gill is correct, "Mother Earth" seems largely a product of sustained contact; ethnologists and white travelers collapsed numerous earth-connected goddess figures into a single construct. Scholarly arguments for her ubiquity in North America reflect "the needs of the storytellers," in this case, scholars who assume such a figure must exist. "Mother Earth has seemed so primordial, so archetypical, so fundamental as to be herself a key element in the demonstration of a variety of theories concerning the nature and development of religion and culture." And Indian peoples, for their own reasons, have in the twentieth century largely accepted her existence and made allegiance to her part of modern pan-Indian identity. For Indian peoples, the very acceptance of "Mother Earth" by whites and their insistence on ascribing it to all Indians both gives Indian claims to particular authority when speaking of the land credence and creates a unifying element among diverse traditions. Mother Earth gives Indians "identity, purpose, responsibility, and even a sense of superiority over very powerful adversaries."[25]

In Gill's formulation, then, "Mother Earth" is a joint cultural creation, a useful project in cross-cultural misunderstanding. It is, he concludes, "an American story." "Mother Earth is the mother of us all."[26] Indians are significant figures in American environmentalism, but as participants in this tangled historical conversation and not as so many Don Juans, founts of ancient exotic wisdom to be incorporated into the twentieth century.

II

Searches for authenticity will, I think, get us nowhere. Indians have not influenced current environmental thought through what amounts to an accurate ethnographic appreciation by white environmentalists of a fully separate set of cultural traditions. Instead, environmentalist Indians, and non-Indian understanding of their knowledge, have been constructed within what amounts to a lengthy conversation that has included Indians. Indians have had an impact on environmentalist thinking, but environmentalist and Western thinking has at the same time helped construct modern Indians' visions of themselves.

Europeans and Anglo-Americans have, of course, long linked Indians with the natural, but this linkage was initially more ambiguous than it became later. Modern environmentalists desire to use Indians as a means of placing all humans back within nature. They seek to counter current conceptions of nature as something "out there," separate from humans. This distancing of the human and the natural was not a problem that confronted the first Europeans to see Indian peoples. In the *Diario* of Columbus, for example, the landscape and "nature" of the Indies were not wild but humanized and domesticated. Columbus never presented the Indies as a wilderness. In the landscape of Hispaniola, Columbus found the expected marks of humans. Hispaniola was "in appearance all very intensively cultivated."[27] Columbus was, in a sense, cognitively closer to the Indians than we are now. He no more than they perceived a nature separate and "out there."

Columbus expected nature to display the marks of humanity, but he did not expect humanity to display so openly the marks of nature. The Indians went naked. Nudity became, in a sense, the central fact about Indians for Columbus, and nudity became the human expression of nature in the diary. The word *natura*, as used by Columbus, referred not to nature but to human genitalia, which the Indians barely covered or openly displayed. Columbus struggled to place Indian society within parallel European forms, but he repeatedly was dismayed and frustrated by their nakedness, by their open display of nature/*natura*. He reacted to Indian nudity by partially naturalizing Indians in a way that set them off from Europeans. Prepared for a linkage between humans and nature displayed in the body of the land, the landscape, Columbus was disconcerted by a linkage displayed in the human body.[28]

Columbus's *Diario* begins the dialogue with Indians and nature, but in the nature Columbus "discovered," humans were part of an organic nature more because nature was humanized than humans were naturalized. Nature was not yet "decisively seen as separate from men." Columbus discerned an organic landscape that incorporated both humans and other forms of life. This connected Indians and the land in a manner little different from how Europeans connected peasants and the land. But because the Indians were naked, because they displayed their *natura*, they became associated with the natural in a way that distanced them from Europeans.[29]

Perhaps the most important decision Europeans made about American nature in the centuries following Columbus was that they were not part of it but Indians were. Europeans removed themselves from the natural world and claimed technical mastery over it. They defined Indians as "natural men" distinct, for better or worse, from the "civil" men of Europe. America came to stand for nature, and Indians stood for the natural. An American voyage became by definition a voyage into nature. The humanized landscapes of Columbus did not disappear, but they increasingly receded into the unseen interior of the continent. The actual Indians encountered and naturalized became inferior to those people who had risen above nature, those who were civil—that is, to Europeans.[30]

The equation of nudity, nature, and savagery marked Indians as both natural and inferior. So compelling was the linkage that the discoverers of New Mexico disregarded the obvious marks of Indian civility their own senses proclaimed. In 1540, for example, Coronado doubted whether the Zuñi could "have the judgment and intelligence needed to be able to build these houses in the way in which they are built, for most of them are entirely naked." Such assertions—"fantasmatic representation[s] of authoritative certainty in the face of spectacular ignorance"— became typical of much of the colonial discourse about Indians and nature. By definition, naturalized humans could not have produced a civil society and a human landscape.[31]

Once established, this linkage by Europeans of Indian peoples with the natural had a full and oft-cited history of its own.[32] Nature and human nature overlapped and existed in considerable analytical confusion. A benign nature yielded a benign human nature and noble savages. A "howling wilderness" full of wild beasts and wilder men yielded a vicious human nature and ignoble savages. Indians could be associated with the wild and its need to be conquered or, by the nineteenth century, with a romantic nature that yielded benefits to human beings.

Representations of Indians were, however, not a *necessary* element in Euro-American discourses of nature. For example, though they were basic to Thoreau, John Muir could largely (but not entirely) ignore them.[33] They could as easily be threats to the "wild" that whites valued as they could be a part of it. George Catlin might want to make Indians, like buffalo and antelope, an element in national parks, while John Wesley Powell blamed their burning for the destruction of vast reaches of the Rocky Mountain forests.[34]

The emergence of a culturally dominant view of Indians first as conservationists and later as environmentalists came only in the twentieth century. It initially was the work of a group, almost entirely male, who were prominent sportsmen and leaders in the nature appreciation and conservation movement and who, in distinction to later environmentalists, could claim firsthand experiences with Indian peoples. There was George Bird Grinnell, for twenty years the editor of *Forest and Stream*, the founder of the Audubon Society, and the cofounder of the Boone and Crockett Club. Ernest Thompson Seton, the nature writer who founded "Seton's Indians," a forerunner of the Boy Scouts of America, was perhaps even more influential. There was also James Willard Schultz, who married a Blackfoot woman, and Archie Belaney (Grey Owl), who also married an Indian woman and posed as an Indian.[35] Finally, and in the long run probably more influential than any of them, there was Charles

Eastman, a Dakota, or Eastern Sioux, who graduated from Dartmouth in 1887 and received a medical degree from Boston University three years later.

In the writings of these men, Indians graduated from mere exemplars of woodcraft to models of a proper relation with the natural world. Seton used Indians as "the model for outdoor life in this country" in his *Book of Woodcraft and Indian Lore*, and Seton himself began to copy "little rituals and ceremonies" borrowed from various Indian peoples. Grinnell used his position as editor of *Forest and Stream* to include at least one article on Indian "traditions" and practices in virtually every issue after 1889. For years, the editorial page of the magazine was entitled "In the Council Lodge" and decorated with an Indian motif.[36]

What distinguishes this from earlier episodes of noble savagery is not only the actual experience with Indians of the men promoting the conservationist Indian, but the willingness of Eastman—in this regard a forerunner of other Native American intellectuals—to enter the discourse on these terms. Eastman had lived with his grandmother and uncle in Canada until he was fifteen, before being reclaimed and educated by his father, whom the family had thought dead. Although his particular experience as a Dakota validated him as a "real Indian" to whites, Eastman usually chose to write about generic Indians, the "typical American Indian." This typical American Indian admittedly often seemed to resemble Dakotas quite closely in specific traits, but nonetheless Eastman presented general attitudes supposedly shared across the continent.[37]

Eastman's typical Indian was a noble savage tutored by nature and nature's God. "There were no temples or shrines among us save those of nature." The Indian was "a natural man," and nature revealed God to him. The Indian loved "to come into sympathy and spiritual communion with his brothers of the animal kingdom, whose inarticulate souls had for him something of the sinless purity that we attributed to the innocent and irresponsible child."[38]

Nature served, however, not as an alternative to Christianity but as its parallel. The vision quest "may be compared to that of confirmation or conversion in the Christian experience." The sweat lodge was the equivalent of baptism; the incense of the pipe was the Indian version of communion. The virtues nature yielded turned out to be the same as the Christian virtues; the dangers Indians avoided through their contact with nature were the seven deadly sins. Eastman's "natural man," whose religion was a natural religion, turned out in his telling to be the real

Christian, whose life was a reproof to the professed Christians. Eastman accepts the identification of Indians with nature both to critique (within the long tradition of noble savagery) white society and to argue that there is no real difference between Indian beliefs and Christian beliefs, only different roads to knowledge. "I believe that Christianity and modern civilization are opposed and irreconcilable and that the spirit of Christianity and of our ancient religion is essentially the same."[39]

Eastman did not create the connections between Indians and nature, and he was not the first to emphasize the spiritual and sacred, but in his very popular books Eastman validated these constructions and put them at the center of an "Indian" identity. He did this for his own purposes. He was a progressive Indian in a dual sense of the term, arguing for the integration of Indians into the larger society and stressing the contributions they could make. He entered the discourse about Indians and nature at a particular point and for particular reasons. He created a Victorian Indian. Nature taught restraint, chastity, generosity, respect for property. Eastman had no desire to return to the older, purer life. Like most whites who used Indians as models, he place that life irrevocably in the past. Whites, Eastman contended, had corrupted the modern Indian. Rituals like the Sun Dance had become "a horrible exhibition of barbarism."[40]

Just as Eastman took the discourse about Indians and the natural world and put it to his own purposes, so white readers could take his works and put them to their own. Today, Eastman's books are still in print and selling well. *Soul of the Indian*, in particular, can be read as a validation not only of the environmental Indian but of the spiritual Indian. "The Indian," Eastman proclaimed, "was a religious man from his mother's womb." His mother instilled "into the receptive soul of the unborn child the love of the "Great Mystery" and a sense of brotherhood with all creatures." What makes this Indian spirituality so accessible to white environmental audiences is not only its parallels with Christianity that Eastman underlines but also its similarity to so much white nature writing. Eastman was contemporary of John Muir and Ernest Thompson Seton, and their similarities are everywhere apparent. He writes, for example that the Indian mother:

> humbly seeks to learn a lesson from ants, bees, spiders, beavers, and badgers. She studies the family life of the birds, so exquisite in its emotional intensity and its patient devotion, until she seems to feel the universal mother-heart beating in her own breast.[41]

Eastman clearly shares much with Seton, but Seton in turn borrowed from Eastman. Eastman was one of the those, along with Grinnell, whom Seton consulted in his writing of *The Gospel of the Redman*. "The culture of the Redman," Seton wrote, "is fundamentally spiritual." Grey Owl, who attained his greatest popularity in Canada and Great Britain in the 1930s, wrote in the same terms. He reported that in his 1935–1936 appearances in England, "I did not need to think. I merely spoke of the life and the animals I had known all my days. I was only the mouth, but Nature spoke." The fake Indian had become the voice of nature.[42]

Grinnell, Eastman, Grey Owl, Seton, and others moved the old connections between Indians and nature into an image of the Indians as conservationists and made the connection a fundamentally spiritual one. All of these men were aware of various Indian beliefs and practices, but they took elements of a variety of practices and beliefs and wove them into a generic Indian, and then used that Indian to support conservation: a set of policies, practices, and ideology created within the larger society. This was a process in which Indian intellectuals such as Eastman could participate, albeit for different reasons, as well as whites.

The writings of Seton and others created the context in which other Indian texts, both the fabricated and the genuine, would be read. In a 1976 article, "Inspired by Indians," in *Conservationist*, a journal published by the New York State Department of Environmental Conservation, William Carr remembered that at twelve he was "an ersatz Indian," and he "never quite recovered from the experience." He credited his reading of Seton, among others, for his learning "that Indians had a high regard for wildlife and only killed animals they needed for food or other purposes, and that they also had regard for certain creatures in a religious sort of way." He credited his desire to "think and behave like an Indian" for his eventual interest in conservation. Even as an adult, when alone in the woods, he wondered "whether there is some spirit of an Indian near me, seeing what I see, hearing what I hear, and keeping out of sight. After all, they were there first."[43]

What Carr derived from Seton's generic conservationist Indians could, however, be put back into the mouth of historic Indians. The most obvious example of this is, of course, the speech Chief Seattle never gave and will keep on giving long after historians like myself have carped our way to the grave. In the early renderings of this speech, Seattle told his listeners, "And when your children's children think they are alone in the fields, the forests, the ships, the highways, or the quiet of the wood, they will not be alone . . . they will throng with the returning spirits that once thronged them, and that still love these places. The white man will never be alone."[44] The latest incarnation of Chief Seattle's speech—*Brother Eagle, Sister Sky: A Message from Chief Seattle*—spent a lengthy period on the *New York Times* best-seller list.

This speech (or sometimes a letter) has taken on a life of its own as one of the most magnificently self-serving of American documents. It praises Indians in order to bury them. Sentimental and, in its most recent versions, environmentalist, it chides whites but only, in the end, to forgive them. Here is a man dispossessed; his land taken up by strangers, saying it's all okay: Just remember our ghosts and respect the land. No wonder everybody likes this Indian.

Seattle has become not just an American, but an international environmental hero. He has appeared in pamphlets published by German youth organizations and in the texts of the Women's World Day of Prayer. His speech/letter has been translated into Swedish, Dutch, Portuguese, Italian, and Danish. Friends of the Earth has used the text; so has the Sierra Club, but so, too, for that matter has the magazine of Northwest Airlines.[45]

Rudolf Kaiser, who has examined the genealogy of Seattle's speech, has detailed how it went from being a very dubious rendering of an 1854 speech by Seattle (produced nearly forty years after the original was given) to an environmentalist version created from virtual whole cloth for a film script in 1969 and 1970.[46] This version, however, has proved useful not only for white environmentalists: Indian peoples have, in turn, adopted it as their own and used it in appeals for white support. The Snoqualmie Tribe, while battling for recognition in the Pacific Northwest, for example, featured Seattle's speech prominently in a brochure produced as part of their campaign for tribal status.[47]

The creation, quite successfully, of Indian environmental texts has gone on apace over the last few years. *The Education of Little Tree*, a bestseller written by Forrest Carter, a racist and, by some accounts, neo-Nazi, is particularly interesting. It is an account of a boyhood spent with Cherokee grandparents in the 1930s, during which Little Tree receives an education in "The Way." While much of it is the kind of nature lore predictable in fictions of this sort, other sections show how Indian fictions can sanction a set of essentially neo-Spencerian ideas. According to "The Way," the hunter takes only what he needs, which is a conventional enough environmental Indianism. But also according to "The Way," the hunter takes the slow and the weak and spares the strong. The inferior must die and the strong live. Given Carter's racism and how he would select the strong and weak, these become uncomfortable lessons in eugenics. Indian environmental wisdom can serve as a cloak for some frightening as well as appealing ideas.[48]

In terms of actual influence on popular environmentalism, these written texts are far less influential than movie versions. *Dances with Wolves* was the first environmentalist western, a sort of *Rin Tin Tin* in Ecotopia. In the movie, Kevin Costner meets a whole village of environmentalist Lakotas and learns to dance with wolves and talk to horses. *Dances with Wolves* conflated Indian religion with the kind of popular American sentimentality toward animals usually found in cat food commercials. The result had tremendous appeal, for it united, no matter how tenuously, two very different versions of animal personhood. That Indians remain hunters who kill animals could be temporarily forgotten by audiences weeping over the death of the wolf. Like Seattle's speech, this connection of Indians and popular environmentalism could be used by both sides. The Lakotas were quick to adopt Costner, and no one was so impolite as to remind him that he joined such other worthies as Calvin Coolidge and Herbert Hoover in this particular Indian ritual.

III

Seattle's speech, *The Education of Little Tree*, and *Dances with Wolves* are culturally important, but they are clearly fabrications. What is interesting about them is that they pose as authentic. Their ideas seemingly cannot stand alone; they need to be propped up by Indian traditions.

Native American intellectuals and some tribal leaders have recognized that this non-Indian need to validate environmental attitudes by claims on Indian precedents can give Indians a ready-made audience and political standing. They have not hesitated to take advantage of this. N. Scott Momaday, the Kiowa novelist, and Oren Lyons of the Six Nations have been able to speak in a way that serves both environmental and Indian purposes. Lyons has realized that the environmentalist Indian allows him to speak not just for a relatively small group of Indian people in New York and Canada, but for the Earth itself.[49]

Lyons and Momaday do not speak so much on the level of Indian practices as on the level of Indian spirituality and belief. And here they have recently found an interesting ally who demonstrates how tangled these conversations can become. In an article published a decade ago, one I have already cited, Calvin Martin, a prominent, if hardly mainstream, ethnohistorian, denied that the cognitive worlds of native peoples could be a model for modern environmentalism. Martin argued that the Indian could not "function as our spiritual leader in teaching us wise land use—a new land ethic—because his traditional interpretation of the world beyond him is profoundly different from Western cosmology."[50] In a new book, *In the Spirit of the Earth*, Martin has partially changed his mind. He finds models in paleolithic hunter-gatherers, a group that, admittedly, at once stretches beyond Indians and excludes most historical Indians. Nonetheless, most of his examples are taken from Indian peoples, and precisely those peoples whose cosmologies are most distinct from our own.

Martin has reversed his opinion in large part because he is persuaded that Western cosmology must now be discarded. Of all the elements that must go, the chief one is history.

> Historical consciousness itself has become perhaps our greatest enemy to true progress, the greatest obstacle to imagining ourselves and recalibrating our affairs in line with the new environmental consciousness. The latter is a primal consciousness, actually, for it marks the essential mind and speech and artifice of hunter-gather societies.[51]

Martin begins his book in a canyon near Shiprock on the Navajo Reservation, among Anasazi ruins; he was there seeking something that might "speak or sing to me." He was on a vision quest of sorts. He found a projectile point. A dust devil swirled around him, and he saw a "whirlwind being," slightly taller than himself. Later, the head nurse of the clinic—"witchy, remote, and wildly sensual. A dawn runner, a woman who could dance stripped to the waist at night before a fire"—validates the vision.[52]

Martin's account is interesting because the conventions of its cultural markers—given among Indians and validated by the witchy, wildly sensual woman—are so thoroughly of the late twentieth century that he wishes to escape. However, this is only part of my concern here, because there are also actual, modern Indian people on the margins of Martin's account. Scattered through the canyon are empty, broken bottles of Garden Deluxe, a cheap wine that the Navajos drink. Modern shards of glass left by drunken Indians are counterposed against the stone arrowheads left by ancient hunters that Martin seeks. There is no artifice or knowledge in a broken bottle, but in the lithic remains, the artifice of an ancient craftsman, is "a key [that] unlocks the door to the realm of another being." There is also a sign that a spark of the older communion with the natural world remains. Counterposed against the bottles are carefully piled stones, signs of "recent human meditation." Left by a "Navajo, no doubt . . . still fluent in these powers."[53]

Martin is selective in his Indians, both ancient and modern, and he does not attribute exclusive knowledge to them. He puts true knowledge in the hands of the neolithic. But ancient Indians for Martin are still the primary guides to environmental salvation. Martin, the environmentalist, is not interested in stewardship or management. He is interested in a "biosphere of communion, of transformation, of translation." The earth is not ours to care for; as the hunters and gatherers knew, "this place and its processes . . . always takes care of us."[54]

Martin's book will, I imagine, become a New Age classic. That it is a history written by an academic historian accessorized with all the scholarly accoutrements—footnotes, a bibliographic essay, professional resentments—that demands the rejection of history will probably be ignored by most of its readers. It is a book that is not a return to primal knowledge but a part of an ongoing environmentalist dialogue. It is a book praised by N. Scott Momaday on its back cover. And it is interesting, and important, for how it reveals the complications of the appeal to Indianness in environmentalist discourse, what it reveals of the conversation between Indians and non-Indians.

While I can understand the book on this level, I remain, frankly, uncomfortable with Indians as a route to transcendence, communion, and transformation. Indeed, I am uncomfortable with these things as a destination at all. I am inclined to think that the connections that environmentalists should forge with Indian peoples are not links to shamans and ancient traditions but links to the Navajos who left the bottles in the canyon near Shiprock.

The Indians who have the most to say to environmentalists are those whose reservations are the leading sites for proposed temporary low-level nuclear storage sites and many toxic waste disposal plants. The Navajos dead and dying after working in unventilated uranium mines, the Navajos who work well-paying jobs strip-mining Black Mesa, the Navajo tribal officials who have decided that they must log off their last old-growth forests to maintain the timber cuts that keep tribal sawmills open in an economy with huge unemployment rates—these compromised Indians interest me more than shamans. I am, in short, inclined to think that it is better to recognize that we are all caught up in a single, interrelated set of environmental problems in a triumphant capitalist economy and not to think that answers are secreted away in remnants of authentic or separate traditions.

I am also inclined to think that understanding Indian practices in the historic past is critical to understanding real environmental dilemmas in at least the modern West. As the direction of academic ecology has changed, as ideas of balance, succession, and stability have yielded to a view of natural systems as historically derived and contingent, what Indians did matters more and more.[55] Many current environmental problems can be understood only in terms of earlier Indian practices—particularly burning—which left us human-manipulated landscapes that we have tried to pretend were natural—often with disastrous results in places like Yellowstone in Wyoming and the Blue Mountains of Oregon.

There are going to remain Indian voices within the current environmental dialogue. And I admire the way that Lyons and Momaday use the larger society's craving for authenticity for their own purposes. And I know they might very well not admire my admiration. But I would feel more comfortable and optimistic if Indians who speak not for ancient traditions but for compromised, contemporary Indian communities had a greater role in this cultural conversation.

NOTES

1. Bill Devall and George Sessions, *Deep Ecology* (Salt Lake City, Utah: Peregrine Smith Books, 1985), 96–97.

2. Barry Lopez, *Crossing Open Ground* (New York: Charles Scribner's Sons, 1988), 88.

3. See, for example, Richard K. Nelson, *Make Prayers to the Raven: A Koyukon View of the Northern Forest* (Chicago: University of Chicago Press, 1983), xiii-xiv.

4. This sense of other-than-human persons composing the natural world was most fully analyzed by A. Irving Hallowell, *Culture and Experience* (Philadelphia: University of Pennsylvania Press, 1955), 177–182. It has become a staple of descriptions of northern hunters. See, for example, Adrian Tanner, *Bringing Home Animals: Religious Ideology and Mode of Production of the Mistassini Cree Hunters* (New York: St. Martin's Press, 1979). For a general discussion of Indian environmental religions, see Christopher Vecsey, "American Indian Environmental Religions," in *American Indian Environments: Ecological Issues in Native American History*, ed. Christopher Vecsey and Robert W. Venables (Syracuse, N.Y.: Syracuse University Press, 1980).

5. Alfonso Ortiz, *The Tewa World: Space, Time, Being and Becoming in a Pueblo Society* (Chicago: University of Chicago Press, 1969), 18–20.

6. For the first suit to employ these tactics, see *Northwest Indian Cemetery Protective Association v. Peterson*, 565, F. Suppl. 586 (N.D. California 1983). The Sierra Club filed an important brief in support of the Indian plaintiffs.

7. Donald Huges and Jim Swan, "How Much of the Earth Is Sacred Space?" *Environmental Review* 10 (Winter 1986): 247–259.

8. Raymond Williams, *The Country and the City* (New York: Oxford University Press, 1973), 8–10.

9. Tanner, *Bringing Home Animals*, 1–13, 25–26.

10. Nelson, *Make Prayers to the Raven*, xvi.

11. See William Cronon, *Changes in the Land* (New York: Hill and Wang, 1983); Timothy Silver, *A New Face on the Countryside: Indians, Colonists, and Slaves in South Atlantic Forests, 1500–1800* (New York: Cambridge University Press, 1990); Stephen Pyne, *Fire in America* (Princeton, N.J.: Princeton University Press, 1982); Calvin Martin, *Keepers of the Game* (Berkeley: University of California Press, 1978); Richard White, *Land Use, Environment, and Social Change: The Shaping of Island County, Washington* (Seattle: University of Washington Press, 1980); Carolyn Merchant, *Ecological Revolutions: Nature, Gender, and Science in New England* (Chapel Hill: University of North Carolina Press, 1989).

12. Substance of a Talk delivered at Le Macouitonong . . . by the Indian Chief Le Maigouis or the Trout Coming from the First Man Created . . . May 4, 1807, M. G. 19, F 16, Public Archives of Canada, Ottawa.

13. F.B. O'Callaghan, ed., *Documents Relative to the Colonial History of the State of New York* (Albany: Weed, Parsons, and Company, Printers, 1853–87) 9: 162–163.

14. Kirkpatrick Sale, *The Conquest of Paradise: Christopher Columbus and the Columbian Legacy* (New York: Knopf, 1990), 87.

15. Tanner, *Bringing Home Animals*, 136–181; Nelson, *Make Prayers to the Raven*, 220.

16. In *Keepers of the Game*, Calvin Martin argues that the fur trade amounted to a war against beavers prompted by the Indians' belief that beavers had purposefully spread epidemics and diseases among Indians.

17. Reuben Gold Thwaites, *The Jesuit Relations and Allied Documents: Travels and Explorations of the Jesuit Missionaries in New France, 1610–1791* (Cleveland: Burrows Bros., 1896–1901).

18. This was the belief of early-twentieth-century Shawnees (C.F. Voegelin, *The Shawnee Female Deity* Yale University Publications in Anthropology No. 10 [reprint, New Haven: Human Relations Area Files Press, 1970], 20), but it corresponds to earlier Shawnee cosmology. See also Jeanne Kay, "Native Americans in the Fur Trade and Wildlife Depletion," *Environmental Review* 9 (1985): 118–130.

19. Archer Hulbert and William M. Schwarze, eds., *David Zeisberger's History of the Northern American Indians* (Ohio State Archaeological and Historical Society, 1910), 84.

20. Dan Flores, "Bison Ecology and Bison Diplomacy: The Southern Plains from 1800 to 1850," *Journal of American History* 78 (1991): 465–485.

21. Raymond DeMallie, "The Lakota Ghost Dance: An Ethnohistorical Account," *Pacific Historical Review* 51 (1982): 390–392.

22. Joseph E. Taylor, "Making Salmon: Economy, Culture, and Science in the Oregon Fisheries, Precontact to 1960. (Ph.D., University of Washington, Seattle, 1996), 12–67.

23. Calvin Martin, "The American Indian as Miscast Ecologist," *The History Teacher* 14 (1981): 243–252.

24. Sam D. Gill, *Mother Earth* (Chicago: University of Chicago Press, 1987).

25. *Ibid.*, 120–122, 151–152, 157–158.

26. *Ibid.*, 158.

27. Oliver Dunn and James Kelley, Jr., *The Diario of Christopher Columbus's First Voyage to America, 1492–1493* (Norman: University of Oklahoma Press, 1989): 54–

55, 104–105; mountains, etc., 152–155, 172–173; Hispaniola, 208–215, especially, 212–213. Antonello Gerbi *Nature in the World* (Pittsburgh, PA: University of Pittsburgh Press, 1985); Stephen Greenblatt, *Marvelous Possessions: The Wonder of the New World* (Chicago: University of Chicago Press, 1991).

28. Columbus, *The Diario*, 212–213. For Columbus's use of *natura*, etc., see concordance, p. 453.

29. Raymond Williams, *Problems in Materialism and Culture* (London: Verso, 1980), 79; for organic view, see Carolyn Merchant, *Death of Nature* (San Francisco: Harper & Row, 1983), 1–41.

30. Tzvetan Todorov makes this formulation, but he goes out of his way to point out that Rousseau and Chateaubriand, authors usually cited as the major exponents of "natural men" and "le bon sauvage," held much more complicated views. Todorov, *Nous et les autres la réflexion française sur la diversité humaine* (Paris, Seuil, 1989), 310–329.

31. Ramón A. Gutiérrez, *When Jesus Came, the Corn Mothers Went Away: Marriage, Sexuality, and Power in New Mexico, 1500–1846* (Stanford, Calif.: Stanford University Press, 1991), 44–45. The second quote is from Greenblatt, *Marvelous Possessions*, 90.

32. Robert Berkhofer, *The White Man's Indian: Images of the American Indian from Columbus to the Present* (New York: Knopf, 1978), 71–96.

33. Robert E. Sayre, *Thoreau and the American Indians* (Princeton, NJ: Princeton University Press, 1977). See John Muir on Indians in John Muir, "The Mountains of California," in *Nature Writings* (New York: The Library of America, 1997), 372.

34. Stephen Pyne, *Fire in America: A Cultural History of Wildland and Rural Fire* (Princeton, N.J.: Princeton University Press, 1982), 80.

35. George L. Cornell, "Native American Contributions to the Formation of the Contemporary Conservation Ethic" (Ph.D. diss., Michigan State University, 1982), 4–5, 99–105, 110, 199–204.

36. Cornell, "Native American Contributions," 111, 112, 174, 177–178.

37. Charles Alexander Eastman, *From the Deep Woods to Civilization: Chapters in the Autobiography of an Indian*, 2d ed. (Lincoln: University of Nebraska Press, 1977); idem., *The Soul of the Indian*, 2d ed. (Lincoln: University of Nebraska Press, 1985), 3.

38. Eastman, *Soul of the Indian*, 5, 15.

39. *Ibid.*, 7, 83–84, 9–14, 23–24, 45–46, 119–144.

40. *Ibid.*, 88–89, 91–109, 54–55.

41. *Ibid.*, 28, 33.

42. Cornell, "Native American Contributions," 119, 203. Ernest Thompson Seton, *The Gospel of the Redman* (Santa Fe, N.M.: Seton Village, 1936), 1.

43. William H. Carr, "Inspired by Indians," *Conservationist* 30 (1976): 38–39.

44. Rudolf Kaiser, "Chief Seattle's Speech(es): American Origins and European Reception," in *Recovering the Word: Essays on Native American Literature*, ed. Brian Swann and Arnold Krupat (Berkeley: University of California Press, 1987), 524–525.

45. *Ibid.*, 497–502.

46. *Ibid.*, 497–536. Stories about the spuriousness of the speech or letter are fairly common and nearly totally ineffective; see "Myth-quoted: Words of Chief Seattle Were Eloquent—But Not His," *Seattle Times*, July 1, 1991.

47. "In the Beginning." Pamphlet, Snoqualmie Tribe, Redmond, Washington.

48. Forrest Carter, *The Education of Little Tree* (Albuquerque: University of New Mexico Press, 1991), 8–11.

49. For Oren Lyons, see, for example, "Our Mother Earth," in *I Become Part of It: Sacred Dimensions in Native American Life*, ed. D. M. Dooling and Paul Jordan-Smith (Parabola Books: New York, 1990), 270–274, and "An Iroquois Perspective," in *American Indian Environments: Ecological Issues in Native American History*, ed. Christopher Vecsey and Robert W. Venables (Syracuse, N.Y.: Syracuse University Press, 1980), 171–174. For N. Scott Momaday, see "Confronting Columbus Again," in *Native American Testimony: A Chronicle of Indian White Relations from Prophecy to the Present, 1492–1992*, ed. Peter Nabokov (New York: Viking, 1991), 436–439; "Native American Attitudes toward the Environment," in *Seeing with a Native Eye: Essays on Native American Religion*, ed. Walter Capps (New York: Harper and Row, 1976), 79–85.

50. Martin, "The American Indian as Miscast Ecologist," 244.

51. Calvin Martin, *In the Spirit of the Earth: Rethinking History and Time*, (Baltimore: Johns Hopkins University Press, 1992), 120.

52. *Ibid.*, 23–25, 28.

53. *Ibid.*, 11, 24.

54. *Ibid.*, 120, 71–130.

55. There is a large literature on this. For a history of twentieth-century ecology, see Anna Bramwell, *Ecology in the 20th Century* (New Haven: Yale University Press, 1989); R. P. McIntosh, *The Background of Ecology* (New York: Cambridge University Press, 1985). For interpretations of recent changes, see Daniel Botkin, *Discordant Harmonies* (New York: Oxford University Press, 1990), and Donald Worster, "Ecology of Order and Chaos," *Environmental History Review* 14 (1990): 1–18.

TERENCE TURNER

Indigenous Rights, Environmental Protection, and the Struggle over Forest Resources in the Amazon

The Case of the Brazilian Kayapo

THE FATE of the tropical forest of the Amazon has been a central concern of modern environmentalism since its inception as one of the new social movements of the 1960s. The struggle to save the "rain forest" (as the Amazon forest is popularly, if largely inaccurately, designated) has converged at important points with the struggle for the physical and cultural survival of the indigenous peoples of the forest, as waged by indigenous support groups, human rights organizations, government agencies, and the indigenous peoples themselves. There are obvious reasons for this convergence. The forces menacing the physical and cultural survival of the indigenous peoples of the region are the same as those threatening the destruction of the forest ecosystem. These include, above all, massive "development" schemes, financed by governments and international financial institutions, and Brazil's road-building and settlement programs. There is also the point that native Amazonian peoples are themselves an integral part of the forest ecosystem. Their traditional forms of subsistence and environmental adaptation have evolved over millennia in sustained coexistence with it. Objectively, the indigenous areas of the Amazon are those in which the forest survives, if not intact, then at least as a viably self-reproducing ecosystem, while in areas of nonindigenous settlement it has tended to disappear. Indigenous peoples' struggles to defend their traditional lands and resources thus appear to converge with the environmentalist defense of the Amazonian ecosystem. Many environmentalists have therefore supported the struggle for indigenous land rights and cultural survival, and organizations supporting indigenous land rights have identified themselves with environmentalist causes.

The alliance of environmentalists and indigenous forest peoples, however, has been unstable and fraught with difficulty on both sides.

145

Indigenous groups and their supporters have resisted efforts on the part of some environmentalist organizations to exclude indigenous people from tracts of Amazon forest they acquire for preservation or for which they seek government protection as ecological reserves. Some environmentalists, for their part, have oscillated between unrealistic extremes in their attitudes to indigenous peoples. Some have romanticized Amazonian Indian peoples as primitive ecologists, with a spiritual feeling of kinship for everything natural. A hypertrophic form of this romantic view is the fictitious claim disseminated by certain anthropologists that Amazonian Indians, or at least one important group of them, the Kayapo, possess and practice a highly sophisticated "science" of forest "management", through which they have virtually created the forest through a kind of ecological gardening (Posey 1982, 1985, 1993).

At the opposite extreme, some environmentalists and ecologists have reacted to recent cases of indigenous groups who have permitted ecologically destructive practices such as mining or logging on their land in exchange for fees. The Kayapo again have served as the leading example, this time as villains rather than heroes. As news reports of the Kayapo's complicity in mining and logging operations on their newly recognized reserves accumulated, and more careful scientific studies exposed the inaccuracy of the enthusiastic claims for Kayapo ecological science and forest management (Parker 1992, 1993), environmentalists were forced to realize that Amazonian cultures, at least as represented by the Kayapo and others who had engaged in similar extractivist activities, do not share Western ecological values. Specifically, they are not committed to the preservation of nature for its own sake, have no spiritual reverence for individual trees and animals, and have no idea of "managing" their environments to create or preserve them. Some environmental activists have overreacted to such revelations, taking them to imply that the Kayapo and other indigenous Amazonians lack *any* ecological notions or values and look on their environments merely as stocks of resources to be exploited with no concern for their sustainability or conservation. The same writers have further argued, plausibly enough, that indigenous peoples' immemorial record of sustained coexistence with the forest ecosystem is a mere incidental by-product of the weakness of their technologies, coupled with a lack of capitalist incentives for surplus accumulation, which have unfortunately now arrived (Redford 1990; Redford and Stearman 1993; Stearman 1994). The logical conclusion from these premises, explicitly drawn by certain advocates of these propositions, is that native peoples must be seen by the environmentalist movement as enemies, rather than as allies.

Variants of these two extreme views have dominated much of the increasingly intense debate over the relationship between environmentalist and indigenous peoples' interests in the Amazon. There is much at stake in this debate for both environmentalists and indigenous peoples: in Amazonia, neither are likely to achieve their goals without at least tacit support from the other, and the alternative to some form of alliance, namely, overt hostility and opposition, would likely result in their separate defeat. There are also more general issues involved. Is environmentalism, and the ecological principles on which its values are founded, after all to be understood as an ethnocentric product of Western culture, or can common ground be found between even widely different cultures in the struggle to protect their environments from the newly intensified threats brought on by the expansion of global capitalism? The tendency of all sides in the debate to conduct it purely on ideological or cultural grounds raises a second basic issue. For both the contemporary West and simple indigenous societies like the Kayapo, ecological views, values, and practices assume their cultural forms, change, and exercise their effects through social processes subject to political and economic factors. Scholarly research directed at understanding ecological notions and activist attempts to organize collaborative projects and alliances alike must take such pragmatic dynamics and interdependencies into account if misunderstandings, ineffective policies, and project failures are to be avoided.

The Kayapo

The Kayapo, an indigenous people of the southern fringe of the Amazon forest, have been the objects, and even for many the prototypes, of both extreme positions in the debate over the compatibility of environmentalist and indigenous interests. The Kayapo case is important because it demonstrates, in dramatically compelling terms, that indigenous political processes and cultural values have important contributions to make to environmentalist causes and that indigenous societies, like our own, may constantly learn from experience and change their views and policies toward their relations with the environment. Indeed, what we call their "culture" turns out to be more fundamentally concerned with the social mechanisms and ideological dynamics of forming and changing their views of and material relations with their environment and other matters than with any particular set of ideas or customary patterns considered to be their "traditional" or "authentic" cultural repertoire. As we shall see, the Kayapo have, in the past few decades, gone through a series of struggles to defend or transform their relations with their natural environment, together with their relations with the encompassing regional, national, and global political economic systems. This historical process has specifically included reformulating and transforming their environmental relations and values. In this paper, I attempt to explain how this process and the reorientation of Kayapo economic relations with the natural environment arose out of the internal dynamics of Kayapo society as it came into new and intensified interaction with external political, economic, and ecological factors.

Kayapo Relations with the Environment: Resource Use, Cultural Values, and Political Economy

The Kayapo currently number about five thousand persons. They live in relatively large, mutually autonomous village communities ranging from one hundred to twelve hundred in population, much as they did in the time before the inception of peaceful relations with Brazilian society in the late 1950s. There are presently fifteen villages, scattered over a large area on both sides of the Xingú River, a major southern tributary of the Amazon. Kayapo country is a mixture of forest and savannah. There is a rainy season and a dry season: the forest in the Kayapo area is thus not "rainforest" in the strictest sense, and includes many deciduous species. Their territory is contained mostly in six reserves, recognized by the federal government in the 1980s and 1990s, that cover a combined area of some 100,000 square kilometers.

Past and Present Kayapo Relations with the Natural Environment

The appropriate place to begin to attempt to understand Kayapo attitudes to "nature" and the environment is an examination of the activities comprising their traditional subsistence base. For most Kayapo today, these activities continue to be the main business of their lives, and they continue to provide the great bulk of Kayapo subsistence needs. In the past, and for the great majority of Kayapo, presently, the Kayapo produce their means of subsistence by a combination of slash-and-burn horticulture, hunting, fishing, and gathering. According to the division of labor by gender and generation, men engage in all productive pursuits incompatible with the care of young children, while women perform those which can be carried out while caring for children. This means that men hunt; fish; do the heavy and dangerous work of clearing gardens; and gather certain products that grow at great distances, requiring overnight journeys. Women do most of the planting, weeding, and harvesting of gardens; cut firewood; cook the food; build traditional shelters (now done almost exclusively in trekking camps); and care for children. Girls begin to help their mothers with household and garden chores while still children, but boys do no productive labor until they reach the men's house, and relatively little even then until they are initiated in their middle teens.

In the course of a year, Kayapo communities spend part of the time in their base village near their gardens and part of the time on collective treks, moving through the country and hunting and foraging as they go. Trekking expeditions typically last from one to three months and take several forms. Most frequently, the whole village goes on trek together, to gather food for a ceremonial feast. The ceremonies in question are rites of passage, initiations, or the bestowing of honorific names on children. A community may go on two or three such treks per year. In trekking camps, age sets, defined as comprising members of the same gender and stage of the social life cycle, act collectively as units in the production of food and shelter, the clearing of new camps, and the transportation of produce from the gardens of the base village to the camps. In the base village, the nuclear family is the main unit of cooperation in the production and consumption of material subsistence, but as a social unit it owes its form primarily to its role in producing new social persons, rather than its functions in organizing subsistence activities.

In both trekking camps and the base village, the production of material subsistence thus forms an integral part of the process of producing human beings as social persons identified in terms of collectively recognized categories of gender and social age. It is not the dominant aspect of this process in the sense of determining its social form; nor does it appear as a distinct sphere of activity (i.e., "the economy"), much less as the primary or exclusive type of "production" in our sense of the term. There is no "economy" in this limited sense in Kayapo society. Rather, the division of labor in subsistence production is determined by the division of labor in the production of human beings, including the "socialization" of children but extending to the reproduction of the whole social life cycle as the basic pattern of social identity for adults of both genders.

In general terms, then, Kayapo society is constituted as an integral process of social production, which subsumes the production of material subsistence, the production of social persons, the production of the nuclear family and extended families (in their capacities as social units of production relations), and the communal institutions that regulate the reproduction of these units. This global notion of production is the basis of the Kayapo perspective on the natural environment. The production of food, shelter, and utensils from natural materials forms a direct and continuous part of the production of human persons, families, and (thus) society as a whole, rather than being made the object of a separate sphere of "economic" activities. The well-being and productivity of the ecosystem and that of society and its members are thus implicitly felt to be interdependent.

This sense of interdependence is emphatically not the same thing as the spiritual reverence for nature that some environmentalists and indigenous spokespersons have claimed as the characteristic attitude of indigenous peoples to the natural world. It does not entail any general idea or category of "nature"; nor does it preclude an instrumental attitude toward natural entities and their destruction in the furtherance of pragmatic individual or collective interests. The Kayapo do indeed impute certain forms of spiritual force or magical power to natural beings, but they tend to conceive of such powers in practical terms as sources of danger to be guarded against or of power to be appropriated by those with the special knowledge and skills to do so, not as grounds for mystical identification, protection, or respect.

This whole set of attitudes and ideas presupposes an experience of the natural world as a system far beyond the powers of human techniques to alter in any general way, and thus collectively impervious to human exploitative activities. Gardens may be slashed and burned, animals may be killed, and fish may be caught, but such limited activities never put in question the renewal and abundance of forest and animals as such. This perspective has obviously arisen, in the case of the Kayapo and other indigenous Amazonian peoples, from their experience as a society of relatively low population density, with a relatively rudimentary technology and no incentives for material production beyond subsistence, inhabiting a vast area of abundant resources that were, from their point of view, inexhaustible.

The Cultural Roots of Kayapo Environmental Activism: The Altamira Demonstration as Ritual Drama

This same experience, however, when filtered through the Kayapo notion of the society-environment relation as a single great process of the production of human, social, and natural life, might also be expected to promote a profound collective identification between human beings and the natural environment (seen as a vast and inexhaustible source of energy and self-renewing power) that, while different from Western environmentalist values or scientific ecological understandings, would nevertheless constitute a positive basis for a kind of indigenous "environmentalism." This is a possibility that has been overlooked both by those who have argued that the indigenous experience of the Amazonian environment as an inexhaustible resource has led to an absence of any ecological understanding or concern and by those who have imagined that indigenous peoples have a spiritual reverence for the ecosystem or a "scientific" concern for its conservation. It is a possibility that finds rich ethnographic confirmation in the case of the Kayapo. The Kayapo sense of human-ecosystemic interdependence receives expression in ritual, myth, and cosmology. It is a perspective that implicitly makes a threat to the continued existence of the one tantamount to a threat of destruction of the other. This was not an eventuality that the Kayapo had to confront until their recent encounters with Brazilian projects for the economic exploitation and development of their territory. Once these encounters were perceived to pose just such a threat, however, Kayapo ideas and attitudes about the forest and riverain environment became a potent basis for political mobilization and resistance. A case in point was the Kayapos' campaign against the Brazilian government's secret plan to build, with World Bank support, a series of hydroelectric dams on the Xingú River, which runs through their territory. This effort culminated in the great rally of indigenous Amazonian peoples organized by the Kayapo at Altamira, near the site of the proposed dam on the lower Xingú, in early 1989.

For five days in February 1989, some six hundred Kayapo Indians, together with contingents form forty other indigenous nations of Amazonia, gathered at the small river town of Altamira in the state of Para, Brazil. Also present were more than four hundred representatives of the Brazilian and world news media, documentary filmmakers, photographers, and diverse nongovernmental organizations (NGOs) that had rarely, if ever, cooperated with one another on any common cause: environmentalists, human rights advocates, critics of multilateral development banks and the sorts of development projects they tend to fund, especially big dams; and indigenous support groups. The announced purpose of the gathering was to force representatives of the Brazilian government, the World Bank, and Brazilian construction and electric power companies to give a public account of their hitherto secret pans for a huge hydroelectric dam scheme in the Xingú River valley. The Kayapo, who had called the meeting and invited the media and NGO representatives, as well as the other indigenous peoples, had made it clear that they intended to call upon the representatives of the government and construction companies to disclose their projections of the environmental and human impact of the scheme and to explain why they had attempted to keep all plans for it secret from those who would ultimately be most affected by it: the native peoples and rural Brazilian population of the region. The government and World Bank had been understandably reluctant to accept the Kayapo invitation. Only when it became clear that hundreds of national and international journalists, filmmakers, and opinion leaders would attend the gathering did President Sarney of Brazil agree to send a personal representative, as well as the chief engineer of Eletronorte, the state power company in charge of the dam scheme, to present the case for the government project.

To the hundreds of Brazilian, European, and North American journalists and activists who attended the event, and the millions who saw news clips of it on their televisions or read of it in their newspapers and magazines, the Altamira gathering appeared to be a combined Earth Day demonstration and environmentalist panel discussion, put on by painted Indians wielding primitive weapons and performing exotic ritual dances. It was easy for a Western observer to miss the ritual framework that to the Kayapo participants constituted its essential organizational nexus and code of cultural meaning.

The Altamira gathering was planned by Kayapo chiefs who met at Gorotire (the largest Kayapo village) in October 1988. Their common conceptual vocabulary (for some of them, their only vocabulary) for organizing a collective Kayapo social action was their shared cultural background of Kayapo ritual forms, cosmological concepts, and social principles. The Altamira rally was planned primarily in these terms, and only secondarily as a political demonstration and discussion of ecological policy, which were the terms in which it was exclusively represented in the Western news media. It was in these Kayapo cultural terms that its Kayapo participants understood its message and the meaning of their own participation. The Portuguese speeches and press releases of the bilingual Kayapo leaders to the assembled government delegates and reporters were couched in the rhetoric of international First World environmentalism and international Fourth World ethnic nationalism, but the basic ideas they expressed were Kayapo concepts and values embodied in the cultural forms the Kayapo leaders drew upon to organize the meeting. These forms included, most importantly, the construction of the Altamira encampment as a complete Kayapo village, with family households and children as well as a men's house; the New Corn Ceremony with its associated myth, employed as the organizational and ideological matrix of the event; and the more general use of traditional ritual, oratory, and song in the Kayapo language throughout the public sessions as well as in the encampment.

The Kayapo leaders who planned the Altamira gathering faced a dual problem. They wished, on the one hand, to make the encounter at Altamira a demonstration to the Brazilians and the world at large of the vitality and richness of their own culture. They shrewdly realized that from the standpoint of the impact on Brazilian and world public opinion, to dramatize the viability and value of their society and culture as the human reality threatened by the dam scheme would be as important as any arguments they might present for their political rights to be consulted about the dams or their legal claims to the land and its resources. The Altamira encounter, on the other hand, was essentially planned as a confrontation and dialogue with Brazilian government authorities in the presence of Brazilian and international media personnel and support group representatives. This meant that the proceedings would have to take place in Portuguese, with occasional English translations. How then to motivate the mass of monolingual (and monocultural) Kayapo to make the arduous journey to be present at such an event, and how, once they arrived, to provide ways for them to participate in the proceedings in terms meaningful to themselves?

These questions posed themselves, at the simplest level, as the problem of coordinating the actions of members of more than a dozen mutually autonomous communities, who not only acknowledge no common leadership but whose traditional culture lacks a calendar in terms of which common action could be precisely coordinated in time. How to get the different Kayapo villages to synchronize their organization for the great journey and then actually converge at Altamira on the same date?

Non-Indian supporters of the Kayapo were puzzled and not a little disturbed when the Kayapo chiefs announced their decision to hold the Altamira meeting at the end of February, in the middle of the rainy season, when road travel throughout the area becomes difficult. From the standpoint of organizational efficiency, it seemed an irrational choice. From the standpoint of the Kayapo leaders struggling with the above questions, however, the decision was the outcome of rationally compelling considerations. Late February is the time when the final rites of the New Corn Ceremony are performed, and the New Corn Ceremony is the only major Kayapo ritual that is celebrated at the same time of year by all the Kayapo communities, being tied to the actual time of ripening of the new maize crop, which is the same over the whole region. Like all major Kayapo ceremonies, it takes the form of a series of initial rites, followed by a collective hunting and foraging trek lasting several weeks, followed in turn by a final climactic ceremony. The Kayapo leaders therefore couched their appeal for attendance at Altamira in terms of the New Corn Ceremony: all Kayapo were exhorted to come together for a collective celebration of the final rites of the New Corn Ceremony at Altamira; the initial rites were to be performed in each village, as usual, but in place of the seminomadic hunting trek to collect food for the final ceremony, there would be the great trek to Altamira.

For the Kayapo participants, then, the basic organizational and calendrical framework for the Altamira event was not an abstract date on the Western calendar or a list of arbitrary assignments to chartered buses, but the New Corn Ceremony with its collective trek simply redirected toward the common final ritual celebration site at Altamira. The New Corn Ceremony served not only as an organizational structure but as a structure of meaning, informing the Altamira event with a series of messages and associations that were drawn from traditional Kayapo culture but nevertheless closely approximated the official Kayapo stance, as articulated in Western political and environmentalist terms adopted by the Kayapo leadership to deal with the Brazilians and the world press in the public sessions at the Altamira municipal sports hall.

The New Corn Ceremony is, among other things, an agricultural first-fruits ritual. This celebration of the growth and maturation of the garden crop is coupled with parallel rites of passage for human children of both sexes. Before the ears of corn ripen, when the new corn plants reach the height of about one meter, the young men of the village clip the hair of the unmarried girls (this short hairstyle being the main public badge of their membership in the young girls' age set). Later, after the women of the village present the first newly ripened ears of corn to their brothers, adult sons, or brothers' sons in the men's house, young boys of around eight years of age are ritually removed from their mothers' houses, where they have resided up to that time, and inducted into the men's house, which henceforth serves as their collective dormitory until they marry and move in with their wives. The final rite of the ceremony, that which was celebrated as the final act of the Altamira encounter, is the culmination of this collective initiation of youths.

Both the agricultural and human rites of passage that make up the New Corn Ceremony revolve around yet another, more fundamental rite of passage, which forms the dramatic centerpiece of the whole ceremony and occurs at its midpoint, just before the celebrants depart on their collective monthlong hunting and foraging trek. This involves the cutting down of a great tree in the forest. After the trunk is trimmed of branches and a ten-foot section decorated at one or both ends with red paint representing fire, it is carried by the men into the center of the village plaza and thrown down in front of the men's house. There, in its transformed and domesticated form, it thenceforth serves as a bench.

The significance of this rite is clarified by the myth associated with the ceremony, which explains that the ancestors of the Indians first discovered maize growing like fruit on the branches of a single great tree in the forest. They cut down the tree and thus obtained the ears and kernels, which they planted. The result was the multiple reproduction, in miniature, of the original corn tree by the cornstalks that sprang up in their gardens. This was the origin of agriculture. The multiplication, however, did not stop with the corn. As they gathered and planted the kernels, the ancestral Indians, who up until then had been a single society speaking a common language, began to speak the different, mutually unintelligible languages of the Indian societies of the contemporary Amazon, and scattered into the mutually dissociated native groups found in the region today.

The great tree cut down for the ceremony represents the ancestral maize tree of the myth. Like its mythical precursor, it embodies the natural powers of reproduction and growth that society must tap and channel into the domesticated forms of garden horticulture and the socialization of human individuals. The cutting, symbolic decoration, and bringing in of the tree to the center of the village represent this process of domestication and the renewal of society and the garden crops it produces. The same act expresses the dependence of society and its reproduction upon the natural forest environment, and the powers of reproduction and growth of which the forest is the embodiment and source. The New

Corn Ceremony, in sum, expresses the Kayapo conception of the inter-dependence of society and nature, a relationship focused in the productive processes entailed in the reproduction and growth of both. By making the New Corn Ceremony the organizing schema for the Altamira encounter, the Kayapo leaders implicitly communicated to the mass of Kayapo participants the essence of what they were asking them to defend and what was threatened by the destruction of the regional environment that would be caused by the hydroelectric dam scheme.

The myth of the ancestral maize tree associated with the New Corn Ceremony simultaneously provided an ideological charter for the common front that the Kayapo leaders attempted to forge among the other native societies of the region. The Kayapo were historically a raiding people who had been on hostile terms with a number of these neighboring groups and had regarded non-Kayapo peoples with a measure of disdain. The Kayapo leaders, however, fully realized the importance of bringing all the indigenous nations of the region together for a common stand in defense of their threatened forest world. The myth of the maize tree and the identification of the Altamira gathering with the New Corn Ceremony became their rhetorical basis for conveying this message to their people. Although Kayapo tellers usually construe the ancestral group who chopped down the maize tree as the ancestors of the Ge-speaking groups like themselves and shrug off the question of the ancestry of the non-Ge groups, in rallying support for the Altamira demonstration Kayapo leaders and orators expanded this interpretation to identify the group who had cut down the maize tree as the ancestors of all the indigenous peoples of the area. The myth thus became a rhetorical charter for bringing together all native Amazonian societies as related offspring of the same ancestral stock, kinsmen who were merely being invited to a long overdue family reunion. The New Corn Ceremony thus served not only to organize the participation of the mass of rank-and-file Kayapo at Altamira, but to rationalize their coparticipation and solidarity at the Altamira gathering with non-Kayapo peoples for whom many of them had little regard.

The construction of the Kayapo encampment at Altamira as a total Kayapo village community, complete with families, children, and households pursuing domestic activities including cooking, child care, and the manufacture of artifacts, was another deliberate symbolic gesture calculated by the Kayapo planners of the Altamira event to send many of the same messages to the outside world as the use of the New Corn Ceremony would convey to the Kayapo participants. The Kayapo might have chosen to send only a small delegation of prominent men, or perhaps a larger group of male "warriors," to confront the Brazilians at the meeting. That they did not was the result of a conscious decision about self-presentation, a crucial consideration for what they conceived from the outset as essentially a media event. By building a total social community at Altamira directly in the path of the dam, the Kayapo concretely presented themselves not as a particularly aggressive and litigious horde of ex-warriors, but as a total society with a rich and vital culture, whose way of life was being threatened by the economic development projects of distant governmental planners, politicians, and bankers (on Kayapo self-representation and use of media, see Turner 1991a, 1992b). The original plans for the encampment (not followed through in the event) actually called for the village to be maintained indefinitely, as a sort of permanent "live-in" at the dam site, consistent with the intentions of the planners to make the Altamira encampment a total presentation of Kayapo culture and society to the world at large.

An aspect of the Kayapos' presentation of themselves and their cause at Altamira that struck many observers was the prominent role of women. In the part of the program devoted to speeches by representatives of indigenous nations, the Kayapo alone chose to be represented by a female as well as a male speaker. The woman (a senior woman of A'ukre village) spoke first and at equal length with the man, who was one of the most prestigious Kayapo chiefs and a famous war leader. Several women also made spontaneous interventions as orators. The most electrifyingly aggressive rhetorical confrontation of the session was by a woman who interrupted a speech by the Eletronorte engineer by brandishing her machete in his face, while telling him, in no uncertain (Kayapo) terms, how she would use that machete on him if he were to show up in her village with a team of dam construction workers.

Such performances were not merely for display to non-Kayapo spectators. The participation of women was a manifestation of the same themes embodied in the construction of the Altamira encampment. Both in their public speeches and in talking among themselves, women articulated concerns specific to their own social roles and perspectives about the threat to their children and families posed by the environmental destruction that would be caused by the Brazilian projects. They asserted that whereas the Brazilians acted out of a need for money, which they required to feed their own families, the Kayapo, lacking the abstract medium of money, can meet the same need only through the direct exploitation of forest resources, and thus require a healthy environment. They thus adjured Brazilian policymakers to understand and accede to Kayapo objections to the dams and the destruction of the forest out of their love for their own children (or, in a negative vein, charged them with acting like kinless orphans without families to care for, who "had not been held enough" by their own mothers, the most grievous Kayapo insult). This culturally gendered female voice thus expressed, in the strongest and most vivid terms, the idea of the interdependence of a healthy environment and the reproduction of human beings and families, and thus of society, that I have suggested is the fundamental "environmentalist" value of Kayapo culture.

The Altamira demonstration, in sum, was a dramatization of the environmental values of Kayapo culture in the service of a Kayapo version of environmental activism. Supported and attended by many Western environmental activists and organizations, it nevertheless asserted and defended a cultural and political vision of the relation of human society and the natural environment different from theirs in fundamental respects. The Kayapo at Altamira were not fighting to defend "nature" for its own sake, but to protect their environment as a source of resources indispensable to the survival of their own society, understood to entail their existing relations of productive interdependence with the natural world. The maintenance of these relations requires the survival of the forest and riverain ecosystems in more or less their existing form; thus any threat to their survival must be resisted. This principle, however, does not preclude any forms of exploitation that may involve the destruction of individual trees or animals but do not threaten the reproduction of the ecosystem as such. That the symbolic framework of the Altamira rally was the ritual chopping down of a great tree, evocative of the reproduction of Kayapo society through the appropriation of the powers of the natural environment, may stand as an ironic comment on the conceptual divergence between Kayapo and Western versions of environmentalism, even as it made possible their pragmatic convergence.

The success of the Altamira rally and a whole series of similarly effective Kayapo actions succeeded in guaranteeing continuing Kayapo control over the area occupied by their far-flung villages and its environmental resources. It did not, however, answer the questions of how the Kayapo could or should exploit these resources in the context of their intensifying articulation with the Brazilian and world political and economic systems. These issues had become pressing as gold miners and loggers increasingly penetrated Kayapo areas beginning in the middle 1970s, at first as invaders and then under the protection of Kayapo leaders in exchange for a small cut of the proceeds. Some argued that piece-meal collaboration of this sort was in the long run the only realistic alternative to government-sponsored invasions by massive development projects (like the dam at Altamira) or waves of landless settlers.

The Kayapos' Complicity in Environmentally Destructive Extractive Enterprise: Gold Mining and Mahogany Logging

The Altamira meeting marked the high point of the alliance between the Kayapo and the loose confederacy of environmentalist, human rights, and indigenous advocacy organizations, supported by large sectors of First World public opinion, who had rallied to support their resistance to the Xingú dams. Some saw in this emergent alliance of hitherto divergent and uncoordinated single-issue groups a portent of a new political conjuncture with implications far beyond the Amazon. Just as the Kayapo and their brilliant and charismatic Kayapo leader, Payakan, had played a leading role in precipitating this hopeful coalescence, however, they soon found themselves playing a leading role in its breakup.

The honeymoon between the Kayapo and the environmentalists, indigenous support organizations, journalists, and public opinion leaders who had done so much to help (and hype) them broke up over two issues. The more serious was that of the Kayapo's complicity in mining and logging on their own recently won reserves and the misuse of the communal funds derived from the extractive concessions by some Kayapo leaders. Beginning in the late 1980s, there was a steady trickle of reports in the Brazilian and international media that Kayapo leaders were entering into contracts with logging and mining companies, granting them concessions to operate on Kayapo lands in return for a percentage of the proceeds. To some outsiders it appeared that the Kayapo had become collaborators in the destruction of their own forests and rivers for the sake of short-term monetary profit. Some of the media accounts reported that some Kayapo leaders were using the income from these contracts to maintain lavish personal lifestyles in Brazilian towns far from their home villages, complete with townhouses, Brazilian servants and bodyguards, cars, airplanes, drinking binges, and Brazilian mistresses and prostitutes. The biased impression conveyed by the stories of Kayapo "wealth" based on the lifestyles of a few Kayapo leaders with houses in Brazilian towns was magnified by omission of any description of the poverty of the 99 percent of the Kayapo population who remained in the villages. Some of these reports included Payakan among the Kayapo leaders supporting an urban lifestyle with money from logging contracts, in direct contradiction of his symbolic persona as an ecowarrior *sans peur et sans reproche.*

Payakan, indeed, became the second major issue that shook the alliance of environmentalists, indigenous advocates, and the Kayapo. The immediate cause was a media-promoted scandal over the charges of rape brought against Payakan and his wife by a Brazilian girl. The story broke in the media in the same week as the Rio United Nations Conference on Environment and Development (UNCED) meeting in August 1992, at which Payakan had been expected to speak. Instead, Payakan, who had become for his North American and European supporters the incarnation of the myth of indigenous peoples as defenders of the biosphere, found himself, together with his wife, facing charges of raping a Brazilian girl after a boozy party in the Brazilian town of Redenção, where they had been living in preference to their own village of A'ukre. The charges were found, upon investigation, to be exaggerated, and Payakan was eventually (December 1994) acquitted by a Brazilian court, which also declined to sentence his wife on grounds that she was insufficiently acculturated to understand Brazilian law. The decision, however, came too late to undo much of the damage caused by the inflammatory media attacks. Meanwhile, the legal evidence and journalistic reports elicited by the charges and accompanying scandal had made all too clear that while the specific charge of rape was unfounded, it was true enough that Payakan had for long been secretly involved in logging contracts; nor was he unique among the town-dwelling Kayapo leaders in this respect.

All of this aroused widespread dismay among supporters of the Kayapo and other rain forest peoples, many of whom had imagined the Kayapo only as guardians of the forest, living in pristine harmony with their environment, in conformity with their village family system and their ancient culture. The revelation that they were actually aiding and abetting the logging of their own forest and the pollution of their own rivers with mercury and mud by gold miners was for many the end of the heady conjuncture of indigenous resistance and green activism at Altamira and after.

The Gorotire Revolt and the Expulsion of Miners and Loggers from Kayapo Territory in September–December 1994

Even as Payakan and others among the younger Kayapo leaders in their townhouses became the objects of controversy among nonindigenous critics and supporters alike, they and the mining and logging concessions they had negotiated were arousing steadily mounting resistance by a conjuncture of social and political forces in their home communities. Resentment was building up in the villages of Gorotire, Kikretum, A'ukre, and Catete, with the longest history of logging or mining concessions, against the way the new generation of leaders appeared to be exploiting their role as intercultural mediators to set themselves up as an embryonic new class, using their control of the new economic dealings with the Brazilians to divert to themselves the lion's share of the benefits accruing from them, leaving relatively little for their fellow villagers.

To many, "the Kayapo" seemed to have succumbed to the appeal of easy money and abandoned whatever environmentalist principles they might ever have had. Not only were logging activities damaging their forests (more through the construction of logging roads than the actual cutting of the widely dispersed mahogany trees), but the proliferation of gold mining was polluting their rivers both with mud and, far more seriously, mercury. Suspicious miscarriages and other birth defects began to appear at the most seriously affected community, Gorotire, where tests showed many individuals with shockingly high levels of mercury poisoning. Agents of the National Foundation for the Indian (FUNAI), NGO representatives, anthropologists, public health workers, and their own senior chiefs repeatedly explained to the residents of Gorotire and the other villages involved in the gold concessions the nature of the danger and warned against the cumulative effects of continued gold mining using the mercury-based Placer technique, all to no apparent avail (Gonçalves et al. 1993).

Appearances turned out to be deceptive, however. In the villages, the sense that the concessions to Brazilian extractivists were cumulatively threatening irreversible damage to the life-sustaining ecosystem and the health and social well-being of the affected communities was spreading and intensifying. Awareness of the extent and seriousness of the environmental damage and the associated social distortions was gradually building to a level of concern for the survival of the society and its environmental resource base analogous to that which had motivated the demonstration against the great Xingú dam scheme at Altamira. The mounting tension finally exploded in an event as spectacular in its way as the Altamira rally itself.

This upheaval took place in the autumn of 1994 in the key village of Gorotire, the largest of the Kayapo communities. Gorotire, it will be recalled, is both the largest Kayapo village, with a population of slightly over one thousand, and the one that was most heavily involved in gold mining and logging concessions. It was therefore also the Kayapo community with the longest and most intense experience of the environmental costs of logging and mining: disturbed forest areas resulting in diminished hunting and foraging, fouled water unsuitable for fishing and drinking, and mounting levels of mercury poisoning. For ten years, the people of the village had put up with these problems as the price of their entry into the Brazilian commodity economy. The symptoms of mercury poisoning were slow to appear, large areas of forest remained intact, hunting was still good in some places, and there were still fish to be caught and good water to be found in small streams and springs, even if the main river and its fish were no longer usable. The environmental effects, however, were cumulative and increasingly affected the daily lives of the average villager in ways that could no longer be ignored.

The young leaders in their townhouses in Redenção, of course, were insulated to a great extent from these effects. This difference was added to the increasingly obvious fact that most of the economic benefits of the timber and mining concessions were going directly to support the lifestyle of these leaders in the towns, and to a less extent of the senior chiefs in the village, but few were trickling down to the ordinary villager. Even the medical and educational services, to pay for which the concessions were ostensibly given in the first place, were often neglected, with medicines and school supplies sporadically out of stock and teachers' salaries sometimes unpaid. The villagers saw themselves increasingly as standing to lose, while the leaders who dealt with the Brazilian loggers and miners stood to gain, by these arrangements. They gradually became aware that this situation was not merely a temporary or contingent state of affairs but a new social order based on the system of logging and mining concessions mediated by their leaders, an order that was likely to continue as long as the prevailing arrangements remained in force. They became, in so many words, class-conscious; it was a short step to class struggle.

That step was finally taken with the influential urging of a new element in village society: members of the younger generation just reaching full adulthood, many of whom had some education and acquired the same basic language, literacy, and numerical skills that had hitherto been the monopoly of the first generation of young leaders. Enough young people have now acquired these skills, either at the sporadically functioning village school or in some cases at Brazilian schools in the town, to neutralize the monopoly of the current young leaders. As ordinary villagers themselves, they tend to identify with the disfranchised common villagers, to whom they are able to explain the logging and mining contracts with the Brazilians and the distribution of the theoretically communal revenues those arrangements provide. They are the organic intellectuals of the new class politics of the Kayapo, a politics that in the crisis of autumn 1994 at Gorotire found its synthetic expression in a revolt of the commons backed at the crucial moment by traditional chiefly authority.

In August 1994, the men of Gorotire embarked on a collective hunting trek, as a normal part of a communal naming ceremony. As is customary on such occasions, the bachelor youths and recently married fathers acted as collective units, functioning separately from the older men. While the senior men were off hunting, the younger men met together to discuss the logging and mining contracts. Most had become bitterly opposed to both forms of extractive activity, partly because of their effects on the environment and health and partly because they, and the community as a whole, were seeing too little of the benefits the contracts had been supposed to bring. Those who had actually worked in the gold mines alongside Brazilian miners and had seen with their own eyes how much wealth the Brazilians were taking out of their land, and how little they were giving back, took the lead in challenging the concessions.

Those calling for an end to the concessions carried the day in the young men's meeting. The young men as a group then confronted the senior men when they returned to the camp. One of the senior chiefs was present, and he and the rest of the older men resolved to support the younger men. All the men left the hunting camp and marched directly to the main Gorotire gold mine of Santidio, where almost three thousand Brazilian miners were then working. The Kayapo assaulted the mine, burned down the miners' shelters, broke their machines and threw them in the flooded pits they had dug into the landscape. They then drove off the unresisting miners, who were obliged to walk seventy-five kilometers through the forest to the nearest road.

Most of the angry and bewildered miners made their way back to Redenção, where two thousand of them camped on the central boulevard of the town and commenced intimidating the citizens, threatening to burn down the offices of FUNAI, and calling on the mayor to help them get back into the reserve if he valued the peace of his city. Meanwhile, the Gorotire returned to their village, where they immediately obtained the support of the other senior chief for their collective demand that the mining of gold should cease and all miners be expelled from Gorotire

territory. The senior chiefs ordered the younger leaders, their own sons
and nephews, to repudiate the contracts they had negotiated and to confirm the order to expel the miners. The young leaders, who might have
resisted either the authority of the senior chiefs, if unsupported by an
aroused populace, or the demands of a mob of villagers unsanctioned by
chiefly authority bowed to their combined political pressure and authority and canceled the contracts (Turner 1992c, 1995a).

The federal prosecutor's office also played an influential role in the
intensifying opposition to the logging and mining contracts. Logging
and mining on indigenous lands are illegal under the terms of the Brazilian federal constitution, which specifies that mining operations on indigenous land require a special vote of the federal congress. There had been
no such votes on any of the mining activities in Kayapo territory. In early
1964, an activist federal prosecutor began calling for the expulsion of
miners and loggers by federal police. Kayapo leaders blocked this move
by threatening armed resistance to any police attempting to enter the
reserves to enforce the order. After the Gorotire revolt, however, the
prosecutor called again for enforcement of his order, this time obtaining
the acquiescence of Kayapo leaders at a meeting in Redenção on December 10th. On December 15th, federal police began expelling miners and
loggers from the Kayapo Indigenous Territory, the reserve that includes
Gorotire and the other villages then involved in extractive concessions.

Both logging and mining contracts were included in the general sus-
pension of extractive activity. The ban on mining in officially demarcated Kayapo reserves has remained in effect until the present, with Kayapo support. As of this writing (September 1998), an attempt to revive
production at the Santidio mine under Kayapo management, combining
mining operations carried out by Kayapo labor with reclamation of the
ecological damage, approved after much controversy by the Gorotire villagers in 1996, has remained blocked since that time by FUNAI and
other political forces in the Brazilian government, which has shown itself
fiercely opposed to projects granting indigenous communities control
of economically significant production and income, even on recognized
indigenous territory (Turner n.d.). Logging has been selectively renewed in some areas, especially the Menkragnoti reserve to the west of
the Xingú River. Logging in the Kayapo area has been almost exclusively
for mahogany, the most valuable of the tropical hardwoods, but one for
which the Kayapo have no indigenous uses. Mahogany grows sparsely,
with individual trees scattered at wide intervals through the forest. The
logging roads that must be cleared to take out the huge trunks cause

more damage to the forest than the cutting of the individual trees, but even so, damage to the forest ecosystem from mahogany logging operations is minimal. One can fly over forest areas that have been logged for mahogany without seeing breaks in the canopy. As in the case of Kayapo garden clearings, the forest quickly closes over and reclaims cutting sites and logging roads. Except for a few areas, such as part of the Catete reserve, there has been no clearcutting or harvesting of other species. Most Kayapo therefore do not regard the mahogany logging operations of their Brazilian concessionaires as serious threats to their environment, despite their temporary damage and negative effects such as driving off game. By Kayapo cultural standards, they are not ecologically deleterious to anything like the extent of the massive dams or gold mines against which they have mobilized militant actions. Kayapo opposition to logging contracts has therefore been less consistent and more ambivalent. It has focused more on issues of underpayment and cheating on contracts by loggers and inequitable distribution of the income from the contracts by Kayapo leaders than on any perception of serious ecological damage. The issue is rapidly becoming moot, as commercially interesting mahogany trees are logged out. The level of logging activity in the eastern Kayapo territories around Gorotire has already fallen to a level far below its height of five to ten years ago, as a result of the combination of contract cancellations and opposition by Kayapo communities, sporadic enforcement by federal police and other government agencies such as Instituto Brasileiro da Amazônia (IBAMA) (the agency charged with protecting the environment of Amazonia), and dwindling stocks of loggable mahogany.

Conclusions: The Indigenous Political Struggle for Popular Empowerment and Environmental Conservation

The Gorotire crisis, which marked the collective repudiation by an important indigenous community of what had become the leading example of collusion by an indigenous Amazonian people in environmentally destructive extractive activity, is a portentous event with potentially general implications. It represents a further example of the Kayapo's ability to develop new and different conceptual perspectives, political tactics, and economic policies toward their environment in response to changing political, economic, and ecological circumstances. This record of political struggle and accomplishment stands as a salutary corrective to the still prevalent view of indigenous societies as monolithic "cultures" lacking the political capacity to adapt to new situations or to change their collective ideas and perspectives on subjects such as their relations with their environment.

As at Altamira five years before, the Kayapo acted to protect their environment from the effects of destructive extractivism, despite the large income it was bringing their communities. Once again, the precipitating factor was neither a spiritual reverence for nature (which the Kayapo lack) nor a "scientific" grasp of ecological systematics (which they also lack), nor yet environmentalist principles like those of Western "greens," but rather a sense that the damage being inflicted on the environment by commercial extractive activities had reached a point that threatened the reproduction of their own society, both by eroding its natural resource base *and* by distorting its social relations and political processes to an unacceptable degree. The fundamental cultural presupposition underlying this sense is the notion explained above that human society and the natural world form interdependent components of a single great process of reproduction, which is sustained and renewed through myriad productive activities. These acts of production involve destructive transformations of natural entities, but for their own replication require that the natural world as a whole remain able to reproduce itself, and thus replace the entities and resources thus transformed.

Just as at Altamira, Kayapo cultural attitudes toward the environment were important in the movement to expel the miners and loggers. Again, it was not a concern for nature in the abstract or the value of natural beings as such, but the sense that the continuity of the ecosystem as a whole is essential to the reproduction of Kayapo society that underlay the collective resolve to act to prevent further environmental damage. The general implication appears to be that when a point is reached where the ecosystem is seriously threatened, as by the destruction and pollution of rivers and their fish by mining, or when so much of the forest is damaged by logging that hunting and gathering become increasingly unproductive, the indigenous people will resist if they can, even if the policies responsible for the destruction have been sanctioned by their own leaders and produce a modicum of wealth in Brazilian goods and devices such as cars, motorboats, and airplanes.

The Kayapo example also serves to suggest that the notions of indigenous peoples as "primitive ecologists" with environmentalist values analogous to, if more "spiritual" than, our own and the opposite conception of them as destructive exploiters, culturally indifferent to environmental values and ready to seize any opportunity for short-term profit from any form of destructive assault on the ecosystem are false alternatives and do not exhaust the real spectrum of indigenous environmental relations. The context of human-environment relations in the Amazon, as in many parts of the world, is one of rapid change between a regime of subsistence and a capitalist economy dominated by extractive forms of production for profit. In this historic context of fundamental change, "culture" does not stand still, an inert body of "traditional" concepts and attitudes, like a cookbook of some bygone era no longer able to deal with today's ingredients but incapable of change. Nor are "culture" and "politics" mutually exclusive ways of relating to environmental problems, such that indigenous peoples operate only with "culture" and we with "politics." As the example of the Kayapo shows, indigenous communities, like ourselves, are fully capable, given the opportunity, of transforming their relationship to their environments through processes of internal political struggle, combined with active relations of resistance and accommodation to the ambient society and economic system.

That indigenous communities like the Kayapo may arrive, as a result of these processes, at environmentally sound policies and practices for reasons other than environmentalist principles in our sense should give us no cause for concern. On the contrary, it seems to me that the Kayapo experience may point to a moral for us. If environmentalist causes are to succeed politically in our own society, it will be because they become progressively more integrated into our consciousness of how environmental exploitation and despoliation are inseparable from the degradation of people in class society and how the way we live with nature is an integral part of the way we reproduce ourselves and our social world.

NOTE

This paper is based on fieldwork carried out on numerous visits to the Kayapo between 1989 and 1998. These trips have included extensive cooperation with Brazilian indigenous support organizations, chiefly the Centro Ecumenico de Documentação e Informação and the Nucleo de Direitos Indigenas and its successor, the Instituto Socio-Ambiental (ISA); the Fundação Mata Virgem and its successor, the Associação pela Vida e Ambiente, which merged with ISA in 1995; the Centro de Trabalho Indigenista and agencies of the federal government of Brazil, including the National Foundation for the Indian and the Secretaria da Amazonia Legal of the Ministerio do Meio Ambiente. The trips were funded by the Center for Latin American Studies of the University of Chicago, the Lichtstern Fund of the Department of Anthropology of the University of Chicago, the Spencer Foundation, Granada Television International, ISA, Conservation International, the Harry Frank Guggen-

168 SOCIAL STUDIES

heim Foundation, and the Cornell International Institute for Food, Agriculture and
Development.

REFERENCES

Cleary, David. 1990. *The anatomy of an Amazon gold rush.* Oxford: Basingstoke in
association with St. Anthony's College.

Destro Junior, Nelson Cesar. 1995a. Relatório de Viagem a ADR Redenção. Report
to Fundação Nacional do Indio, Brasília, Brazil.

———. 1995b. Relatório de levantamento efetuado nos PV's Purure e Nhakim na
Área Indígena Kayapo. Report to Fundação Nacional do Indio. Brasilia, Brazil.

Espirito Santo, Marco Antonio do. 1995a. Relatório sobre o momento Kayapo e
perspectivas. Report to Fundação Nacional do Indio, Brasilia, Brazil.

———. 1995b. Relatório de Viagem ADR de Redenção-Operação Kayapo. Report
to Fundação Nacional do Indio, Brasilia, Brazil.

Ferrari, I., et al. 1993a. Saude, garimpo e mercurio entre os Kayapo: Estudo explora-
torio. *Salusvita,* Bauri 12:113–126.

Ferrari, I., et al. 1993b. Mutagenicity, health, goldmining and mercury: An epidemi-
ological study of chromosomic aberrations in miners, Indians and controls in Brazilian Amazon regions. Paper presented at the Sixth International Conference on
Environmental Mutagens, Melbourne, Australia.

Fisher, William. 1994. Megadevelopment, environmentalism and resistance: The institutional context of Kayapo indigenous politics in Central Brazil. *Human Organization* 53:220–232.

Gianninni, Isabelle Vidal. 1994. Indigenous peoples and forestry management alternatives in the Brazilian Amazon. Report of an experiment in progress to develop
a management plan for the natural renewable resources in the indigenous area
of the Xikrin do Cateté. "Traditional Peoples and Biodiversity Conservation
in Large Tropical Landscapes." Case study. Institution Sócio–Ambiental. São
Paulo, Brazil.

Gonçalves, Aguinaldo. 1994. Resultados dosimetricos e genotoxicos Kayapo—
aldeias Gorotire e Djudjetuktire [Kikretum]. Relatório do Projeto fomentado pela
Fundação Mata Virgem. Rain Forest Foundation. Archive of the Socio–
Environmental Institute. São Paulo, Brazil.

Gonçalves, Aguinaldo et al. 1992. Garimpos, mercurio e contaminação ambiental.
In *Saude e desenvolvimento: Vol. II. Processos e consequencias sobre as condições de vida,*
edited by N. C. Leal, São Paulo, Brazil: Hucitec-Abrasco.

———. 1993. Garimpo, mercúrio, saude e atividade física em aldeias Kayapo: Pecu-
liaridades observadas entre gestantes. Paper presented at VIII Congresso Brasileiro de Ciencias. Brazilian Congress of Sciences. Belem, Brazil.

Parker, Eugene. 1992. Forest islands and Kayapo resource management in Amazonia: A reappraisal of the apete. *American Anthropologist* 94:406–427.

———. 1993. Fact and fiction in Amazonia: The case of the *apete. American Anthropologist* 95:715–723.

Posey, Darrell A. 1982. The keepers of the forest. *Garden* 6:18–24.

————. 1985. Indigenous management of tropical forest ecosystems: The case of the Kayapo Indians of the Brazilian Amazon. *Agroforestry Systems* 3:139–158.

————. 1993. Reply to Parker. *American Anthropologist* 94:441–443.

Redford, Kent H. 1990. The ecologically noble savage. *Orion Nature Quarterly* 9(3):25–29. Republished *Cultural Survival Quarterly* 15(1):46–48.

Redford, Kent H., and Allyn MacLean Stearman. 1993. Forest dwelling native Amazonians and the conservation of biodiversity: Interests in common or in collision? *Conservation Biology* 7:248–255.

Stearman, Allyn MacLean. 1994. Revisiting the myth of the ecologically noble savage in Amazonia: Implications for indigenous land rights. *Bulletin of the Culture and Agriculture Group* 49 (Spring):1–6 (American Anthropological Association).

Schmink, Marianne, and Charles H. Wood. 1992. *Contested frontiers in Amazonia* New York: Columbia University Press.

Turner, Terence. 1989. Altamira: Paradigm for a new politics? Paper presented at symposium, The Kayapo Offensive, Annual Meeting of the American Anthropological Association, Washington, D.C., November 1989.

————. 1991a. Representing, resisting, rethinking: Historical transformations of Kayapo culture and anthropological consciousness. In *Post-colonial situations: The history of anthropology*, edited by G. Stocking, Madison: University of Wisconsin.

————. 1991b. Baridjumoko em Altamira, *Provos Indígenas de Brazil: Aconteceu*, Editora CEDI, São Paulo, Brazil. 337–338.

————. 1992a. Os Mebengokre Kayapo: De comunidades autónomas à sistema inter-étnica. In *Historia dos Indios do Brasil*, edited by M. Carneiro da Cunha. São Paulo, Brazil: Companhia das Letras. 311–338.

————. 1992b. Defiant images: The Kayapo appropriation of video. *Anthropology Today* 8: Also appeared as "Imagens desafiantes: A apropriação Kaiapo do video. In *Revista de Antropologia*, 36, São Paulo. 1993: 81–122.

————. 1992c. Viagem aos Kayapo, 11–24 Julho 1992. Report to Centro Ecuménico de Documentação e Informação, São Paulo, Brazil.

————. 1993b. The role of indigenous peoples in the environmental crisis: The example of the Kayapo of the Brazilian Amazon. *Perspectives in Biology and Medicine* 36:526–545.

————. 1995a. A sociedade Kayapo. Memorandum presented at Seminário Kayapo, Instituto Socio-Ambiental, 13–14 March, São Paulo, Brazil.

————. 1995b. Relatório: Aspectos da atualidade Kayapo, Setembro 1995. Report to Fundação Nacional do Indio, Brasília, and Instituto Socio-Ambiental, and São Paulo, Brazil.

————. 1995c. An indigenous people's struggle for socially equitable and ecologically sustainable production: The Kayapo revolt against extractivism. *Journal of Latin American Anthropology* 1:98–121.

————. 1999. Extrativismo mineral por e para comunidades indígenas da Amazonia: A experiencia de garimpo entre os Waiãpi e os Kayapo do Sul do Pará. Forthcoming in *Cadernos do Campo* (São Paulo, Brazil: Editor Universidade Federal de São Paulo) and forthcoming as "Mineral Extraction by and for Indigenous Amazonian Communities: Gold Mining by the Waiãpi and Kayapó," In *Mining, Oil, Environment, People, and Rights in the Amazon*, edited by Leslie Sponsel.

BARBARA EPSTEIN

Grassroots Environmental Activism

The Toxics Movement and
Directions for Social Change

SEVERAL YEARS ago I wrote a book on one phase of radical grassroots environmental activism, namely, the nonviolent direct-action movement of the late 1970s and early 1980s, revolving around mass civil disobedience mobilizations against nuclear power and nuclear arms. The movement was initiated by young people, mostly white, mostly of middle-class origin, who were part of a radical subculture with ties to the women's movement and roots in the movements of the 1960s. Many of the activists in the nuclear power/weapons movement hoped that it would develop into a broader movement for a nonviolent, egalitarian, ecologically oriented society. But the movement was unable to sustain itself past the decline of its two major targets—the nuclear power industry and the arms race. In the mid-1980s, the movement that had formed around these issues faded.

While the direct-action movement was declining, another grassroots environmental movement, this one based on a very different constituency, was gaining strength. This movement (which might be described more accurately as an arena of activism, or a series of overlapping arenas) is concerned with toxic hazards and the threats they pose to health. It includes local groups opposing hazardous waste sites in their neighborhoods and environmental justice groups that stress the disproportionate exposure of communities of color to such hazards. It also includes groups concerned with toxics at the workplace and their effect on workers; farmworkers concerned with pesticides; and women's groups concerned with the link between environmental hazards and women's health, especially breast cancer. This constellation of movements has become quite extensive. Starting from local, particularized concerns about hazardous chemicals, antitoxics activists are now linking many issues and developing an analysis that has radical social implications. In this essay, I summarize the development of grassroots activism around toxics, describe the diversity of constituencies that it draws on, and comment on the movement's strengths and weaknesses. I also address the relationship of humanist academics to grassroots environmental activism, specifically, why this relationship has thus far been minimal and how it might become more significant.

What was at first called the hazardous waste movement began to emerge in the late 1970s out of isolated, spontaneous protests in places where groups of people became aware that they and their communities were being affected by chemical waste dumps. The issue of chemical waste was a product of World War II and its aftermath. The development of military technology during the war and the industrial expansion that took place after the war were accompanied by an enormous expansion in the production of toxic substances. These were disposed of, during and after the war, more or less the way other industrial wastes were disposed of—in any site that was available and cheap.

Though the environmental movement of the late 1960s and 1970s made the question of pollution a priority, and some federal legislation was achieved on this issue during the seventies, the issue of hazardous waste did not become a focus of activism or public concern until 1978. What brought hazardous waste to the attention of the public was Love Canal, in Niagara Falls, New York. Over several decades, highly toxic chemicals had been dumped in an abandoned canal by the Hooker Chemical Company, which, when the canal's capacity as a dump was used up, leased the area for $1 for the construction of a school. In 1978 Lois Gibbs, a Niagara Falls resident, began investigating the high rate of health problems among children attending the school, including her son. When she and others confronted the state health department and other government officials, they met denial and resistance. They organized, and were able to force the government to acknowledge the problem, close down the school, and provide some assistance toward relocation. The Love Canal Homeowners' Association, which led this struggle, consisted mostly of women from the middle- and lower-middle-class families that made up the community. These were people with no previous political experience, people who had not thought of themselves as environmentalists until forced into a battle over environmental issues.[1]

Covered by the national media, Love Canal increased people's awareness of the connections between toxics and health and no doubt gave some people the sense that they had the right to protest such conditions and might win something. Before Love Canal, the press had reported a dozen or so local protests against toxics, but none had created much of a stir. After Love Canal, protests proliferated. In 1981 Gibbs established the Citizens' Clearinghouse on Hazardous Wastes (CCHW) to assist community groups engaged in struggles against toxics. By the end of 1984 the CCHW had served six hundred community groups; by the end of 1985, more than a thousand; and by 1988, more than forty-five hundred. Meanwhile, in 1984, the National Toxics Campaign was formed to address the issue of toxic wastes; along with other activities, it served community groups. Greenpeace also became involved in helping community groups addressing issues of toxics. These organizations gave assistance to several thousand local groups.[2]

In the early 1980s struggles against toxic waste sites began to emerge among communities of color; over the last decade, this has become the most rapidly growing part of the toxics movement. In 1982 North Carolina decided to build a PCB disposal site in Warren County, a rural area with a high rate of poverty, a large African American population, and a sizable Native American population. Residents, mostly from these two groups, organized a series of protests and marches, culminating in a demonstration in which they attempted to block the trucks carrying wastes to the site. Five hundred twenty people were arrested. This protest failed to stop the building of the disposal site, but it did inspire other protests in similar communities. Some efforts were successful: in 1985 the Concerned Citizens of South Central (Los Angeles), an African American environmental group, defeated efforts to build a toxic wastes incinerator in their neighborhood. An attempt to transfer this site to a nearby Latino community was also defeated by the joint effort of the Concerned Citizens and the largely Latino Mothers of East Los Angeles. With the support of a number of other community and environmental organizations, they sued the government and won.

The coalescence of what was coming to be called the environmental justice movement was encouraged by the publication, in 1987, of a document titled *Toxic Wastes and Race in the US*, by the Commission for Racial Justice of the Church of Christ. This document described the racial and class compositions of the immediate environs of toxic dumps throughout the United States, and reported that the location of these dumps was associated even more strongly with race than with class. This report gave further legitimacy to the claim that disposal of toxic wastes is, among other things, linked to racism. Meanwhile, environmental activists of color were raising the issue of the whiteness of the mainstream environmental groups and their lack of interest in issues affecting people of color. In 1990, environmental activists of color sent letters to the Big Ten, the mainstream environmental organizations, raising these issues and demanding that the numbers of people of color on the staffs and boards of these organizations be increased. This led to widespread debate throughout the environmental movement. The most extensive changes took place in progressive organizations such as Greenpeace and the National Toxics campaign. In the mainstream environmental organizations, the response was more limited. In some cases, people of color were hired as staff or invited onto boards, and there were acknowledgements that the agenda of environmentalism must go beyond wilderness issues to include environmental justice and the effect of toxics on health. But on the whole these issues have remained low priorities for the mainstream environmental organizations.[3]

In 1991 many leading environmental activists of color, a number of them based in a series of regional networks, gathered together in the People of Color Environmental Leadership Summit in Washington, D.C., and ratified a set of principles concerning environmental racism, control of the production of toxic substances, disposal of toxic wastes, and other sources of pollution. There was discussion of whether or not to form a national organization, but many people felt that it was important to maintain the local and regional character of the movement. For several years antitoxics activists had been using the phrase "we speak for ourselves" to indicate their unwillingness to be dominated by experts or other outsiders; at the Washington conference, the phrase was used to mean refusing to be represented by people unfamiliar with the experience of a particular community. The conference decided to develop nationwide networks but to avoid forming a national organization. These networks (including the Southwest Network for Environmental and Economic Justice, the Gulf Coast Tenants' Organization, and others) have come to play a major role in defining the agenda for the environmental justice movement and grassroots environmentalism as a whole.

The decision by leading environmentalists of color not to form a national organization reflected a commitment to localism—to the importance and the autonomy of local groups—that is widespread among grassroots environmental activists, whites as well as activists of color. However, this conception of a local focus is seen as entirely compatible with the growth of informal national networks and a consciousness, on the part of local activists, that they are part of a national movement. This consciousness encourages a breadth of perspective that has not always been apparent. In the early stages of the movement, many people involved in struggles against toxic facilities simply wanted to get them out of their own communities and were not very concerned about whether they might be located elsewhere instead. Since white communities were the most likely to be able to drive toxic dumps out of their communities, a purely localist orientation had racist implications: dumps could be sited in communities of color.

By the mid-1980s this insularity was being replaced by a broader perspective. Activists increasingly understood themselves as members not just of local groups but of a movement, concerned with keeping toxics out of everyone's neighborhood. Local groups began to help each other in opposing toxics; the national networks, such as the CCHW, worked to oppose toxic hazards anywhere in the nation; the acronym NOPE (Not on Planet Earth) expressed the movement's commitment to doing away with toxics altogether. As the movement has developed, the perspective of activists has broadened in other ways as well. Though the vast majority of local efforts are defensive in that they focus on preventing the siting of toxic facilities, it would be difficult to find an experienced activist who is not aware that solving the problem of toxics requires changes in the production process. Though some businesses might voluntarily change over to environmentally sound technologies, it would be unrealistic to expect most to do so. Massive government intervention would be required, and it would have to be accompanied by a shift in values, giving public welfare priority over the freedom of the marketplace. This would involve a dramatic departure from the current celebration of private enterprise, its elevation over the public sector.

It is debatable whether the goal of an environmentally clean economy could be achieved under capitalism. An environmentally sound capitalism might be theoretically possible, just as a capitalist system free of racism and sexism might be. But in fact, U.S. capitalism has come to rely extensively on practices that are environmentally unsound. Some industries would have to make extensive changes, and others, such as the petrochemical industry, which rests on the exploitation of a nonrenewable resource, presumably would cease to exist in their present form. The degree of government intervention that this would require would be difficult to reconcile with the ideology of the free market. Though aware of the potentially radical implications of demanding that the production of toxics be curtailed, most activists avoid condemning capitalism or production for profit and argue instead that production processes could be changed to accord with environmental needs without serious economic losses. This is no doubt tactically wise: in the current climate, calls for socialism or even social democracy would be likely to seem utopian and tend to marginalize the movement. But failing to address the issue means avoiding the question of what it will require to achieve a relatively toxic-free environment.

The movement has also broadened its concerns from hazardous waste dumps to the problems posed by toxic contamination of the environment more generally. In the early 1980s, the efforts of local groups were focused more or less exclusively on the problem of hazardous waste sites: getting rid of them or preventing their construction. Over the eighties, many groups began to address broader issues of industrial pollution; they began to develop alliances with groups concerned with farmworkers protesting the use of pesticides and with occupational health and safety groups. National networks around military toxics and toxics in high-technology industries have encouraged the growth of local groups dealing with these issues. In recent years, links have begun to be developed between the women's health movement, particularly women's groups concerned with the impact of environmental toxins on health, and toxics and environmental justice groups.

Going beyond a focus on neighborhood issues to broader questions often provides a basis for bringing together white activists and activists of color, as the two national networks concerned with military toxics show. Developing ties between activists of color and white activists has been an explicit priority among women activists addressing the link between health and the environment. Over the past few years links have begun to be developed between activists in the women's health movement who see their work as connected to environmental issues and women from the grassroots environmental movement concerned with environmental dangers to health. In April 1994, a conference was held in Austin, Texas, sponsored by Greenpeace and the Women's Environmental and Development Organization on breast cancer, toxics, and nuclear radiation. The conference brought together an interracial group of women activists from around the country, from the women's health, environmental justice, toxics, and antinuclear movements. In a number of cities, umbrella organizations have formed to coordinate the efforts of local groups concerned with this range of issues. In May 1995, Greenpeace sponsored a meeting of leading women activists from the environmental justice, toxics, women's health, and antinuclear movements. This resulted in the formation of the Coalition for Health and Environmental Justice, which is encouraging the formation of local networks within which activists from these movements can work together and support each other's efforts. On the local as well as the national level, activists see it as a priority to maintain racial balance in these efforts and to build an interracial leadership.

The toxics/environmental justice movement is largely made up of people who are poor, working class, or lower middle class, many of whom live in rural communities or poor neighborhoods of cities. It is overwhelmingly a movement of women; this is particularly true among whites. (Most networks of organizations of people of color are led by men, many of whom have roots in the movements of the 1960s; virtually all of the leadership of white grassroots groups is female.) It is impossible to count the members of this movement, and in a certain sense there is no such thing as membership: this is a movement made up of local grass-roots groups, most of which come together around a particular issue and then disband when that struggle comes to an end. The CCHW claims that 80 percent of the leadership of the movement is female; it may be that female membership of local groups is even higher.

There has been considerable speculation about why women are more likely than men to become involved in this movement. Some argue that women are more concerned than men with health issues, particularly the health of their own children. My sense is that local toxics groups are likely to be formed by women rather than men because women are often at the center of already-existing community networks. Groups that take it upon themselves to address the issue of toxics are likely to be made up of people who already consider themselves responsible for community welfare and have the networks that make organizing possible. In the United States at least, women are particularly likely to occupy this role. If a local group is formed by women, it is likely to recruit women more readily than men. Furthermore, in the communities in which these groups are formed, men are more likely to define their identity around their work. Women, whether they have jobs or not, are likely to be described by the press and local officials as housewives. Even if they have jobs and also the primary responsibility for children and housework, the women in these communities seem to be more likely than men to think they have time for community work such as this.

The toxics/environmental justice movement has had an impact on several levels. Public consciousness of the health implications of toxics has been raised; it is much more difficult to build a new chemical dump now than it was ten or fifteen years ago. There is increased public consciousness of racial disparities in exposure to toxics. The movement has had some success in relation to local campaigns: it has been particularly effective in preventing the siting of new hazardous facilities. The movement has also changed the consciousness of its own participants in important ways.

Particularly among whites (meaning mostly white women), involvement in the toxics movement is likely to be the first experience of political activity. Previously apolitical women who become involved can find themselves changing their views on a range of issues. The degree of impact varies. Women who participate in a local struggle and end their involvement when that issue is resolved may gain a more skeptical view of the government and the corporations; greater respect for dissent, and perhaps more awareness of the unequal treatment of women, people of color, and poor people. Women who become leaders of local groups, especially those who continue their involvement in the movement after the local struggle that drew them into the movement has been won or lost, are likely to find their outlooks changing in profound ways. An example would be Penny Newman, who lives in the small rural community of Glen Avon, in a canyon in the mountains east of Los Angeles, California. Penny and her former husband, a fireman, moved to Glen Avon in the early 1970s because they thought it would be a good place for their children to grow up. In the mid-seventies, while Penny was the chair of the local PTA, she became aware of the existence of chemical dumps—the Stringfellow Acid Pits—in the hills above the town: a woman asked her to allow someone to speak about this issue at a PTA meeting; Penny refused. She didn't think about the issue again until the winter of 1978–1979, when there were heavy rains. As later became clear, the management of the pits became concerned that the walls of the dumps might collapse; they decided to pump out a million gallons of chemicals and allow the stream to run down the side of a road, through the town. This stream of chemicals passed in front of the elementary school. For several days the children were playing in the foam that formed in front of their school. The teachers had been informed about the chemical release: they had been cautioned not to tell parents about it and had been told that in case of an emergency—that is, if one of the walls were to collapse, releasing a large flow of chemicals—a bell would sound. If there were one bell, they should put the children in a bus that would be waiting outside. If there were two bells, they should put the children on the desks and hope for the best. A few teachers disobeyed their orders and told the parents, and Penny and others—women whom she knew through the PTA and other community groups—organized the Concerned Neighbors in Action. Through vigorous organizing they were eventually able to get the dumps shut down and to put the community on the Superfund list as a priority for a federal cleanup.

Penny told me that she and the other women involved became very angry at the government agencies that had allowed the acid pits to function. The corporation that ran the pits, she said, was not there to protect the community; but the health department was, and it was very difficult to understand why government officials were so anxious to protect the corporation. "Realizing this was hard for people," she said. "It was like everything you were ever taught was a lie. Once you crossed the line of realizing that, there was no going back. It changed the way you looked at everything." When I asked, "What, for instance?" Penny said, "You began to question everything. I began looking at other issues. I saw farmworkers complaining about being sprayed, and it began to make sense to me. I reregistered from being a Republican to being a Democrat. I started looking at feminism, because of the way I was being treated by the agencies. However much information we had, we were still dismissed as hysterical housewives. I began to see, where people raised questions about injustice, there was probably a root to that." Penny added that although she had become comfortable thinking of herself as both a feminist and an environmentalist, many women in the movement were uncomfortable with these terms, thinking of them as referring to people of the upper middle class. She added that many more women than men had become involved in the movement and, in some cases, where husbands had not supported their wives' activities, there had been divorces.

The process of becoming involved in the movement and the impact it has on people's thinking seem often to be different in communities of color than in white communities. (Though these terms can be misleading. With the exception of Indian tribes living on reservations, there are relatively few communities affected by hazardous wastes that consist of people of only one race; it is more accurate to speak of predominantly white, or African American, or Latino communities.) The organizations that form in communities of color are more likely to be made up of people who either have some previous experience of activism themselves or have friends or relatives who have been activists. Memories of the civil rights movement are much stronger among African Americans, and among people of color generally, than among whites. For many whites, political protest carries an association of deviance, of abandonment of respectability. Coming to see the government as not on one's side can be a shock. People of color are less likely than whites to be surprised when government agencies fail to protect them, and becoming involved in

protest activities is not as likely to be experienced as a wrenching process. People of color are less likely to regard the authorities as their allies
and more likely to be comfortable with dissent and protest. For instance,
an African American activist from Rahway, New Jersey, recounted:

> When they sited the incinerator for Rahway, I wasn't surprised. All you have
> to do is look around my community to know that we are a dumping ground
> for all kinds of urban industrial projects that no one else wants. I knew this was
> about environmental racism the moment that they proposed the incinerator.[4]

The toxics/environmental justice movement has not received much
attention from humanists. This is not due to a lack of interest in social
movements: particularly in areas such as cultural studies, American studies, and women's studies, but also in the more conventional disciplines,
there is widespread interest in social movements and a great deal of writing in a spirit of solidarity with social movements. The social movements
that are invoked, however, tend to be those that revolve around identity
politics, especially those with a strong countercultural orientation. The
toxics/environmental justice movement does not fit this model. Intellectuals help social movements in many ways: by analyzing the conditions
that the movements are concerned with and suggesting ways of dealing
with them, by describing social movements sympathetically and increasing their public visibility, and by introducing students to activism. These
things cannot happen if a movement goes unnoticed because it does not
fit prevailing definitions or theories.

During the 1980s, many academics who both study and have some
links to progressive social movements became interested in the body of
writing that is broadly described as new social movement theory. Developed largely by a group of European scholars (Alain Touraine, Alberto
Melucci, and others) new social movement theory argued that the movements of the 1960s and beyond represented a sharp break from the "old"
movements, centered around class issues and the vision of socialism, that
had held sway from the mid-nineteenth to the mid-twentieth century.
Though a range of views have been expressed under the broad rubric
"new social movement theory," on the whole this theory argued that new
movements were, and would continue to be, concerned more with opposing the state and constructing community and cultural identity than
with economic issues and that they were based more on groups defined
in terms of identity than on those defined in terms of class. New social
movement theorists argued that the new social movements called for a
post-Marxist theory, one that would privilege issues of culture, identity,

and community and would be influenced more by an anarchist sensibility of opposition to centralized authority than by a Marxist vision of socialism.

Many of the movements that emerged in the late 1970s and early 1980s, especially in western Europe and to some extent in the United States as well, seemed to bear this approach out at least in some ways. The feminist, gay, and lesbian movements all emphasized identity and the defense and construction of alternative communities. These movements and the antinuclear movement were to a large extent infused by an antiauthoritarian, antistate sensibility; more activists identified with anarchism than with Marxism. These movements were also largely white and based on people of middle-class origins, even if their current circumstances were outside what is ordinarily thought of as the middle class. New social movement theory did not fit movements of people of color, even in Europe and the United States, quite as well. Movements of people of color are concerned with issues of identity and community, but the strong anarchist sensibility characteristic of movements of young, middle-class white people has not been nearly as prevalent in movements of people of color. To be sure, these movements have been critical of the state, but they have not generally put forward a vision of a decentralized or stateless society. And large sections of these movements have continued to be concerned with issues of class and economic justice.

In some respects, the toxics/environmental justice movement confirms new social movement theory. This is not a class movement; though it is made up mostly of people on the lower economic rungs of society, it reaches out to anyone who is affected by or concerned about toxics. Issues of identity, particularly gender and race, play an important role in the movement, both in its internal politics and in its understanding of issues. The focus on the issue of racism exemplifies this. Furthermore, the toxics/environmental justice movement is concerned with cultural change in the broad sense that it opposes a range of current environmental practices (and the racism often implicit in them) and calls for something different. But unlike activists in the direct-action movement of the late 1970s and early 1980s—and also the youth movements of the 1960s—activists in this movement are not particularly interested in constructing an alternative culture or developing a general critique of U.S. culture. The lack of explicit cultural radicalism in the toxics/environmental justice movement is both a strength and a weakness. The absence of cultural radicalism in the toxics/environmental justice movement allows the movement to speak easily to large numbers of people; the fact that its members are seen as ordinary people rather than hippies, radicals, or intellectuals makes it easier for many people to listen to their arguments or imagine themselves becoming part of the movement. On the other hand, the movement's lack of explicit radicalism, cultural or otherwise, also means that it lacks a clear vision of a better society. Visions can sustain movements through difficult times; too little vision can be a problem.

In addition, the toxics/environmental justice movement's approach to the state is different from that which new social movement theory associates with new movements. Of the two culprits involved in the creation and perpetuation of toxics, the state and the corporations, it is the state that tends to draw most of the movement's fire. Toxics groups often find themselves confronting state agencies rather than corporations, since state agencies are responsible for regulating hazardous wastes. One of the aims of the toxics/environmental justice movement is to bring about more extensive and effective state regulation, to force the state to regulate corporate behavior on behalf of the public good. The demands of the movement—that the state be responsible to the people rather than to corporations and that there be greater regulation of production for the public welfare—suggest a kind of radicalism different from the anarchism associated with many countercultural movements and with new social movement theory.

The toxics/environmental justice movement is more removed from academia than many movements of the recent past. Many of the movements of the 1960s and early 1970s, especially the antiwar movement and the women's movement, were based in and around universities; the civil rights and black power movement also had major student constituencies. The antinuclear movement of the late seventies and eighties was based mostly outside the universities—but many of its activists were university trained, and many returned to the university when the movement collapsed. The constituency of the toxics/environmental justice movement comes largely from poor, working-class and lower-middle-class communities that have little connection with university communities. Though the movement includes students and other young people, it is not dominated by youth or students in the way that many of the movements of the sixties were. In the sixties it was easy for professors to find the antiwar movement: they could see it from their office windows. The toxics/environmental justice movement is further removed.

It is the humanities that are, of all the academic disciplines, least connected to this movement. There are scientists actively involved in the movement as technical advisors and supporters. There are social scientists who study, write about, and publicize the movement. Robert Bullard, a sociologist, is both a leader of the environmental justice movement and the author of two collections of articles from within and about the movement; Andrew Szasz's *Ecopopulism* is a study of the toxics movement; Robert Gottlieb's *Forcing the Spring*, a revisionist history of U.S. environmentalism, includes a chapter on the movement; a collection edited by Richard Hofrichter includes voices from the movement and discussions of it; and at least two graduate students are writing dissertations on the movement.[5]

On the whole humanists have payed little attention to this movement. The partial exception to this statement is Laura Pulido, who has written an excellent book on the environmental justice movement, *Environmentalism and Economic Justice: Two Chicano Struggles in the Southwest*.[6] Pulido is a geographer; she draws upon social science methodology and also perspectives more associated with the humanities, in particular new social movement theory. She describes the environmental justice movement as lying somewhere between old and new social movements, emphasizing the marginality or "subalternity" of the groups that she describes. Pulido's work is an important first step toward a cultural analysis of environmental activism, though in other arenas in the movement marginality might not be the key term.

It would be a good thing for both the academy and the toxics/environmental justice movement if there were more connection between the two. It would be good for the academy because it would help to provide an alternative to the rather precious and self-referential version of radicalism that flourishes in some quarters. It would be good for the movement, which is weak in areas where intellectuals can be helpful—particularly in beginning to develop a strategy, a sense of how particular struggles fit into the effort to construct a better society. This is an arena in which humanists and social scientists who are concerned with how social change takes place and may have some historical and theoretical perspective on these issues might be able to contribute something. The absence of strategy has an impact on the day-to-day effectiveness of a movement and on its ability to maintain its strength and credibility over time. Though there has been considerable research into alternative, environmentally sound methods of production, there has been relatively little discussion of what sort of political pressure or government controls would be necessary to bring about such changes, and what the ramifications for society generally would be. Nor has research addressed the question of how to build an international movement that could prevent corporations from simply shifting toxic hazards out of nations with strong environmental movements into more vulnerable nations.

I am not advocating that academics or other intellectuals simply walk into the toxics movement and announce that they are there to provide a strategy for the movement. In fact, given the fact that the movement is organized around local struggles, it is not easy for outsiders to "join" the movement at all. But as the toxics/environmental justice movement begins to construct ties with other movements that have overlapping concerns, the possibility emerges of a broad environmental and socially progressive movement, in which intellectuals will have to play an important part. Such a movement—with its interracial character, the prominence of women within it, its basc in poor and working-class communities—would have a better chance of developing a progressive agenda than any other contemporary movement that I know of.

NOTES

1. For a firsthand account, see Lois Marie Gibbs, as told to Murray Levine, *Love Canal: My Story* (New York: Grove Press, 1982).

2. Andrew Szasz, *Ecopopulism: Toxic Waste and the Movement for Environmental Justice* (Minneapolis: University of Minnesota Press, 1994), 72.

3. For a scathing critique of the record of the mainstream environmental organizations on the issue of environmental justice, see Mark Dowie, *Losing Ground: American Environmentalism at the Close of the Twentieth Century* (Cambridge, MA: MIT Press, 1995), chap. 5, "Environmental Justice."

4. Celine Krauss, "Women of Color on the Front Line," in *Unequal Protection: Environmental Justice and Communities of Color*, ed. Robert D. Bullard (San Francisco: Sierra Club Books, 1994), 264.

5. In addition to the collection cited above, Bullard is the editor of *Confronting Environmental Racism: Voices from the Grassroots* (Boston: South End Press, 1993); Szasz, *Ecopopulism*; Robert Gottlieb, *Forcing the Spring: The Transformation of the American Environmental Movement* (Washington, DC: Island Press, 1993). See especially chap. 5, "Grassroots and Direct Action: Alternative Movements"; Richard Hofrichter, ed., *Toxic Struggles: The Theory and Practice of Environmental Justice* (Philadelphia: New Society, 1993). Dissertations on the toxics/environmental justice movement are being written by Hal Aronson, Sociology Department, University of California, Santa Cruz; and Patrick Novotny, Political Science Department, University of Wisconsin, Madison.

6. Laura Pulido, *Environmentalism and Economic Justice: Two Chicano Struggles in the Southwest* (Tucson: University of Arizona Press, 1996).

OLEG N. YANITSKY

Russian Environmental Movements

THE ENVIRONMENTAL movement in the former Soviet Union and Russia has existed for about forty years. It is one of the most stable, professionalized, and scientified social movements; it has exerted a significant impact on other new social movements and on the emergence of civil society in Russia at large. My recent estimates put the total membership of ecological nongovernmental organizations (NGOs) and pressure groups in Russia at between 30,000 and 50,000. According to some estimates, in 1992 in 287 Russian cities there were 840 environmental organizations, including 580 local, 233 regional, 51 interregional, and 38 with an international scale of activity (Analiticheskii Vestnik 1993). In dollars, the annual budget of a Russian ecological NGO of average size is about $20,000 to $25,000. For Russia this is quite a large amount of money.

Nowadays the movement is undergoing deep transformations. These transformations are the result of a rapidly changing societal context (e.g., the alignment of social forces and the state-movement relationships) and the movement's internal evolution, conditioned by its differentiation, ideological and political diversification, changing strategy, and means of resource mobilization.

The Movement's Emergence and Development: A Brief History

The establishment of the first student nature protection organizations in 1958 at Tartu University, Estonia, and in 1960 at Moscow State University is usually seen as the beginning of the recent environmental movement in the USSR. These two events initiated the movement of the Student Nature Protection Corps (SNPC), which still exists today. The year 1958 also marked the beginning of the movement in defense of Lake Baikal, which included a set of mass campaigns, collective appeals, and protests and which has also extended for more than thirty years.

By 1970, there were more than thirty SNPCs in the USSR. In 1970, the SNPC of the Faculty of Biology at Moscow State University began to coordinate systematically all the SNPCs' work, contacting newly established groups, organizing training seminars, and collecting and publishing news of their activities. From 1972 on, the Youth Council on Nature Protection of Moscow University began to play a leading role in coordinating the activities of all SNPCs throughout the country.

During the next fifteen years, this movement continued to develop despite ups and downs. It worked out some intergroup programs like "Flora," "Fauna," "Reserves," "Recreation," and "Gunshot" (the program aimed at the struggle against poaching). It conducted a number of regional and national training schools for SNPC activists. The foci of SNPC activities include the constant exchange of documentation, the establishment of databases, printing of newsletters and guides, and steady efforts to gain the movement's independence from Komsomol (Communist Youth) patronage. In 1985, there were more than 150 SNPCs across the USSR.

Starting in the early 1980s, numerous other ecologically oriented civil initiatives emerged outside the student movement. First, there emerged ecological groups, sections, and commissions within the so-called creative unions of architects, artists, cinematographers, and other intellectuals. Second, there were ecological initiatives of other social movements, both new and well established. Third, there were the initiatives and mass campaigns in defense of Lake Baikal and the Aral Sea and against the Northern Rivers Project, campaigns launched by prominent writers and public opinion leaders such as Valentin Rasputin, Sergei Zalygin, and Vasilii Belov. Fourth, 1987 saw the rise of numerous city-based ecological clubs (usually established around newspapers). Fifth, newspapers became the nests of proecological activities. Sixth, many official public organizations, such as the Soviet Peace Committee, began to initiate various quasi-independent ecological associations and funds. Finally, the most significant phenomenon was the emergence of thousands of urban grassroots movements and groups. They fulfilled different functions—educational, recreational, amateur activity, self-help, and mutual assistance—but their shared aim was to exercise self-government with the aim of protecting their immediate environments.

The period of 1987 to 1990 could be labeled *"informal green politics,"* which marked the beginning of the overall *green movement's politicization.* Mass campaigns, rallys, blockades, and other forms of direct action, taken to place environmental issues on the political agenda, were characteristic of this stage of the movement's development. Among the most important stimuli for the radicalization and massification of the Soviet environmental movement were governmental plans for building more hydropower plants in Siberia and the Northern Rivers Project. Radicalization and massification did not result from adopting a radical (i.e., ecoanarchist) ideology, but rather from the great influence on public opinion of a few widely known ecological leaders, including Sergei Zalygin and Alexei Yablokov.

These years were also marked by the emergence of numerous local, regional, and republican *Popular Fronts in Defense of Perestroika*, mass protest movements of a political character with a strong ecological coloring. In their actions, Popular Fronts involved both branches of the environmental movement—"conservationists" and "civil initiatives" (see descriptions in "*Ideological Aspects of Environmental Movements*"). Another form of politicization was the establishment of *Public Committees of Self-Management* that were set up in urban neighborhoods across the country.

Subsequently, green movements were established in republics of the former USSR, such as Estonia, Latvia, Ukraine, and Belorussia. The most radical were Estonian and Latvian movements. Among their goals were to promote alternative values; to raise national consciousness and protect national culture; gradually to replace nuclear power plants by safe, renewable sources of energy; and to oppose state actions that destroyed regional ecological balance and aggravated the social and economical crises in these republics. The peak of the movement's politicization occurred in 1989 and 1990, during the election campaigns to the national and republican parliaments of the former USSR, as well as to the local soviets all over the country.[1]

The scope of activity of all these new social groups was impressive, especially considering their challenges to routine policymaking structures. Remarkably, these groups stood outside the official policymaking process. But during that period they were relatively successful in mobilizing mass support.

By the end of 1990, however, the phase of informal politics had come to an end, and the environmental movement began to split into nonpoliticians and politicians. By "nonpoliticians" I mean the numerous local protest groups that represented the movement's major social resource. These local groups—which gave their leaders away to the democratic and later to the nationalistic movements, as well as to the parliamentary and other power structures—remained without a unifying ideology or shared program. The chance to set up an adequately mass-based and politicized environmental movement in Russia was lost.

The political behavior of the more politicized faction became more sophisticated. Its members started to shape their own platforms, promoted particular candidates to central and local parliaments and began to develop specific tactics and offer different perspectives on the development of environmental/political reforms as well as of the movement itself.

The most recent period of the movement's development is difficult to label: too many processes, within the movement and outside, overlap and interrelate. Nevertheless, the major trends of this period may be summarized as diversification, increased reformism in both ideology and tactics, and bureaucratization and professionalization.[2]

1. Diversification. There were several reasons for the movement's differentiation. When the period of informal politics ended, it became clear that various groups in the movement pursued quite different goals. It also became obvious that large organizations like the Socio-Ecological Union (SEU), Moscow Ecological Federation (MEF), and others were umbrellas that sheltered many divergent groups. The split between the movement's radical minority and its reformist majority also deepened.

But the crucial factor in diversification was the changing character of resource mobilization: each NGO began to look for its own resources. The collapse of the USSR and the sovereignization of its republics intensified the processes of decentralization and autonomization of various branches of the movement and its reorientation toward local environmental issues. In sum, intergroup solidarity diminished, while intergroup competition increased.

2. Reformism. With the exception of ecoanarchists, the movement was always inherently reformist in character.[3] Conservationists, its leading force, were never inclined to radicalism, but tried to work within the system. When in 1991 the adherents of democratic reforms gained victory over the communists, many ecological NGOs began to present their organizations as "respectable" and "responsible." The majority of these NGOs began to collaborate with central and local administrations.

3. Bureaucratization. The causes of the movement's bureaucratization were both subjective and objective. By subjective I mean the mentality shaped under the rule of several generations of a well-trained, specialized, and forceful technobureaucracy, both in the apparatus of the Communist Party of the Soviet Union and in industry. In recent times, dismissed or retired technobureaucrats began to establish agencies or set up public actions and NGOs in order to defend their own social status. In one way or another, Soviet bureaucratic culture still has a great impact on all kinds of social activity.

Among objective causes are a constant search for resources, a growing amount of routine work, and research and development that require a well-organized collaboration with experts and other professionals. Many protest groups evolved into formalized ones (coordination and research centers, foundations, summer schools, and the like) staffed by the activists-turned-bureaucrats. Many NGOs' leaders were coopted by the federal, regional, and local administrations to set up new environmental agencies. These developments parallel the evolution of the U.S. environmental movement from 1970 to 1990 (Dunlap and Mertig 1992, 8).

4. Professionalization. The reasons for the movement's professionalization are quite obvious. During thirty years, the conservationists, adherents of bioscientism, have been and still are the leading stratum in the movement. Furthermore, the overwhelming majority of ecological NGOs engage in research and development, educational, or publishing activities. Participation in parliament hearings, litigation, maintenance of international contacts and fundraising activities, as well as involvement in discussions and decisions concerning global issues, rapidly transform amateurism into professionalism.

The Movement's Organizational Structure

It is impossible to outline with precision the organizational structure of the current environmental movement: current Russian statistics are scant and dubious. However, some generalizations can be made.

The recent organizational structure of the movement is complex and changing. There are a few large umbrella organizations such as the SEU, MEF, and some others that embrace anywhere from 10 to 150 NGOs. They have no official membership lists or membership dues. The majority of these NGOs are multifunctional, dealing with nature protection, research and development, ecological education, and propaganda.[4] The longer such organizations exist, the more functions they embrace, since plurality of function favors keeping them afloat (Yanitsky 1994).

Another major component of this organizational structure are SNPCs, which are based at universities and supported by the SEU. These corps are mainly aimed at nature protection and ecological education. There are also a few radical organizations, such as Khraniteli Radugi (Rainbow Keepers) and the Russian division of Greenpeace, which stand somewhat apart from the mainstream organizations. Nevertheless, overlapping individual and collective membership is common, with the exception of the nationalistic organizations, with which ecological ones have rather tense relationships (for details, see Khalyi 1994). Also, there are many marginal organizations that sometimes label themselves "eco-logical." Finally, many organizations have the prefix "eco" in their names, without, however, having anything in common with genuine environmental NGOs.

To clarify the interdependent evolution of the movement in context, let us compare two figures that illustrate the disposition of social forces vis-à-vis environmental issues. Figure 1 represents this alignment in 1989, at the time of massive protest actions. Figure 2 represents the situation in 1995.

In 1989 there were four major social forces at work: the state administrative-command system (the System), environmentalists, workers, and residents. The main axis of conflict was between environmentalists and the System. The System's members in the Soviet case were potentially

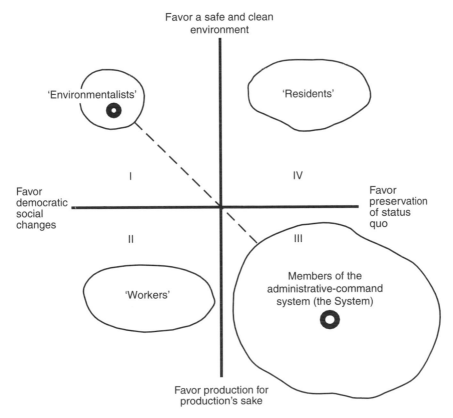

FIGURE 1. A spatial representation of the disposition of social forces in relation to environmental issues in the USSR, 1989.

consistent opponents of the aims and values of the environmental movement (Fig. 1). The former saw nature only as a means for ensuring the System's existence, while for the latter nature was valuable in itself, something that must be defended and revitalized. For the System's members, relations within and outside its boundaries were typically "organizational." There was no place for pluralism and concern for others. In contrast, environmentalists were not only more humane and tolerant, but also more socially active. The first group were "organization men"; the second, "citizens." While the System members saw technology and science as instruments for strengthening their power, environmentalists considered science as a source from which a new value system was to be formed.

These two camps also held opposite views on the organization and development of the Soviet society. Over the decades, the System had created a powerful infrastructure for exploiting the country's natural and human resources. The ecological concept of social reproduction and its relation to the nonhuman environment was alien to the System's ideologists, because if they adopted it, they would have had to show concern not for generic "masses" and "labor power," but rather for specific individuals and groups.

The most burdensome legacy of the System was probably the type of personality it created: an uprooted individual, a dealer without any feeling of responsibility or attachments, an exploiter, a quick profit-seeker. Environmentalists tried to oppose this rootless and nomadic psychology by stressing the values of creative work, mutual aid, and solidarity. They advocated forms of moral incentive and reward such as self-estimation and social reflexivity. In the former Soviet society, environmentalists painstakingly tried to establish oases of proecological thinking and ways of life.

At the core of the System were those elites who held crucial positions in the distribution of resources in the former USSR. At the rather extensive periphery of the System were two camps. The first included individuals and groups who served the core zealously, because it provided them with material benefits and vertical mobility. The second was made up of those who depended on the System either legally (e.g., military personnel) or because they had no choice. Among the latter were those who were engaged in nomadic professions such as building power stations: for decades such people had been involved in the taming of nature.

The nucleus of the opposing group (shown as quadrant I) was made up of leading environmental organizations, including the SEU and local protest groups. Environmentalists usually had at least average income and as a rule were individuals with the personal qualities typical of the Russian intelligentsia: broadly educated, humane in outlook, with a developed civic self-awareness. "Professionals, citizens, and activists" sums up this core group of the environmental vanguard.

The periphery around the core of the leading environmental organizations was made up of various types of urban civil initiatives: youth clubs, neighborhood public committees for self-management, and numerous groups for the defense of their immediate environments. I estimate that in 1988 these environmental groups together comprised about 8 percent of the urban population over age fourteen, and that 1 percent were constantly involved in proenvironmental activity (Yanitsky 1991, 536).

Quadrant II of Figure 1 is occupied by workers, who were directly involved in the system of material production, while quadrant IV is the niche inhabited by residents, namely, those who were primarily involved in the sphere of social reproduction. These two groups had much in common, since they were members of the same families and inhabitants of the same neighborhoods. But their attitudes toward the environment were typically different. Workers were more System dependent, with more rationalist and technocratic habits of mind, whereas residents were more oriented to the human scale. Like environmentalists, they valued self-reliance and mutual support.

Residents and workers are, of course, social concepts. The former denotes the link between the city dweller and his or her immediate environment. The more residents invest their time and skills in the local environment, the more they will value it and the more their input into its protection will grow. There is an even stronger social link between the worker and his or her place of work, which provides salary and other social benefits. Because work as such is not an aim in itself, the worker's priority becomes obedience to the laws of labor collectives as a large family.

Residents comprise two different groups: the nonmobile—young people, mothers with little children, and persons who are retired or have a disability—and those employed in education, culture, and modern computerized spheres of production and services that make it possible to work at home. Workers are mainly those employed in large state enterprises; rural immigrants to cities, especially in the first generation; and, of course, tens of millions of office staff. State socialism created a powerful stratum of "Soviet clerks," who used to perform low-skill, routine paperwork. This last group was one of the strongest adherents to the values of the System.

The cleavages between workers and residents were particularly obvious in their attitudes toward science and research. Workers were largely indifferent and even hostile to science, because it created the danger of technological innovation. Residents had a positive attitude to science: independent expertise and alternative scientific projects were among the limited resources they could use against the System.

Changing Alignments of Social Forces

By 1995, the alignment just outlined had changed dramatically. Under the pressure of inner social, economic, and political conflicts and cleavages and outer forces, the environmental movement experienced deep transformations, as shown in Figure 2.

In the camp of environmentalists we observe disintegration and differentiation. Initially, disintegration was caused by the collapse of the USSR, which deepened the sovereignization of Russian republics and regions. Disintegration was intensified by ideological cleavages and tactical differences among environmentalists, as well as by the growing competition among ecological groups for access to foreign aid. Disintegration also resulted from growing state pressure, which tried to compromise the movement by every possible means, thus forcing it out of the political arena.

In addition, the movement's disintegration was probably an inevitable outcome of its maturation once the period of mass protest was over and various branches started looking at their individual identities. My investigations in 1992 through 1994 revealed seven typical branches within the Russian environmental movement: conservationists, alternativists, traditionalists, civil initiatives (local protest groups), ecopoliticians, ecopatriots, and ecotechnocrats. (The ideological stands and political preferences of each group are discussed below.)

In sum, today we have no united environmental movement, and therefore the previous axis of political struggle has disappeared.

Let us look at quadrant III in Figure 2. The System proved to be stable and resistant to external pressures. Despite some serious transformations, it continues to maintain the dominant social paradigm of the former USSR. This means that the System's orientation toward restoring its geopolitical status quo and rehabilitating the old resource-consuming and highly exploitative industrial system remains unchanged. Indeed, in some ways, the System has even extended and strengthened its influence and domination.

Despite the apparent political differences between the new and the old elites that populate quadrant III, its core members remain those who occupy ruling positions in the distribution of financial and natural resources. As a result, workers and residents have become more dependent on the System than ever before.

I call this dependence *negative solidarity* for two reasons. First, it is negative because it is coerced. In fact, there are many contradictions and

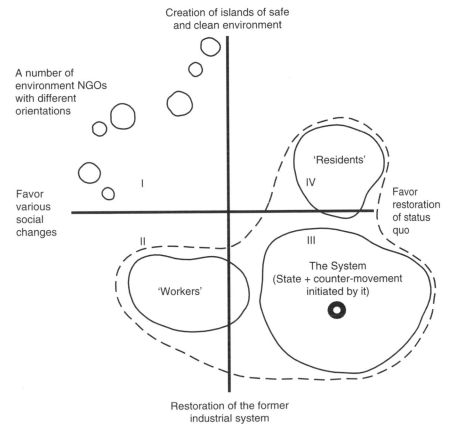

FIGURE 2. A spatial representation of the disposition social forces in relation to environmental issues in Russia, 1995.

conflicts between workers and residents on one side and the System on the other. Second, it is negative because all groups are surviving at the expense of the overexploitation of natural resources.

Let us have a closer look. During recent years, workers have become divided into two social strata. A minority have become owners of small businesses, and as such, they are totally dependent on the distributing elite and numerous bureaucrats in the state and local administrations, since it is the latter who may give or refuse them permission to run their businesses. But the overwhelming majority of workers remain poor and without rights. They survive only from the sale of stolen state property, or thanks to the harvests from their private kitchen gardens. They raise only short-term economic demands and resist any attempt at ecological modernization of their factories. It is symptomatic that in late 1993, workers, small-business owners, and some trade unions, with the direct support of the state bureaucracy, formed the so-called Constructive-Ecological movement, which is in fact a countermovement aimed at pre-empting the leadership of the SEU and other genuine environmental Russian NGOs.

Residents have also split into two strata. The mobile young minority have become shopkeepers, hawkers, profiteers, and racketeers. The immobile majority, as well as teachers, scientific workers, and so on have rather quickly become poor, disoriented, uprooted, and psychologically exhausted. The point is that their immediate environments have lost their capacity for social reproduction, mutual support, and safeguarding of security.

The proecological power of residents was once concentrated in their protest and civil initiative groups, which later were transformed into public committees of local self-management. But now protest groups are dying out: most of them were disbanded after the introduction of the posts of elected mayors and district prefects in the end of 1991, which meant the total economic and political dominance of executive bodies.

This raises an important question about the place of the Russian intelligentsia in the relationships of the above-mentioned social forces. The intelligentsia originally played a key role in raising and resolving environmental issues. During the early stages of the movement's development, it was the intelligentsia who encouraged the emergence and development of environmental initiatives and groups and the raising of environmental consciousness of both workers and residents.

Nowadays the environmental stands developed by the intelligentsia have lost their principled character. In the programs of most of the social groups concentrated in quadrant III, concern for environmental well-being has today become no more than a rhetorical cover for conservative, nationalistic, and even fascist positions. Meanwhile, the intelligentsia has become directly involved in political struggle. For this thin layer of the environmental vanguard, political involvement has had catastrophic consequences. The democratic intelligentsia have dissolved in numerous politicized movements; political conflicts are exhausting the intelligentsia's resources and lowering its members' interest in environmental issues.

Today, the environmental movement in the former Soviet Union has no mass support among the country's urban and rural populations. From 1991 to 1995, the movement's leaders did not pay attention to the critical problems of workers and residents—declining standards of living, unemployment, growth of crime, the problems of refugees and migrants, and so on. They did not organize or even take part in any mass campaign for the defense of social and economic rights of workers or residents. Having spent a huge amount of time and money to build e-mail and other communication networks so as to consolidate the environmental movement itself, they have not spent a penny for the creation of a "feedback system" with workers and residents.

Ideological Aspects of Environmental Movements

For decades under the pressure of communist ideology, Russian researchers of social movements hardly paid any attention to the movements' ideological foundations. But Western sociologists are correct in stressing the significance of ideological processes aimed at producing and maintaining the movement's mobilization potential (Klandermans and Tarrow 1988; Neidhardt and Rucht 1991).[5]

Today, all groups that make up the environmental movement pay growing attention to the ideological underpinning of their activities. This search for ideological foundations seems quite natural—indeed, it seems an indication that the movement has become more mature. The less the movement is reactive and the more it is proactive, creative, and professional, the deeper ideological differentiation has to be. There are many reasons for this process. Some are historical and cultural, since different environmental groups have emerged historically from different subcultures. Conservationists, for example, came from the subculture of university intelligentsia, a rather specific subculture of Russian naturalists. Grassroots ecologists represent quite another sort of subculture, that of residential communities, householders, and house-keepers.

But the major reason for these diversified intellectual attempts is the desire to have a coherent ideology. Under the conditions of our post-communist society, we no longer have a unifying official ideology, but the desire to have a definite ideological base is strong among the people. The deeper the social, economic, and political changes and the more ambiguous they are, the more people strive for a secure ideological guide and justification for their daily activities. What they look for is not so much a new official doctrine as a set of values that can assist them in restoring the identity lost after the collapse of the Soviet regime.

Let us consider the ideologies and values of seven branches within the movement.

The *conservationists*, who form the core of the Russian ecological movement, have never articulated their ideology systematically. Never-theless, today (as thirty years ago) their central points are bioscientism ("nature knows best") and the idea that an ecological catastrophe is inevi-table. Their ultimate goals are the creation of a world brotherhood of greens and the construction of a "parallel society" that makes modest material demands.

The *alternativists* are the most ideologically oriented branch within the movement. Its founders and leaders were proponents of ecoanarch-ism. Many alternativists are members of green parties and the radical ecological groups. This is a unique group, which combines sociopolitical activities with constant ideological reflection. The alternativists are de-termined adversaries of the state as a political institution. In their opin-ion, radical ecological modernization can be carried out only by means of an alternative project for the society at large.

The *traditionalists* have no specific ideological doctrine. They typi-cally belong to the Russian intelligentsia, with its traditional ideals of tolerance, nonviolence, and the desire "to understand and to help." Tra-ditionalists are oriented toward the past in the sense that they value nineteenth-century Russian culture, with its ideals of serving the people and their enlightenment. Traditionalists oppose "Russification" and the Westernization of the way of life of the ethnic minorities that populate Russia. The individual ideological inclinations of the group members are various. But despite their spiritual heterogeneity, traditionalists are united by their reflective disposition and their capability to assess the world and their own deeds critically. The traditionalists are therefore the group that is most ideologically stable and resistant to external effects.

The *"civil initiatives"* have no ideology of their own. The majority of this group, the largest in the movement, are still carriers of traditional socialist values such as egalitarianism, collectivism, mutual support, and self-government. They are opponents of the idea of market economy. Unfortunately, the main aim of the branch—public self-government— was not reached during the years of *perestroika* and subsequent reforms. Today the civil initiatives, having exhausted the potential of anticommunist protest and losing their faith in democratic leaders, are a group at high "ideological risk."

The *ecopoliticians* as a branch include the movement's founders, who, although not its formal members (like Valentin Rasputin), have had a great ideological impact on it; theorists and ideologues who adapted the already well-developed doctrines to the movement; professional ecopoliticians; the leaders of numerous Russian green parties; activists who first became politicians (people's deputies) and then joined the movement; and, finally, practical politicians who came from civil initiatives, combining the roles of professional politicians and movement activists. The common feature of this branch's ideology is "Policy first!" and "Policy decides everything." Strategically, this means that political reforms should precede economic and environmental changes; tactically, it means justifying the use of any political means, parliamentary or non-conventional.

The ideology of Russian *ecopatriots* is marked by right radicalism, betting on the forceful ecologization of society and explicit sympathy for socialism. Its key points are state patriotism, a stiffly regulated market, restriction of private ownership of large plants, an ecologically oriented socialist and private economy, and social justice. Some ecopatriots even call for the establishment of an "ecological dictatorship" in Russia.

The *ecotechnocrats* see the solution of ecological problems in a wide use of ecologically sound technologies. Strictly speaking, they do not espouse a technocratic ideology in the common sense of the term. I call them "technocrats" because they display a kind of naïve technocratism, specifically, the belief that technological innovations created by them will be adopted by society. The ecotechnocrats are the least ideologized and politicized part of the movement (Yanitsky 1996).

Environmentalists' Participation in Policymaking

The development of the Soviet/Russian environmental movement raises questions about interrelationships between the internal evolution of the movement and its impact on its social and cultural context. This well-organized, resourceful, and tactically experienced movement contrasts with the low standing of environmental issues on the agenda of recent Russian politics.

In part this may be explained by the fact that the movement remained more moderate in its demands and actions than did many of its allies in central Europe and the Baltics. But it seems more important that Russian environmental NGOs have tried to avoid confrontational tactics, preferring to act within the System and concentrating on pragmatic issues. Under the Soviet regime, no environmental groups could operate outside the framework of a totalitarian political system. For decades, the political opportunity structure for the environmental movement was extremely unfavorable. Neither ten nor five years ago did the Soviet/Russian electoral system give any real chance for the movement to gain significant representation in the parliament and local soviets or to influence actual policymaking.

Even today, when the structure of political opportunities has widened enormously, few environmental NGOs are involved in direct conflicts with the state. The Russian ecoanarchists may be the only group that has devoted time and mental resources to building a comprehensive alternative concept of social order. This lack of radical stands was not a result of the inability of most greens or other ideologists to develop a radical ecological criticism of Soviet industrialism. On the contrary, such a criticism was developed by the leaders of *perestroika* and of the ecological movement. The problem was rather the lack of deep and sharp social-ecological conflicts that could radicalize the movement and society at large.

The majority of the former Soviet and present Russian environmental groups played and still play according to the rules established during the Soviet era: these rules are infiltration, secrecy, and threats. In spite of their latent and open conflicts with the central and local authorities, environmental groups increasingly tend to present themselves as practical, professional, and responsible. Radical ideological expressions and demands, as well as direct actions against the state, become more and more rare. In contrast to the first years of *perestroika*, when the state authorities had to defend themselves from the attacks of environmental protest groups, today state bodies and local authorities have successfully managed to meet these challenges by using tactics of preemption, cooptation, and marginalization. This is ironic, given that the present leaders of many "respectable" Russian environmental groups started as radicals with only one aim: to penetrate into and even to seize the steering wheel of the System.[6] Now both radical and reformist ecologists are at the margins of the Russian political arena.

Russia is a state with enormous concentration of power in the hands of the president and his administration. The crucial factor here, however, is not the structure of political opportunities, but the totally counterecological orientation of all branches and levels of power. Although today there are scores of political parties in the Russian public arena, there is actually only one ruling "party"—the "party" that is in charge of oil, gas, and other natural resources. Needless to say, this "party" is a direct and formidable adversary of the environmental movement.

Another crucial factor in the marginalization of the environmental movement is the uncertainty of the structure of political opportunities, owing to the uncertainty of the ecological policy at large. In the USSR and then in Russia there never was a well-elaborated ecological strategy, or even a set of tactics. Under the pressure of public opinion and of Chernobyl and other accidents, the state bodies were forced to formulate a set of environmental laws and programs; however, nobody follows them.

In essence, state environmental policy in Russia was always ad hoc and highly compartmentalized. This compartmentalization of policy has provoked the further split of the environmental movement into a set of groups and NGOs dealing with specific issues such as biodiversity, nature reserves, radioactive dumps, nuclear power plants, and chemical disarmament. This transformation increases the reformist character of the environmental movement, inasmuch as it creates a way of coping with environmental degradation through specific measures directed by the existing economic and political System itself, skillfully manipulating and coopting protest groups and NGOs that earlier were the System's strong adversaries. A striking example of this transformation is the evolution and gradual cooptation of the most powerful umbrella environmental organization, the SEU. The SEU today tends more and more to play the role of providing central and local administrative bodies with high-skilled staff or consultants and experts.

One plausible explanation of this evolution of the movement is its identity orientation. Since environmentalists were never power oriented, had no political strategy of their own, and had no representatives in the power structures, activists-turned-politicians were forced to accept the preexisting rules of the game and the general policy style peculiar to existing structures. For decades environmental policy in Russia was formulated in secrecy, outside the public eye; at best, the public was "informed" after the fact. Informal networks in the economy and politics were and are still rather strong; informal and latent reciprocal ties continue to play a decisive role in environmental regulation. The respondents in my investigations labeled this latent system the "industrial party."

Another explanation of these transformations, as well a distinction from conditions in the West, is the almost total absence of *green pressure groups*. If by pressure groups one means organized structures shaped for the pursuit of a particular political goal or for creating a favorable public opinion, there are no such groups in Russia. Mainly, there are *protest groups and ecological NGOs* with diverse functions. These NGOs may or may not be involved in campaigning and other activities of the environmental movement.

To follow down the highway of state policy has become a distinguishing feature of Russian environmental groups" behavior. After the Earth Summit in 1992, state environmental policy slowly shifted to global environmental issues and concepts such as sustainable development. In February 1994, President Boris Yeltsin issued the "Federal Action Plan for the Protection of the Environment and Sustainable Development." Later, an open competition was announced for the development of drafts of this plan, but the majority of environmental groups did not want or were unable to draft alternative national environmental strategies, preferring instead to collaborate with state bodies and local administrations in solving particular environmental issues.

Environmental Movement in a Changing International Context

By "changing context" I mean the collapse of the USSR and other changes on the state level, as well as the exploitation of financial resources from Western countries for the environmental movement's maintenance. In the late 1980s and early 1990s, there was an increase of political opportunities for environmental groups across Europe. However, the reasons for the increased opportunities were quite different in the western and in the central/eastern parts of the subcontinent.

Entering the European Union, the majority of western European countries came under the international pressure for stricter environmental standards. In Britain, Germany, Spain, and some other western European countries, environmental groups successfully used this opportunity to increase pressure on their governments and the European Parliament. On the other hand, the formation of the European Union clearly showed the need for international cooperation between environmental groups. "Given the transnational nature of many environmental problems, an approach which moves from their symptoms to the causes must inevitably shift its attention to a supranational level" (Rucht 1993, 77). Thus international cooperation emerged from the inherent nature of environmental problems. In sum, despite obvious constrains on international cooperation, there was clear progress in the activities of western European environmentalists.

The situation in Russia was quite different. The collapse of the Soviet Union, the processes of shaping new nation-states, and the sovereignization of republics and regions within the Russian Federation provoked the disintegration of the previously united environmental movement of the former USSR. In every newly emerged nation-state or republic of Russia, ecological groups became more and more involved in solving particular local and regional problems. In some republics and regions, environmental groups were simply absorbed by the nationalistic movements; in other cases, one could observe contradictions and conflicts between the former and the latter. Another cause of this disintegration on the international level was the unwillingness of the conservationists (who are still the core group of the movement) to enter directly into the political processes and conflicts going on within the Commonwealth of Independent States.

But the critical cause of the weakening of the environmental movement in the former Soviet Union lies in changes in the structure of resource mobilization. The aspirations and demands of residents were a powerful force for the movement's consolidation across the USSR. Up to 1991, the mobilization of humanpower dominated, while financial resources played an insignificant role. With the subsequent breakup of the country into smaller nations, financial resources became critically important. This meant that the interests and attitudes of residents were replaced by the attitudes and the (selfish) interests of the movement's elite, which now became the key holder and distributor of financial resources coming from the West. Both the distributive activity of the movement's elite and the growing competition of local groups for the same pool of resources enhanced the cleavages and conflicts among ecological NGOs. All these NGOs compete for Western financial aid and resources. I call this transformation the movement's overall *Westernization*.

Conclusion: The Future of Russian Environmentalism

In comparison with most other social movements, the environmental movement in the former Soviet Union has been relatively successful during the past forty years. The movement not only survived, but experienced remarkable growth. It contributed to the emergence of numerous local and regional environmental NGOs. It became a significant social force and an integral part of the global environmental community. Without doubt, the movement is now much more professionalized and efficiently organized than ever before. It is well equipped with computers and an e-mail network. It has learned to mobilize technical and financial resources from the West and (to a lesser degree) from domestic foundations. Despite constant internal disputes and disagreements, the movement has developed a set of techniques it could use to influence the public policy.

But the movement has failed in its goals of preventing environmental accidents and protecting the quality of environment. Yet, it is obvious that if the movement had not been around, the situation would have been much worse.

For a number of reasons (such as the emergence of new green parties and quasi-environmental NGOs and the absence of cross-national or even regional mass campaigns), the movement is likely to become still more differentiated in the near future. Given alien and hostile macrosocial and local contexts and chronic resource deficits, this differentiation is primarily a means of self-preservation and survival.

The low level of public concern for environmental quality and the adherence to the ideology of bioscientism by the movement's leaders make it unlikely that the movement will be able to attract new members and more financial support from the West in the near future. The danger that NGOs will lose their role as carriers of environmental values is very real. Besides, as the United States and other Western countries show, the growth of big umbrella organizations does not necessarily mean they can improve the environment. Indeed, accumulation of massive resources seems to be counterproductive.

It is an open question how the ideological cleavages and tactical differences between international and Russian umbrella organizations such as the SEU on the one hand and local nationalistically oriented movements and NGOs, on the other, will be resolved. The answer will strongly define the fate of the Russian environmental movement in the coming years.

The movement's reorientation toward Western financial aid has two consequences. It means further integration into the global network of the environmental community. It also leads to a deepening gap between the movement's elite and its rank-and-file members, to the expenditure of more and more resources on the maintenance of the mainstream organizations, to the growth of distributive bureaucracy, and to the ultimate loss of the movement's contacts with workers and residents.

In the last analysis, while in western Europe we observe the integration of environmental groups on the international level, mutual support and resource exchange among them, in Russia and the Commonwealth of Independent States we see splits and encapsulation, reorientation toward local problems, disintegration, and unwillingness to engage in public political activity. As a result, there is today no strong politicized environmental movement that is able to mobilize a mass of active militants and to influence radically the existing political regime.

NOTES

1. My estimates show that no fewer than three hundred ecologically oriented deputies of the All-Union Parliament were elected. This figure constituted about 15 percent of the total deputy corps. Forty elected deputies were the acknowledged leaders of local, regional, and national environmental groups and movements (Yanitsky 1989, 24).

2. Any social organization that looks for social and economic support will inevitably experience the shift toward a formal, bureaucratic structure, with high-skilled and professional staff, and it will be unable to resist the replacement of radical goals and direct actions with reformist and conventional ones. In other words, three types of change are inevitable: goals transformation, turn to organizational maintenance, and oligarchization (Zald and Ash 1987).

3. Like Friends of the Earth and Greenpeace in western Europe, Russian ecoanarchists and Rainbow Keepers give us excellent examples of the movement's entrepreneurs at works by means of direct actions (protest camps, sit-ins, blockades), they mobilize concerned but passive groups of local residents. At the same time, the Russian movement has never applied professional marketing approaches to the recruitment and rentention of members, since members never were a key resource for gaining financial support for their groups. Herein lies the sharpest difference between the Russian and the Western environmental movements.

4. The Biodiversity Conservation Center of the SEU is rather typical. It serves as a consultational, informational, and fundraising center, coordinating a wide range of projects in the field of biodiversity conservation. Governed by a board of eight conservationists, the Center has eighteen paid staff professionals and thirty to fifty volunteers. The center participates in drafting conservation legislation, collects and analyzes information, and publishes handbooks for professionals and public organi-

zations. Most of the center's programs are implemented in cooperation with governmental bodies, research institutions, NGOs, nature reserves, and international ecological organizations (*Biodiversity Conservation Center of the Socio-Ecological Union*, 1994). On the other types of ecological NGOs, see Yanitsky (1993, chap. 8).

5. I completely agree with McCloskey (1992), who has argued that "History shows that organizations typically enjoy greatest success when they manage to combine elements that are often uncomfortable with one another. Those who have a passion for the mission and are driven by their own visions and those who can manage finances and personnel well" (84).

6. Direct action and other extraparliamentary actions are also used when an organization pursues a goal to take place among existing "established" NGOs. This very tactic was successfully used by some Russian ecoanarchist groups when they decided to seize leadership of campaigns that had been led by the "conservationist faction" for years.

REFERENCES

Analiticheskii Vestnik Informatsionnogo Agentstva Postfactum. 1993. Moskva, 1 July.

Biodiversity Conservation Center of the Socio–Ecological Union. 1994. *Russian Conservation News* N 1. October.

Dunlap, R. E., and A. G. Mertig. 1992. The Evolution of the U.S. Environmental Movement from 1970 to 1990: An Overview. In *American Environmentalism: The U.S. Environmental Movement, 1970–1990*, edited by R. E. Dunlap and A. G. Mertig. Washington, DC: Tayor & Francis.

Khalyi, I. 1994. Environmental and National Movements in Russia: Allies or Adversaries? Paper presented at the session, Ecological Movements, the Committee "Environment and Society" of the International Sociological Association's 13th World Congress of Sociology, 18–23 July, Bielefeld, Germany.

Klandermans, B., and S. Tarrow. 1988. Mobilization into Social Movements: Synthesizing European and American Approaches. In *From Structure to Action: Comparing Social Movement Research Across Cultures*, International Social Movement Research I, Vol. I, edited by B. Klandermans, H. Kriesi, and S. Tarrow. Greenwich, CT: JAI Press.

McCloskey, M. 1992. Twenty Years of Change in the Environmental Movement: An Insider's View. In *American Environmentalism: The U.S. Environmental Movement, 1970–1990*, edited by R. Dunlap and A. Mertig. Washington, DC: Taylor & Francis.

Neidhardt, F., and D. Rucht. 1991. The Analysis of Social Movements: The State of the Art and Some Perspectives for Further Research. In *Research on Social Movements: The State of Art in Western Europe and the USA*, edited by D. Rucht. Frankfurt: Campus Verlag.

Rucht, D. 1993. Think Globally, Act Locally? Needs, Forms and Problems of Cross-National Cooperation Among Environmental Groups. In *European Integration and Environmental Policy*, edited by J. D. Liefferink, P. D. Lowe, and A.P.J. Mol. London and New York: Belhaven Press.

Yanitsky, O. 1989. The Greens in New Parliament? *New Times: A Soviet Weekly of World Affairs* N 20. 16–22 May, Moscow, Russia.

———. 1991. Environmental Movements: Some Conceptual Issues in East–West Comparisons. *International Journal of Urban and Regional Research* 15:524–541.

———. 1993. *Russian Environmentalism. Leading Figures, Facts, Opinions.* Moscow: Mezdunarodnye Otnoshenia.

———. 1994. Dvenadtsat' gipotez ob al'ternativnoi ekopolitike. *Sotsiologicheskii Zhurnal,* N 4: 37–46.

———. 1996. The Ecological Movement in Post–Totalitarian Russia: Some Conceptual Issues. *Society and Natural Resources* 9:65–76.

Zald, M. N., and Ash, R. 1987. "Social Movement Organizations: Growth, Decay and Change." In *Social Movements in Organizational Society,* edited by M. N. Zald and J. D. McCarthy. New Brunswick and London: Transaction.

BINA AGARWAL

Gender and Environmental Action

NONPRIVATIZED LAND resources in the form of forests and village commons (VCs) have always been important sources of livelihood and basic necessities for rural households in developing countries. For many poor households and especially women, who own little private land, they have been critical for survival. Fuel, fodder, fiber, food items, small timber, manure, bamboo, medicinal herbs, oils, materials for building houses and for handicrafts, and resin and gum are just a few of the products obtainable from such sources.[1] Firewood, in particular—the single most important source of domestic fuel in rural South Asia (providing 65 percent or more of the domestic energy in large parts of north India and 95 percent or more in Nepal)—is mostly gathered and not purchased (Agarwal 1987). In the semi-arid regions of India the commons supply over 90 percent of the firewood and provide for much of the grazing needs of landless and land-poor households (Jodha 1986). Access to VCs also reduces income inequalities between poor and nonpoor rural households, and there is a close link between the viability of small farmers' private property resources and the availability of common property resources for grazing or fodder collection (Jodha 1986; Blaikie 1985). Moreover, several million people in India (estimated at about 30 million some fifteen years ago; Kulkarni 1983) depend wholly or substantially on nontimber forest products for a livelihood, a source of survival that proves especially critical during lean agricultural seasons and acute food shortage contexts such as during drought (Agarwal 1990).

At the macro level, forests constitute the "lungs" of the world and many argue forests are a protection against global warming. It is also becoming increasingly clear that forest preservation is not a luxury but essential for sustainable agriculture, since the health of forests affects the health of soils (especially in the hills) and the availability of ground and surface water for irrigation (Rao, Ray, and Subbarao 1988; Agarwal 1986a). Indiscriminate agricultural expansion in the name of development, with little attempt to maintain a balance between fields, forests, and grazing lands, assumes that the relationship between agriculture, forests, and VCs is an antagonistic one. Unstable and even declining crop yields in many parts of the developing world, due to deteriorating soil and water conditions, indicate rather the essential complementarity of the relationship.

However, in India, as in many other regions, first under colonial rule and then after Independence, the availability of forests and VCs to rural communities has declined rapidly. This is due both to degradation and to reduced access to what is available, the latter resulting particularly from the twin processes of statization (appropriation by the State) and privatization (appropriation by individuals). These processes have, in turn, contributed to the further decline of these resources, by eroding traditional systems of resource management and use.

By the late 1970s, deforestation and the degradation of VCs had reached crisis proportions in many regions of India. The alarm sounded by grassroots activists, journalists, and some academics led the government to initiate tree-planting schemes under the banner of "social forestry." Undertaken in a top-down manner, most such schemes succeeded neither in regenerating degraded commons and forests nor in meeting everyday village needs. In particular, they raised serious doubts about the ability of the State or of individuals to develop, without some form of action involving local communities, what was a communal resource. In contrast were emerging success stories of forest protection and management by village communities, including forest protection movements such as Chipko and Appiko, spontaneous initiatives by populations living on the edges of forests, and attempts by some forest officials to involve villagers in the management of degraded forestland as a "joint" venture.

As a result of the lessons learned, today we are seeing small but significant reversals in the earlier processes of statization and privatization toward a re-creation of communal property rights in forests and VCs. Numerous forest protection groups have emerged, some state-initiated under what is termed the Joint Forest Management (JFM) program, others self-initiated by village communities, yet others catalyzed by nongovernmental organizations (NGOs). The JFM program, in particular, reflects the State's long overdue recognition of the failure of top-down, bureaucratic approaches.[2] It is also a recognition that the appropriate property rights arrangement for forest and VC management in most regions might be neither individual-private nor exclusively State ownership, but one in which the *community* has strong vested interests and an assurance of gains.

In one sense, the new initiatives represent a move toward reestablishing some degree of communal property rights. But unlike the old communal property systems, which recognized the usufruct rights of all members of the village, the new ones represent a more formalized system of rights dependent on membership in the emergent institutions. In other words, *membership* rather than *citizenship* has become the defining criterion for access to these resources.

This raises critical questions about participation and equity, especially gender equity. Are the benefits and costs of the new institutional arrangements, for instance, being shared equally by women and men? Or is a new system of property rights being created in communal land which, like existing rights in privatized land, is strongly male centered, thereby depriving village women of the only remaining land resource of significance in which they had rights unmediated through male relatives? What is the extent of women's participation in these initiatives? What constrains or facilitates their participation?

In the existing literature on collective action, that relating to these emergent institutional arrangements is still sparse, and that which takes account of gender even more so. In this paper, I address the above questions in broad terms, based on field visits (especially during 1993–1995) to several sites of environmental action, discussions with villagers and NGOs, and emerging case studies.

I argue here that without women's effective participation in all aspects, the emergent initiatives will have serious adverse consequences for social equity and program efficiency and will further disempower women. Indeed, the twin concerns of efficient environment protection and equity need not be in conflict; quite the contrary. The paper also highlights the problem of treating "communities" as ungendered units and "community participation" as an unambiguous step toward greater equity. In addition, I demonstrate the relevance of the feminist environmentalist perspective (elaborated in Agarwal 1992 and discussed later), as opposed to an ecofeminist perspective, in understanding gendered responses to the environmental crisis.

The paper is divided into nine sections. Section I briefly traces the main processes affecting the availability of forests and VCs. Section II discusses why the impact of these processes has a class and gender specificity and spells out their effects, especially on women of poor rural households. Section III outlines the broad features of the emergent institutional initiatives, and section IV traces the gender gap in participation in these initiatives and the resultant adverse effects on women's and family welfare, program efficiency, and women's empowerment. Section V discusses the constraints underlying women's formal participation in the emergent institutions, section VI highlights contrasting cases especially of women's own informal initiatives for protecting and regenerating local forests and VCs, and section VII outlines ways by which women's involvement in the formal initiatives may be strengthened. Section VIII focuses briefly on what these emergent initiatives tell us about gendered responses to environmental degradation, and section IX contains summary comments.

I. Statization, Privatization, and Institutional Erosion

In the period 1987–89, only 64 million hectares, or 19.5 percent, of India's geographic area was forested (Table 1). Much of this land was highly degraded, with poor tree cover, and the remaining forests were disappearing rapidly. Today, most of the good forestland is concentrated in a few states of central, eastern, and northeastern India. Large tracts of common lands once used by rural communities as pastures or for gathering firewood and other products have disappeared or been severely denuded. Accompanying this degradation has been the decreasing availability of these resources due especially to two major processes: statization and privatization.

Consider first the process of statization. Beginning under colonial rule and continuing after India's independence, State control over forests and VCs grew. British policy involved establishing State monopoly over forests; reserving large tracts for timber extraction; severely curtailing the customary rights of local populations over these resources; and encouraging the replacement of locally used species with commercially profitable ones, under the garb of "scientific" forest management. There was also large-scale felling for building railways, ships, and bridges; establishing tea and coffee plantations; and expanding the area under agriculture in order to increase the government's land revenue base (Guha 1983, 1985). In effect, these policies severely eroded local systems of forest management, legally cut off an important source of sustenance for the poor (even though illegal entries continued) and created a constant source of friction between forest officials and the local people.

Postcolonial policies, at least up to the late 1970s, showed little shift from the colonial view of forests as primarily a resource for commercial use and gain. State monopoly over forests persisted, with its attendant tensions, as did the practice of forestry for profit. Restrictions on local people's access to forest produce actually increased, and the harassment

Table 1. Percent Forest Area by States: 1987–1989

State	Million hectares	% of state's geo-area	% of total forest area in India
India	64.013	19.49	100.00
Northern India			
Northwest			
Haryana	0.056	1.27	neg
Himachal Pradesh	1.338	24.00	2.09
Jammu and Kashmir	2.043	9.20	3.19
Punjab	0.116	2.32	0.18
Rajasthan	1.297	3.80	2.03
Uttar Pradesh	3.384	11.49	5.29
West and Central			
Gujarat	1.167	5.90	1.82
Madhya Pradesh	13.319	30.03	20.81
Maharashtra	4.406	14.32	6.88
Eastern			
Bihar	2.693	15.50	4.21
Orissa	4.714	30.26	7.36
West Bengal	0.839	9.46	1.31
South India			
Andhra Pradesh	4.791	17.40	7.48
Karnataka	3.210	16.80	5.01
Kerala	1.015	26.11	1.58
Tamil Nadu	1.772	13.62	2.77
Northeast India			
Arunachal Pradesh	6.876	81.80	10.74
Assam	2.606	33.10	4.07
Manipur	1.788	80.10	2.79
Meghalaya	1.569	70.98	2.45
Mizoram	1.818	89.47	2.84
Nagaland	1.436	86.12	2.24
Tripura	0.532	50.78	0.83
Other Areas	1.228		1.92

Source: Ministry of Forests and Environment (1991, 23, 28).

and exploitation of forest dwellers by the government's forest guards were widespread.

In the late 1970s and early 1980s there was some State recognition of the need to contain the rate of deforestation and to reclaim degraded forests and VCs. A variety of tree-planting schemes were initiated,[3] some involving direct government management, and others attempting to induce village communities and individual farmers to plant. Many of the government's direct planting ventures had poor tree survival rates and typically did little to alleviate the local fuel and fodder problem; in fact, the species most commonly planted, eucalyptus, provided no fodder and poor fuel. And in some cases even natural mixed forests were replaced by monocultural commercial plantations.

Also the takeover of village land used by villagers for various other purposes, and the failure to elicit community support when the schemes were initiated, led to widespread local hostility and resistance. Women typically did not feature in such schemes, or at best were caretakers in tree nurseries with little say in the choice of species or other aspects of the project. Community forestry schemes also had a high failure rate in the early 1980s, in the absence of effective institutional mechanisms to ensure village participation in decision making and the equitable distribution of costs and benefits.

The real "success" story of that period, with plantings far exceeding targets, was that of "farm forestry" practiced by the richer farmers who in many regions sought to reap quick profits by allotting fertile cropland to fast-growing commercial varieties, eucalyptus again being a great favorite. As a result, in several regions, employment, crop output, and crop residues (also used for fuel) declined, sometimes dramatically (Agarwal 1986a).

Only in the late 1980s (as discussed further below) do we see a noticeable recognition by the State of the *positive* role that local communities could play in the regeneration of forests and VCs.

Parallel to the process of statization has been a growing privatization of community resources in individual (essentially male) hands, especially since the 1950s. Customarily, large parts of VC lands, particularly in northwest India, were what could be termed "community-private": they were private insofar as use rights to them were usually limited to members of the community and were therefore exclusionary, but they were communal in that such rights were often administered by a group rather than by an individual (Baden-Powell 1957; Bromley and Cernea 1989). Over time, these resources have become increasingly "individual-private."

Between 1950 and 1984, VCs declined by 26 to 63 percentage points across seven states (Table 2). Population pressure aside, this can be attributed mainly to State actions (which served to benefit selected groups over others), such as the legalization of illegal encroachments by influential farmers; the auctioning of parts of VCs to private contractors for commercial exploitation; and the distribution of common land to indi-

TABLE 2. Distribution of Village Common (VC) Land to Individual Households in Different Regions

State/District	VCs as a percent of village area, 1982–1984	% decline in VC area, 1950–1984	% of land to		% of recipients among the				Per household area owned (hectares)	
			Poor	Others	Poor		Others		Poor	Others
					Before[1]	After[2]	Before	After		
Andhra Pradesh										
Mahbubnagar	9	43	50	50	76	24	0.3	0.9	3.0	5.1
Medak	11	45	51	49	59	41	1.0	2.2	3.1	4.6
Gujarat										
Banaskantha	9	49	18	82	38	62	0.8	2.0	5.4	8.8
Mehsana	11	37	20	80	36	64	1.0	1.7	8.0	9.8
Sabarkantha	12	46	28	72	55	45	0.5	1.1	7.0	9.8
Karnataka										
Bidar	12	41	39	61	64	36	1.0	2.0	6.4	9.2
Gulbarga	9	43	43	57	60	40	0.8	2.4	4.5	7.7
Mysore	18	32	44	56	67	33	0.9	1.9	4.1	11.6

Madhya Pradesh									
Mandsaur	22	34	45	55	75	25	1.2	2.5	7.7	12.4
Raisen	23	47	42	58	68	32	1.3	2.2	6.2	9.0
Vidisha	28	32	38	62	48	52	1.3	2.5	4.9	6.8
Maharashtra									
Akola	11	42	39	61	58	42	1.0	1.6	3.1	4.6
Aurangabad	15	30	30	70	42	58	1.1	2.2	6.4	6.3
Sholapur	19	26	42	58	53	47	0.7	2.2	3.4	5.6
Rajasthan									
Jalore	18	37	14	86	37	63	0.3	1.7	7.2	12.5
Jodhpur	16	58	24	76	35	65	0.4	1.3	2.3	3.8
Nagaur	15	63	21	79	41	59	1.3	2.5	2.4	5.2
Tamil Nadu									
Coimbatore	9	47	50	50	75	25	0.8	2.5	3.8	5.8
Dharmapuri	12	52	49	51	55	45	1.0	1.9	4.6	7.5

Source: Jodha (1986).
[1]Before the distribution of VC land.
[2]After the distribution of VC land.

viduals under various land reform and antipoverty schemes that were intended to benefit the poor but in practice benefited the well-off (Jodha 1986). For sixteen of the nineteen districts in the seven states studied by Jodha, the share of land obtained by the poor was less than that of the nonpoor (Table 2). Hence the poor lost out collectively while gaining little individually.

The statization and privatization of communal resources not only altered the distribution of available resources in favor of a few, they also systematically undermined traditional institutional arrangements of resource use and management. Many traditional methods of gathering firewood and fodder and the practices of shifting agriculture were typically not destructive of nature.[4] Some religious and folk beliefs also helped preserve trees or orchards that were deemed sacred, and some of these sacred groves are still found in parts of India.[5]

Although much more documentation is needed on the regional spread of these resource management systems, the basic point is that where such systems existed, as they did in many areas, responsibility for resource management was linked to resource use via community institutions. When control over these resources passed from the community to the State or to individuals, this link was effectively broken. In turn, the shift in control increased environmental degradation.[6] In particular, property rights vested in individuals proved no guarantee for environmental protection. Indeed (as noted) individual farmers seeking rapid returns in the early 1980s typically planted the fast-growing eucalyptus, which many argued to be environmentally costly.

Aggravating these trends toward deforestation and VC decline are other factors, including population pressure, agricultural expansion at the expense of forests and pastures, and large hydroelectric schemes.

The adverse implications of these processes have been far from uniform, being felt particularly by poor rural households, and especially by female members of such households, for reasons and in ways discussed below.

II. Class and Gender Effects of Environmental Decline

Class and Gender Specificity

Rural households, as noted earlier, have always depended on VCs and forests for a wide variety of essential items for personal use and sale. Although all rural households use the VCs in some degree, for the poor they are critically important because of the unequal distribution of pri-

TABLE 3. Average Annual Income Derived from Village Commons by Poor and Nonpoor Households: 1982–1985

State/Districts	Poor Households[1]		Other Households[2]	
	Value (rupees)	% of total household income	Value (rupees)	% of total household income
Andhra Pradesh				
Mahbubnagar	534	17	171	1
Gujarat				
Mehsana	730	16	162	1
Sabarkantha	818	21	208	1
Karnataka				
Mysore	649	20	170	3
Madhya Pradesh				
Mandsaur	685	18	303	1
Raisen	780	26	468	4
Maharashtra				
Akola	447	9	134	1
Aurangabad	584	13	163	1
Sholapur	641	20	235	2
Rajasthan				
Jalore	709	21	387	2
Nagaur	831	23	438	3
Tamil Nadu				
Dharmapuri	738	22	164	2

Source: Jodha (1986).
[1]Landless households and those owning less than 2 hectares dryland equivalent.
[2]Those owning more than 2 hectares dryland equivalent.

vate land in the country (Agarwal 1994). Jodha's (1986) study of twelve semi-arid districts in seven states in the 1980s found that VCs accounted for nine to twenty-six percent of total income among poor rural households but only for one to four percent of total incomes among the nonpoor (Table 3). The landless and landpoor are especially dependent on the commons for fuel and fodder: VCs fulfill over 90 percent of their firewood and 69–89 percent (varying by region) of their grazing needs, compared with the relative self-sufficiency (from private land) of landed households (Jodha 1986).

There is, however, a gender specificity to the importance of these communal resources, over and above their class significance.

To begin with, in many parts of India there is a systematic bias against women and female children in the intrahousehold distribution of subsistence income controlled by men, including that used for health care and food, as revealed in anthropometric indices, morbidity and mortality rates, hospital admissions, and especially the sex ratio (which in 1991 was 929 females per 1,000 males for all of India). These differences are acute in northwest India but are found in some degree in most regions.[7] Further, where both women and men control resources, women, especially in poor households, are noted to spend their incomes mainly on the family's basic needs and men in greater part on personal needs (Mencher 1989; Agarwal 1994). Hence resources in the hands of male household heads cannot be assumed to benefit women and children in equal degree, and women's direct access to economic resources (private and communal) assumes particular importance. In the case of female-headed households with little or no male support (estimated to be about 20 percent of households in India), the link between direct resource access and physical well-being needs no emphasis.

There are, however, significant inequalities in men's and women's access to private property resources, leading to women's much greater dependence on common property resources. For instance, the most important productive resource in rural economies—agricultural land—and associated production technology is concentrated largely in male hands (Agarwal 1994). Women are also systematically disadvantaged in the labour market, with fewer employment opportunities, less occupational mobility, lower levels of training, and lower payments for the same or similar work compared with men (Agarwal 1986b, 1984; Bardhan 1977). Due to the greater task specificity of their agricultural work (women are mostly concentrated in transplanting, weeding, and harvesting), they face sharper seasonal fluctuations in employment and earnings than do men and have less chance of finding work in the slack seasons (Agarwal 1984; Ryan and Ghodake 1980).

Because of their limited rights in private property resources and fewer other avenues of livelihood, communal resources such as VCs have been for rural women and children (especially those of tribal, landless, or land-poor households) one of the few independent sources of subsistence. Rights in VCs were customarily linked to residence in the village community, and women were therefore not excluded as they typically were from the ownership of individualized private land. Communal resources acquire additional importance in regions with strong norms of female seclusion (as in northwest India), where women's access to the cash economy, to markets, and to the marketplace itself is constrained and dependent on the mediation of male relatives. While these constraints leave women of poor households particularly vulnerable, those in well-off households are also not immune, since in the absence of personal assets they, too, face the risk of impoverishment in case of widow-hood or marital breakdown (Agarwal 1994).

In addition, and most importantly, there is a preexisting gender division of labor. It is women in poor peasant and tribal households who do most of the gathering and fetching, especially of fuelwood and other nontimber products from forests and VCs. Tribal women, in particular, are major gatherers of nontimber forest items for consumption or sale. An estimated 70 percent of such products are collected in the tribal belts of five states: Bihar, Maharashtra, Madhya Pradesh, Orissa, and Andhra Pradesh (Kaur 1991:43).[8] In contrast, men of small and poor peasant households tend to draw on the commons much more for timber, including small timber to make agricultural implements, and materials for house building.

These preexisting gender differences and inequalities impinge radically on who bears the major burden of deforestation and declining VCs.

Implications for Poor Rural Women

First, because women are the main gatherers of fuel, fodder, and water, it is primarily their working day (already averaging ten to twelve hours) that lengthens with declining forests and VCs. In recent years, there has been a notable increase in firewood collection time, to a small degree in some regions and dramatically in others (Table 4).[9]

Fodder shortages are even more acute. My survey in Rajasthan, Gujarat, and the Kumaon region of the Uttar Pradesh hills during 1993 to 1994 indicates not only an increase in the time spent in fodder collection but a growing dependence on market purchase.[10] Moreover, in regions where grazing is still possible, twenty years ago it was boys or men who usually took the animals out, whereas now (as in the Kumaon village) girls are more often sent for grazing while their brothers attend school. Over time this shift could widen the gender gap in literacy in such areas.

Second, the decline in items gathered from forests and VCs has reduced incomes both directly and indirectly, the latter because the extra time spent in gathering reduces time available to women for crop production. This can adversely affect crop incomes,[11] especially in hill communities where, as a result of high male out-migration, women are often the primary cultivators. Similar negative implications for women's income arise with the decline in grazing land and associated fodder shortage. Also, with the erosion of other sources of livelihood, for many years now selling firewood has been common, especially in eastern and central India. Most "headloaders," as firewood vendors are called, are women,

TABLE 4. Amount of Time Taken and Distance Traveled for Firewood Collection in Different Regions

| State/Region | Year of Data | Firewood collection[1] | | Data Source |
		Time taken	Distance traveled	
Bihar (plains)	c. 1972	NA	1–2 km/day	Bhaduri & Surin (1980)
	1980	NA	8–10 km/day	Bhaduri & Surin (1980)
Gujarat (plains)				
Forested	1980	once every 4 days	NA	Nagbrahman & Sambrani (1983)
Depleted	1980	once every 2 days	4–5 km	Nagbrahman & Sambrani (1983)
Severely depleted	1980	4–5 hr/day	NA	Nagbrahman & Sambrani (1983)
Karnataka (plains)	NA	1 hr/day	5.4 km/trip	Batliwala (1983)
Madhya Pradesh (plains)	1980	1–2 times/week	5 km	Chand & Bezboruah (1980)
Rajasthan				
Alwar plains	1986	5 hr/day (winter)	4 km	Author's observation in 1988
Ajmer plains	1970s	2 hr/journey	1.9 km	Survey by author in 1993
(average for all seasons)	1990s	2 hr/journey	2.1 km	Survey by author in 1993
Uttar Pradesh				
Chamoli (hills)				
Dwing	1982	5 hr/day[2]	over 5 km	Swaminathan (1984)
Pakhi	1982	4 hr/day		Swaminathan (1984)
Garhwal (hills)	NA	5 hr/day	10 km	Agarwal (1983)
Kumaon (hills)	1982	3 days/week	5–7 km	Folger and Dewan (1983)
Kumaon (hills)	1970s	1.6 hrs/journey	1.6 km	Survey by author in 1993
(average for all seasons)	1990s	3–4 hrs/journey	4.5 km	Survey by author in 1993

[1]Firewood collected mainly by women and children.
[2]Average computed from information given in the study.
NA: Information not available.

barely eking out a living (Bhaduri and Surin 1980; Kaur 1991). With thinning forests, however, such activity is becoming increasingly nonsustainable, even as it exacerbates the problem of deforestation.

Third, as the area and productivity of VCs and forests fall, so does the contribution of gathered food in the diets of the rural poor. In addition, nutrition suffers with fuelwood shortages, as households economize on fuelwood by shifting to less nutritious foods that can be eaten raw or need less fuel to cook; by eating partially cooked food, which could prove toxic, or eating leftovers, which could rot in a tropical climate; or by missing meals altogether, as observed in Bangladesh in the early 1980s, where the number of meals as well as cooked meals eaten daily declined in poor households (Howes and Jabbar 1986). A trade-off between the time spent in gathering fuel versus cooking can also adversely affect the meal's nutritional quality. Although these nutritional consequences affect all household members in some degree, women and female children bear the greater burden, because of the noted gender biases in intrafamily distribution of food and health care. Nutritional inadequacies in turn have health consequences.

Fourth, large-scale deforestation disrupts social support networks with kin and other villagers. These networks, built up primarily by women, are important in tiding poor households over periods of scarcity. Such support can include reciprocal labor-sharing arrangements during peak agricultural seasons and loans in cash or kind (small amounts of food, fuel, fodder, etc.), on which many poor women depend (Agarwal 1990).

Fifth, gathering food and medicinal items helps build up an elaborate knowledge of the nutritional and medicinal properties of plants, roots, and trees, including edible plants not normally consumed but critical for surviving prolonged food shortages, as during drought (Agarwal 1990). Such "famine foods" are gathered mainly by women and children. The degradation of forests and VCs and their appropriation by a minority are destroying the material basis on which such indigenous knowledge of natural resources is founded and kept alive, leading to its gradual eclipse. This, in turn, could further undermine the ability of poor households to cope with subsistence crises.

Of course, the implications outlined above vary in strength across India, since there are distinct regional differences in the extent of environmental vulnerability, incidence of poverty, and gender bias. Rural women are worst off in regions where all three forms of disadvantage are strong and reinforce each other, as in parts of northern India. Women are less badly off where all three forms of disadvantage are weak, as in much of southern and northeastern India. However, the effects are likely to be felt in some degree in most regions of the country (Agarwal 1995).

Against this backdrop it becomes especially important to understand the gender implications of recent initiatives for protecting and regenerating forests and VCs and to examine women's own responses to the acute scarcity of firewood, fodder, and other gathered products.

III. Emergent Community Institutions for Forest Management

Initiatives taken in recent years by the State or by village communities can broadly be classified into four categories: the government-initiated JFM program, autonomous forest management initiatives, mixed forest management initiatives (State-cum-autonomous), and people's movements. The main features of each are outlined below.

Government-Initiated JFM Program

The basic idea behind the JFM program is to establish a partnership between the state forest department and village communities, with a sharing of responsibilities and benefits. Although the earliest such initiatives were undertaken in West Bengal in the early 1970s by two district forest officers, these remained isolated cases until the late 1980s, when there was rapid informal expansion.[12] In 1989 a formal policy was approved by the state government, following the proven success of forest protection by villagers in the noted districts (Poffenberger 1990). Subsequently, on June 1, 1990, a central government circular spelled out the government's new national policy for involvement of village communities across the country for reviving degraded forestlands.

To date, sixteen states have passed JFM resolutions; this leaves Kerala in the south and all the states, except Tripura, in the northeast. The resolutions allow the participating villagers free access to most nontimber forest products and to 25 to 50 percent (varying by state) of the mature timber when it is finally harvested. On their part, villagers are responsible for protecting the forests by forming an organization, typically a Forest Protection Committee (FPC). However, eligibility rules for membership in the FPCs vary considerably: some states allow all village residents to be members, others allow membership to only one person per household, and so on (Table 5). These rules have important gender implications that I trace later. From among the FPC members a few are

TABLE 5. Conditions of Membership, by State, for Forest Protection Committees under Joint Forest Management

Membership conditions	States
One person per household	Bihar, Karnataka, Maharashtra, Tripura
One adult per household	Jammu and Kashmir, Uttar Pradesh
One adult male and one adult female per household	Andhra Pradesh, Himachal Pradesh, Madhya Pradesh, Orissa, West Bengal[1]
One male and one female per household	Tamil Nadu
All village adults	Haryana
Anyone interested	Rajasthan[2] and Gujarat
Not clear	Punjab

Sources: Society for Promotion of Wastelands Development (SPWD) (1994); *Wastelands News* (1993–94); 4 Jan 1995 government Order for Madhya Pradesh; personal communication from Sushil Saigal (SPWD Staff) on Uttar Pradesh in 1995.
Note: In some states, the Forest Protection Committees also take the form of cooperative societies or general bodies.
[1]In West Bengal, if the husband is a member, the wife automatically becomes a member.
[2]For plantations on village common lands, however, only one person per household can be a member.

elected to an executive committee (or managing committee), which also usually includes the village council head and some others (varying by state).

No comprehensive figures are available for the area under JFM, but data for five states (Gujarat, Haryana, Jammu,[13] Orissa, and West Bengal) indicate that in 1992, 0.6 million hectares of forestland was being protected by 4,486 FPCs (Society for Promotion of Wastelands Development 1994). In some cases, state governments have worked in conjunction with NGOs (or vice versa) in catalyzing the formation of these committees.

Autonomous Forest Management Initiatives

Parallel to and often prior to the JFM initiatives, numerous self-initiated forest management groups have emerged in several states, catalyzed by local leaders or NGOs. Enormously diverse in form and structure, these autonomous groups have been formed primarily in areas where people are still strongly dependent on forests and have a longstanding tradition of community resource management (Sarin 1995). The groups are present in largest numbers in Bihar and Orissa and to a lesser extent elsewhere, as in Gujarat, Rajasthan, Karnataka and Madhya Pradesh. Their organizational setup varies, taking the form of groups of village elders, village councils, FPCs, village-based voluntary organizations, youth clubs, and so on (Kant, Singh, and Singh 1991). Over time, some of these groups have registered with the forest department, but most remain autonomous, without official standing but with tacit village sanction to punish offenders, including by imposing fines.

There are no exact figures on the number of such groups in different states or their area of coverage. Some close observers estimate that in mid-1992 there were about ten thousand community institutions (including both JFM and autonomous groups) protecting some 1.5 million hectares of forestland in ten states (Singh and Khare, quoted in Sarin 1995).

Mixed Forest Management Initiatives

A diversity of initiatives that are operating in conjunction with the State or as autonomous units could be classified under this category. In particular, I have in mind cases where formal State initiatives have become effectively defunct for various reasons, and a range of protection groups in the form of mahila mandal dals[14] (women's associations) have emerged instead. A good example is van panchayats (VPs; forest councils) established by the colonial government around 1931, some of which were revived in the 1980s by local NGOs.

The original setting up of VPs followed a long period of agitation by local communities against the colonial government's curtailment of their rights to forest use and produce. A committee set up in 1921 to examine people's grievances recommended that forests be reclassified and forest councils be formed to manage parts of them. The government subsequently reclassified forests into two categories. Class I forests were those judged to be of little commercial value but of importance as watersheds and sources of fuel and fodder to local communities. These were placed under the revenue department. Class II forests were those that contained commercially valuable timber species; these were placed under the forest department. In addition, there were the "civil forests," which were forests that fell within village boundaries; these were informally managed by villagers but formally under the revenue department's control. It is essentially from Class I forests and civil forests that VPs were subsequently formed in the 1930s. In 1985 there were 4,058 VPs covering about 0.4 million hectares (or 14 percent) of forest area in five districts of the Uttar Pradesh hills (Ballabh and Singh 1988).

Typically consisting of five to nine members elected from the village (or villages) falling in the VP's jurisdiction, the council is responsible for preventing encroachments and devising rules for forest use (Ballabh and

Singh 1988). The VPs are authorized to collect fees from users and levy fines on offenders. Most hire watchmen and pay them either in cash or kind, but this structure is subject to the administrative and technical control of the revenue and forest departments.

Most VPs have been relatively ineffective, with infractions being high. The typical VP is made up entirely of men. In recent years, however, mahila mandal dals have emerged in some regions as independent bodies that are often neither answerable to nor integrated with the VPs but are doing the effective work of protection and reporting offenders to the formal VP body. In rare instances, all-women VPs can be found.[15]

People's Movements

More loosely structured than any of the above are people's movements for forest protection, the most publicized being the Chipko movement initiated in 1973 in the hills of Garhwal (Uttar Pradesh). It began as an attempt by local people to stop indiscriminate commercial exploitation of the regions' forests, 95 percent of which are owned by the government and managed by the forest department. The specific incident that sparked the movement was the successful resistance by the people of the Chamoli district against the auctioning of three hundred ash trees to a sports goods manufacturer, while the local labor cooperative was refused government permission to cut even a few trees to make agricultural implements for the community. Since its inception the movement has spread within the region, but, equally important, its methods and message have reached many parts of the country and outside, in some cases inspiring less known movements, such as Appiko in Karnataka.

General Features of These Initiatives

Although in most states the emergent community initiatives are too new (five years or under for many JFM schemes) to make generalizations possible, they appear to have typically arisen among communities that are highly dependent on forest resources and are facing considerable scarcities, sometimes due to acute degradation of the resource base. Most involve tribal or hill populations that are relatively less socially and economically differentiated.[16] And whereas some groups, as noted, are formally registered with or formed through the forest department, others have no official standing.

In terms of regeneration, there have been some notable successes. Where the tree rootstock is undamaged, natural regeneration begins at an encouraging pace, often yielding a good harvest of grass within the first year of protection and fuelwood through cutback operations within a few years. Several protected tracts I visited in March 1995 showed impressive natural regeneration. For instance, when Malekpur village (Sabarkantha district, Gujarat) began community forest management in 1990, with the encouragement of an NGO (Vikram Sarabhai Center for Science and Technology [VIKSAT]), and registered a Tree Grower's Cooperative Society, the protected area consisted of little more than barren hillsides, from which it was difficult to obtain much except dry twigs and monsoon grass. However, given the strong rootstock, by 1995 there was a young teak forest there, with trees ten to twelve feet tall, interspersed with other species. In the 1994 monsoon season there had also been a substantial harvest of grass. And in early 1995, thinning and pruning operations yielded enough firewood for domestic use to last every participating village household some five months. Similarly, in the Baruch district of Gujarat, where another NGO (Agha Khan Rural Support Group [AKRSP]) has served as a catalyst, several protected forest tracts that I visited in 1995 showed impressive natural regeneration.[17] Biodiversity was also reported to have increased. A number of other case studies report similar encouraging returns after protection and a decline in seasonal migration.[18]

Undeniably, there are also cases of serious conflict (especially intervillage) and failure, and the factors that account for success or failure need more probing. However, a study of forty-two FPCs in Midnapore district of West Bengal—the state with the longest-standing JFM program—is indicative:

> The most effective FPC is when a single village is involved in the management of the forest, its ethnic composition is tribal, a majority of the households in the village become members of the FPC and . . . forest land is allowed to regenerate rather than afforested with plantations. In contrast, the least effective FPC is one which is managed by several villages, has a mixed population of tribes and castes, only a few households in the villages become FPC members, and the forest land is put under plantation rather than natural regeneration. (Malhotra et al. 1990, 22–23)

IV. The Gender Gap in Emergent Institutions

The question remains: Have the communities that have displayed impressive results in protection and greening been as successful in ensuring gender equity in control over common property resources and in the sharing of benefits? To answer this, we need to examine women's participation in the decision-making forums of the emergent community initiatives, for instance, women's presence and voice in FPCs and executive committees that make the rules about responsibility and benefit sharing. We also need to consider the implications of such participation (or its lack) for the effectiveness of protection and regeneration activities, the distribution of burdens and benefits from them, and women's empowerment.

Women's Effective Participation

Effective participation would involve women's formal membership in management committees, their attending meetings where they are members, and their views' being given weight in the meetings they attend.

In terms of women's formal membership, whether in JFM schemes, autonomous initiatives, or VPs, the overall picture to date is discouraging, with some notable exceptions, discussed later. In several JFM states that allow membership in FPCs to only one person per household, women are effectively excluded, since inevitably a man is the member. Even where they are not so excluded, women's numbers are low. In West Bengal, out of 8,158 members in 72 FPCs in Midnapore district, only 241 (3 percent) are women, mostly widows (Roy, Mukerjee, and Chatterjee 1992). In the villages of Barsole and Lekhiasole (also in Midnapore district), where the FPCs have 44 and 303 members, respectively, the first has only 2 women and the second 17, again most of them widowed heads of household (Guhathakurta and Bhatia 1992). In Tamil Nadu, of the 22,561 members in 2,594 FPCs, only 7 percent are women (Narain 1994). Orissa's self-initiated groups also typically exclude women (Kant, Singh, and Singh 1991; Singh and Kumar 1993), and most VPs in Uttar Pradesh have few or no women members (Ballabh and Singh 1988; Sharma and Sinha 1993).

Membership aside, to participate in decision making women need to attend and be heard in committee meetings. For reasons discussed further below, usually few women attend; those who do rarely speak out; and when they do speak, their views are seldom taken seriously.

Within this rather negative scenario, there are also cases of vocal women being present in notable numbers in some forest protection initiatives. These cases, which are described later, can provide pointers on how women's participation could be increased. But first let us consider why it is important that women participate in their own right.

Importance of Women's Participation

Women's membership in forest protection initiatives and their effective participation (or its lack) in the decision-making forums impinges on at least three crucial aspects: entitlements, efficiency, and empowerment.

Entitlements and welfare considerations. A household's entitlement to a share in the benefits from protection is linked to membership in the forest protection initiatives. To be sure, women could benefit in some degree by virtue of belonging to households where men are members. For instance, where degradation is not acute, member households continue to enjoy the right to collect dry wood or leaves from the protected area.[19] Also some FPCs under JFM have given very poor women special consideration in allowing them to collect leaves for plate making (Arul and Poffenberger 1990). However, for several reasons, the benefits mediated through male membership have welfare disadvantages compared with women's directly being members and participants in the decision-making processes.

First, in many villages in Gujarat, West Bengal, Bihar, and Orissa, when protection began women were barred from any form of collection, even of dry twigs. Where the land was barren this caused no extra hardship. But where women were earlier able to fulfill at least a part of their needs from the protected area, the ban on entry imposed by all-male protection groups has made it necessary for women to travel to neighboring unprotected areas, spending many extra hours and also risking humiliation as intruders.[20] In some protected sites in Gujarat in 1993 and West Bengal in 1994, Sarin (1995) found that women who prior to protection spent one to two hours for a headload of firewood now spend four to five hours and that journeys of half a kilometer have in some cases lengthened to eight to nine kilometers.

Similarly, during my field visit to Gujarat's Sabarkantha district, several women said that they were not allowed even to walk through the protected area to the neighboring one for fuelwood collection, on the grounds that they would break the rules. This forced them to skirt the area and spend several additional hours on their journeys (see box). In the village of Pingot (Baruch district of Gujarat), Shah and Shah (1995) found that since protection started, women have been compelled to take their daughters along to help with collection, spending over six hours a day to walk five times farther than they used to, for the same quality of fuelwood. Over time this could negatively affect the girls' education.

When asked to comment on a recent award for environmental conservation conferred on the village, the women expressed only resentment: "What forest? We don't know anything about it now. We used to go to the forest to pick fuelwood but ever since the men have started protecting it they don't even allow us to look at it!" (Shah and Shah 1995, 80).

These gender-specific hardships have typically surfaced where women are not members of protection committees and therefore did not participate in the initial formulation of rules. The household's everyday requirement of fuel and fodder, which is women's relentless responsibility, was therefore bypassed; what received attention were the sporadic need for small timber to construct and repair houses or make implements (which are men's responsibilities) and the potential cash returns from large timber.

In some instances, interventions by NGOs remedied the situation once women brought it to their notice, leading to cutback operations that yielded considerable firewood per household. For instance, in Malekpur village (Sabarkantha district, Gujarat), where some women say their headload collection time had increased from an hour or two to a whole day, a meeting organized by VIKSAT with the Tree Grower's Cooperative Society (which was doing the protection), led to cutback operations in 1995, which (as mentioned earlier) yielded both firewood and fodder. Households that could afford it also switched to biogas. The hardship in the interim years, however, was borne solely by women.

Also in some of the autonomous initiatives, all-male youth clubs that are protecting the forests have not only banned entry completely, but have been selling (rather than distributing) the forest products obtained from thinning and cleaning operations. Poor households that cannot afford to buy firewood and other forest products are the worst victims of this policy, with the burden again falling disproportionately on women. As one woman commented, "Earlier it was the forest department which controlled the forest, now it is the youth clubs" (Singh and Kumar 1993, 23).

Second, cash benefits from protection, say, through the sale of timber or grass, are often put into a collective fund, rather than being distributed to member households. How that fund gets spent again depends on the male representatives in the protection committees. In the late 1980s, some youth clubs undertaking protection in Orissa made substantial gains (in some cases up to Rs. 25,000) from the sale of forest products obtained during clean-up operations in the protected areas. Although in

Extracts from the Author's Interviews with Village Women from Sabarkantha District (Gujarat)

25 March 1995

Q: On what issues do women and men differ in FPC [Forest Protection Committee] meetings?

A: Women face the problem of firewood. Because women protect the forest they should get some benefit from it. Men can afford to wait for a while because their main concern is timber. But women need fuelwood daily. When we ask for permission to take dry twigs men say: what is the guarantee that you won't cut green branches. You might cut more. The men don't listen to us. We can get some fallen twigs and leaves for only 10 days. The forest is closed for the rest of the year.

Q: What do you do then?

A: At the moment it is closed, so we use crop stalks, cattle dung, kerosene. Some have biogas.

Q: What did you do before the closure?

A: We used to go to the Rajasthan border for fuelwood. The route was through our own forest. On the return journey we would pick up dry wood from our forest.

Q: Do you go to Rajasthan now?

A: No, we can't now, because the route through our forest is blocked. From our forest, we are only allowed to get dry wood for 10 days in the winter. That's all. We collect enough for 2–3 months. But in the monsoon we don't know what we will do. Last year they gave us special permission to collect for 10 additional days. This year we are hoping for permission again.

Q: Will you get permission?

A: At the last parishad meeting they told us they won't give us permission.

Q: If you don't get permission, what will you do?

A: We can only call a women's meeting and talk to the men and put forward our problem. We will say: we have to cook, we have no wood. So what now?

one case the money was reportedly used for a school building, in several others it was spent on constructing a clubhouse or for club functions (Singh and Kumar 1993).

Even if such money were distributed to the participating households through the male members, cash given to men does not guarantee equal sharing within the family. In West Bengal, for instance, the daily wages paid by the forest department during the period of planting and the income from the subsequent sale of trees are usually given to the male household head even if the family works as a group. Guhathakurta and Bhatia (1992) found that the men in one village had used the money from timber sales to buy additional land and those in another village had used it for gambling and liquor, rather than for pressing household needs.

When the question of benefit sharing was discussed in a meeting of FPCs from three villages of West Bengal, in which both women and men were present, all the women unequivocally said that shares should be equal and separate for husbands and wives. "There was no vote for 'joint accounts' or the husband being more eligible as the 'head of the household.' These women are responsible for a major share of household sustenance and they wanted control over their share of the income" (Sarin 1995, 90). Indeed, benefits reaching the women would improve the welfare of the whole household, since poor rural women are noted to spend mostly on family needs (Mencher 1989).

Third, "needs" is only one criterion for the distribution of intrahousehold benefits. Entitlements within the household are also linked to perceptions about women's contribution and notions about rights (Sen 1990; Agarwal 1994). Insofar as "perceived contribution" is an important criterion for the distribution of benefits, women seen to be participating in forest management would be better placed to claim equal benefits. Membership would give women a formal independent right in the new resource and not merely indirect benefits mediated through male members.

Fourth, membership in FPCs can lead to additional financial benefits. For instance, in the villages where the AKRSP works, a part of the daily wage earnings from tree planting goes into savings funds. Where women are not members, the savings have gone into a family account (which is effectively controlled by the male household head). In recent initiatives, however, where female membership is high as a result of the AKRSP's specific attempts to involve women, savings go into separate accounts for women and men, and women often make their own decisions about how they will spend this money.[21]

Efficiency considerations. Women's active involvement also appears necessary for the effective functioning and long-term sustainability of these initiatives. For instance, to prevent infractions, women as the main collectors of firewood and other nontimber forest products need to adhere to the rules. In some cases, male committee members have threatened their wives with beatings if they break the rules, reinforcing existing positions of male power (Sarin 1995). Its reprehensibility aside, this form of control is hardly enforceable in the long run, given that women's collection activities fulfill a basic household need on which men also depend.

Oftentimes both women and men are aware of the importance of women's involvement in protection programs. Britt's (1993) interviews in two villages in the district of Nainital (Uttar Pradesh hills) in 1992, although dealing with VP experience, also have relevance for JFM and other contexts. The VP committees of both villages include only one woman each. Britt notes:

> Males and females generally concur . . . that if more women were to attend meetings, the workings of the forest committee would be improved. When prompted, the majority of the villagers thought that some kind of mechanism necessitating attendance by greater numbers of women, such as a 50 percent reservation policy, would provide for greater information dissemination and better implementation of forest committee rules. (147)

In women's own words:

> It would be good if women went [to forest committee meetings] . . . The men don't seem to realize where fodder and fuelwood come from. (143)

> Women often don't even know what rules the forest committee has decided upon. If more women were on the forest committee then they could pass on the information to other women and the forest would be better protected. (147)

> The male members of the forest committee have difficulties implementing the rules. Women could discuss these problems with the men. Perhaps more "mid-way" rules would be, in the long run, more effective . . . more viable. (148)

Despite this recognition, typically few women participate in the VPs or in the FPCs under JFM. However, as I discuss later, it is not uncommon for them to form informal patrol groups where men's groups are ineffective.

Also women's knowledge as well as preference of plant species often differs from that of men. Involving women in decisions about planting and silviculture practices in the protected areas would be an effective way of ensuring that a larger proportion of household needs from the forest are taken into account and that women's particular knowledge of plants and species enriches the selections made, thus enhancing biodiversity. In the Panchmahals district of Gujarat, women's rich knowledge of medicinal herbs was important in promoting such plants in the protected area (Sarin and Khanna 1993). In the same region, in the village of Muvasa a woman's group replanting a part of the VC land resisted pressure from the men to plant eucalyptus for cash benefits. The women used their considerable knowledge about local trees and shrubs, and their suitability for different uses, to select diverse species instead.

Empowerment considerations. The absence of women's formal participation in the new community initiatives will reinforce preexisting gender inequalities and further reduce women's bargaining power within and outside the household. In contrast, participation in public decision-making forums, such as FPCs, would help reverse rural women's traditional exclusion from such forums and also increase their self-confidence in asserting their rights in relation to public bodies in general.

In Navagaon village (Gujarat), for instance, where women constitute 50 percent of the members in Village Development Associations (promoted by the AKRSP) and are entitled to hold separate savings accounts, they now feel they are treated more respectfully by the village men, not least because they deal with institutions such as banks themselves. Similarly, in the Chipko struggle it is notable that, over time, as participation enhanced women's confidence, they began to demand membership in the village councils and a greater say in their decision-making processes.

More generally, numerous case studies have noted the empowering effect on women of greater control over economic resources, especially land, and of participation in the forums that control these resources, especially via collectivities (Agarwal 1994).

V. Constraints to Women's Formal Participation

What constrains women from formal participation in many of these emergent institutional initiatives? Broadly, the constraints are of five types: the rules governing membership, traditional norms of membership in public bodies, social barriers, logistical factors, and the attitudes of forest department personnel.

First, as noted earlier, in several states the JFM resolutions allow only one member per household. This is inevitably a man, except in widow-headed households. Even where the rules allow one man and one woman per household (as in Andhra Pradesh, Himachal Pradesh, Madhya Pradesh, Orissa, and West Bengal), other household adults remain excluded (including dependent widows and unmarried daughters). The most equitable situation would be to allow JFM membership to all village adults, as is the case in Haryana and Gujarat.

Second, traditional village assemblies and councils customarily excluded women, even among tribal (including matrilineal) communities.[22] In many autonomous forest management initiatives, as in parts of Bihar and Orissa, this longstanding tradition has been replicated in the new institutions. Women are not called to meetings for conflict resolution, even when the dispute directly involves them (Sarin 1995).

Third, women face social constraints that need probing. Most studies attribute women's low attendance in FPCs and executive committees and their not speaking up in JFM and VP meetings to "cultural barriers,"[23] but few explore what these might be. Given that a large majority of the community initiatives discussed herein involve tribal or hill communities where there is no female seclusion and women's participation in economic activities is visibly high, the constraints clearly have little to do with explicit norms of seclusion and much to do with gender ideology, that is, the social constructions of acceptable female behavior, notions about male and female spaces, and assumptions about men's and women's capabilities and appropriate roles in society.

For instance, although many of the women Britt (1993) interviewed in 1992 in the two VP villages recognized that their presence in meetings would improve VP functioning, they did not feel free to attend unless the men invited them, and they felt that the men were not seriously interested in doing so.

> The meetings are considered for men only. Women are never called. The men attend and their opinions or consent are taken as representative of the whole family—it's understood. (148)

> Male committee members are not interested in calling women to meetings even though women ... are the ones who go to the forest and do the cutting. (146–147)

Women's effectiveness is also restricted by their limited experience in public speaking, illiteracy, a lack of recognized authority, and the absence of "a critical mass" of women. As one woman in the VP area said:

> Only I alone cannot change procedures. If I tried to change the rules, people would think what sort of woman is she, that she has these ideas . . . I am not in the habit of speaking publicly, not like other women who have worked with CHIRAG [a local NGO]. (Britt 1993, 143–144)

In Katuual village the sole woman member, although elected to the VP several months prior to Britt's (1993) visit, had yet to attend a meeting. However, she was interested in going to the next one and had requested that meetings be held on Sundays when other family members were home, leaving her free to go. She felt, though, that as the only woman member, and without the acquiescence of other village women, she would not be effective:

> I discuss the forest with other women. Many times I have told outsiders not to go to the forest and cut leaves or trees. I warn them that they will be fined. Sometimes I have lied, telling them that a government officer is coming and that if they are fined in front of him then they must pay a deposit in Nainital. But all this has very little effect on the women. If they intend to collect, then they will. (145)

Women also feel discouraged from attending meetings because their opinions are disregarded. One woman member of a VP committee commented as follows on the attitude of her male colleagues:

> I went to three or four meetings. . . . No one ever listened to my suggestions. I marked my signature in the register. I'm illiterate so I couldn't tell what was written in the meeting minutes. I was told that my recommendations would be considered, but first that the register had to be signed. They were uninterested. (Britt 1993, 146)

There are similar complaints about the functioning of FPCs under JFM from parts of West Bengal (Mukerjee and Roy 1993; Roy et al. 1993). Even women who are executive committee members and attend meetings regularly usually sit at the back as mere observers, while the points raised by male members who sit in front receive priority.

Fourth, age and marital status affect women's participation in meetings. In many of West Bengal's FPCs, the few women members are mostly widows (Narain 1994; Guhathakurta and Bhatia 1992). Sharma, Nautiyal, and Pandey (1987) similarly note for Chipko women, "When one looks at the profiles of a few of these women who have taken active part earlier in the prohibition movement, and later in the 'Chipko' movement on a more sustained basis, they are older women or widows or single women," and "Young married women are more constrained by their family responsibilities and kin-based authority patterns" (50–51).

The burden of work is also usually greater on young married women, especially daughters-in-law (Bahuguna 1984; Britt 1993).

Fifth, women's participation is often impeded by logistical constraints and double work burdens. The timing of meetings (which are often called when women are busy with other work) and women's heavy workload (child care, housework, agricultural activities, and other responsibilities) can be serious barriers. As one village woman said:

> Women are very busy with household work. If they go to the meetings who would watch the children? It is impossible for all women to attend. (Britt 1993, 146)

Most hill women in VP villages told Mansingh (1991) that they did not have time to "sit around for [the] four hours that it took to have a meeting in the middle of the day." As a result, women's attendance tended to thin over time.

Sixth, many male forest department personnel involved with JFM are known to call only men to meetings (Roy, Mukerjee and Chatterjee 1992), while there are few women among the department's personnel. In the Haryana forest department only 15 women village forest workers were appointed as against an official provision for 300 (Narain 1994). In four divisions of Tamil Nadu, only 6 percent of total social forestry workers are women (Venkateshwaran 1992). Some women in parts of West Bengal reported that male officers rarely consult them in preparing the village-level microplans for forest development, others admitted to having heard about the plan only through their husbands, and yet others said they had been consulted initially but not for revisions or updates or on choice of tree species (Guhathakurta and Bhatia 1992). Women interviewed by Narain (1994) in two West Bengal villages complained that male officials discouraged them from coming to the forest office and rebuked them if they came in the evening, even on urgent work. Elsewhere in West Bengal women complain that:

> The forest officers put very little value on what they say and always crosscheck with the men to verify the truth of their words. And if ever there is any conflict or contradiction between the women and the men, the foresters always settle the disputes in favor of the men. (Roy et al. 1993, 15–16)

Many forest department personnel see women's involvement in JFM activities as useful mainly for keeping out other women "offenders," rather than for reasons of gender equity or to take advantage of women's knowledge of plants and trees. Women are seen as better able to catch female culprits, since men doing so are susceptible to being charged with molestation.

VI. Contrasting Examples and Women's Initiatives

Despite these constraints there are contrasting examples where women's presence in forest protection groups is high, sometimes in the formal forums, although more commonly in informal ones. These cases suggest that the "cultural" barriers to women's participation, especially in tribal or hill communities, are not insurmountable.

First consider the formal groups. In parts of Gujarat, 30 percent of the members in the village general body are women, and their presence in the JFM executive committees ranges from 14 to 50 percent (Narain 1994). In a number of other recent initiatives in Gujarat, under AKRSP encouragement female membership in the FPCs has risen to 50 percent. In parts of West Bengal's Bankura district, women are doing most of the protection work: for instance, in Chiligarah village women are the members and men their nominees (Narain 1994), and Korapara village initially had only male members, but now 22 out of 35 FPC members are women (Viegas and Menon 1991). There are also several all-women FPCs in Bankura district (Mukerjee and Roy 1993). Likewise, in the Ranchi district of Bihar, in one village which has a mixed population of Muslims, Hindus, and Scheduled Tribes, women took the initiative of forming a FPC in 1991 when the all-male committee was ineffective in resolving conflicts and in saving the forest. The committee has four hundred to five hundred women members drawn from all sections of the village, covers about 490 hectares of forestland, and has since been given formal recognition by the forest department (Adhikari et al. 1991). In Orissa, women in two villages approached a young forest service probationer, who was also in charge of the forest range, to help them form an all-women FPC (Singh 1993).

More commonly though, it is women's informal groups that are in effect undertaking forest protection and wasteland development. In some villages, it is the failure of men's committees that has led women to form their own. In Machipada village (Baruch district, Gujarat), which falls in AKRSP's ambit, the women started their own protection group in 1994, even though an all-male group already existed. They now patrol the area in rotation with the men. My conversation with some of the women during a field visit in March 1995 threw light on this:

Q: Why don't you leave forest protection to the men?

A: We protect the forest for our children. We have an old relationship with the forest.

Q: Don't men also have such a relationship? Do women have a special relationship?

A: Yes, women do. We go there for firewood, nuts, berries, and many other items.

Q: Don't men protect well?

A: Men don't check carefully for illegal cuttings. Women keep a more careful look out.

Q: Is there any other advantage of your forming a separate patrol group?

A: Our patrolling leads to the feeling that there is continuous protection. Now people feel everyone is taking responsibility for protection.

Q: Since you formed your own patrolling group are you treated better at home?

A: Yes, it makes a difference. Now women can explain why the forest is important. Men listen better. Now it feels like our forest.

In Rajasthan, with the help of the People's Education and Development Organization (PEDO), women in several villages have established plantations and employed watchmen to guard them (Sarin and Sharma 1991). There are also numerous success stories of women's groups reclaiming village wastelands from across India (Singh and Burra 1993).

Again, in the Uttar Pradesh hills where, as noted, most formal VPs have few women, there are numerous cases of women's informal groups guarding the forest. In Buribana village (Nainital district), which I visited in 1993, the mahila mandal dal (women's association), devised its own rules for the collection of forest produce and kept a lookout for offenders, reporting infractions to the VP head. Now they are also invited to VP meetings (although formally the VP members are all men). Elsewhere, women either guard the VP forest themselves or employ a guard (Sharma and Sinha 1993). It is notable that in Sharma and Sinha's (1993) study of twelve VPs, the four they deem "robust" and successful have active mahila mandal dals. In general, they attribute the success of many VPs to the presence of active mahila mandal dals, even though these women's associations have no formal authority for forest protection: "If the condition of the forests has improved in recent years, much of the credit goes to these women's associations" (173). The associations spread awareness among women of the need to conserve forests, exert social pressure on women who violate usage rules, and monitor forest use. The importance of a mahila mandal dal's cooperating with the VP lies especially in the fact that women in this region not only do most of the fuel and fodder collection, they also play a critical and highly visible role in agriculture. If they refuse to follow the rules, the men are unable to effectively enforce them. Moreover, women are also in the best position to apprehend transgressors.

Here Viegas and Menon's (1991) observations for FPCs in West Bengal have wider relevance:

> In complete contrast to their [typically low] representation in the committees, the active contribution of the women to the aims and objectives of the FPCs is . . . much more than that of the male members. This seems to be the irony of the situation in that recognition is given through official membership to the males whose contribution is much less than that of the women. . . . there is no time in the day when a few women are not present in the forest. Hence, in most areas no need is even felt to appoint special patrols to guard the forests from offenders. The women invariably take on this role. Whenever they spot an offender it is they who apprehend him/her directly. (22–23)

Women's participation in movements such as Chipko is again illustrative. Although mobilization and protest in the movement have been typically situation specific, in some Chipko areas women have formed vigilance teams against illegal felling and are monitoring the use of the local forest. Moreover, Chipko women have protested against the commercial exploitation of the Himalayan forests not only jointly with the men of their community, but on occasion even in opposition to the men, revealing different priorities in resource use. On one occasion, women successfully resisted the axing of a tract of the Dongri-Paintoli oak forest for establishing a potato seed farm that the men supported. Cutting the forest would have added five miles to women's fuelwood journeys while they felt cash earned from the project would stay mainly in the men's hands. Again in tree-planting schemes, Chipko women have typically favored trees that provide fuel and fodder rather than commercially profitable varieties often favored by the men.[24]

The above examples clearly highlight women's concerns and organizing abilities. At the same time, the examples are not unambiguously positive, since forming informal groups adds to women's responsibilities and burdens without vesting them with additional authority. Improving women's participation and authority in the formal forums therefore remains vital.

VII. Overcoming the Constraints

What accounts for women's participation in some initiatives and not in others? What factors could make the emergent community institutions for resource management more gender equal and strengthen, in particular, women's formal involvement in the groups?

Gender-Progressive NGO Presence

Typically a major factor facilitating women's participation and also effectiveness in FPC meetings is the presence of a gender-progressive organization. I have noted several examples where such an NGO has explicitly brought women's concerns to the fore and led to those concerns' being addressed to some degree. In the village of Malekpur in Gujarat, I noted that in a community meeting, VIKSAT's focus on the hardships women were facing in fuelwood collection led to cutback operations that yielded substantial fuelwood. In Navagaon village (Gujarat), the AKRSP was able to increase women's membership in FPCs as well as their attendance in meetings considerably. Similarly, in Rajasthan, explicit dialogue with the men of the community, through the intermediation of PEDO, reduced male hostility toward women's efforts at reclaiming VC lands. All three NGOs in question—VIKSAT, AKRSP and PEDO—were not exclusively women's organizations.

All-women organizations, however, can make a particular difference. The contrast between north and southwest Bengal in Mukerjee and Roy's (1993) study is revealing. While in southwest Bengal (Midnapore District) only 3 percent of the 8,158 FPC members are women, in north Bengal female presence in FPCs is marked. In the latter almost all the women members in FPCs are also members of the local woman's organization, the Ganatantrik Mahila Samity. In parts of West Bengal's Bankura district where women's NGO presence is strong, there is also a notable female presence in FPCs. In the village of Korapura, the shift from an initial all-male membership to 63 percent female membership is attributable to the active encouragement of the local women's associations—the mahila samitis (women's groups) of the Nari Bikas Sangh. This organization was initially formed under the leadership of the Center for Women's Development Studies (Delhi) to enable women to develop degraded village lands as an income-generating activity. In 1980, a group of women reclaimed village wasteland and planted Arjun trees for sericulture. By 1988, some fifteen hundred women in thirty-six villages were members of such groups (Mazumdar 1989). Today Nari Bikas Sangh is a registered body, and its members are also among the most active members of FPCs in Bankura district.

The fact that these initiatives primarily involve tribal or hill communities makes it easier to overcome social barriers than would be the case in settings that are more class and caste differentiated, and especially where norms of female seclusion are strong. However, the problem that women's opinions are not given much weight in mixed forums, even when they speak out, is part of a larger issue of the cultural construction of gender and social perceptions about women's capabilities and place in society, from which even hill and tribal communities are not immune. Changing these perceptions will not be easy, since many institutions contribute to the creation of gender ideology, including educational establishments, the media, and religious bodies. At the same time, it is encouraging that many aspects of women's situation are amenable to change over time when women begin to speak out collectively through the facilitating presence of a gender-progressive NGO.

In two meetings that I attended in March 1995 in Gujarat, one convened by the AKRSP, the other by VIKSAT, the personnel from these NGOs helped both in soliciting the opinions of women who were present and in ensuring that those opinions were given due weight. Of course, if the NGO is itself male biased, it can reinforce existing bias within village communities. The gender awareness now being displayed by the AKRSP and VIKSAT is reported by them to be of relatively recent origin—a result both of field experience and of discussions initiated especially by some of their women officeholders.

Gender-Sensitive Forest Officials

A gender-sensitive forest officer can also make a marked difference. For instance women labeled as "offenders" began to be called "defenders" of forests through the intervention and support of forest personnel. Again, in Brindabanpur village (West Bengal), women were forced to trespass into the neighboring village for fuelwood and other necessary forest produce, since they had no forest of their own. When a sympathetic forest officer (who examined the complaints against the women) assured them that they would be allotted some forestland if they formed a FPC and followed its rules, the women constituted an all-women protection group. It is reported that women now monitor the space carefully, sell saplings from a nursery they have developed, and operate a savings account. Some who had earlier depended solely on illicit felling now have part-time employment as a result (Chatterjee 1992). Mansingh (1991) recounts a similar case in a VP area where the matter was similarly resolved by discussions between the villagers, the village government functionary, and a local NGO.

Involvement in the Initial Stages

Experiences in several areas suggest that if women are involved from the beginning when the organization is formed, the chances of their sustained participation is greater. This might be both because they are more motivated and because their presence has greater legitimacy. Mansingh (1991) in her study of women's attendance at VP meetings, found that:

> Women were involved substantially in the voluntary work and always seemed well informed about what had happened. This was in my opinion because they had been present in vast numbers in the first meeting and had understood, agreed to, and participated in the concept of the *van suraksha samiti* [forest protection committee]. (29)

Further:

> It was in the first few meetings that the basic protecting resolutions of imposing a moratorium on the cutting of green wood and [of] leaves for fodder, and stopping the grazing of animals were passed. Women[,] being the main collectors of wood and fodder[,] needed to agree to this unanimously, either themselves, or through their husbands. Though the latter were often the means by which the outcomes of later meetings were communicated to them, it didn't always work with the initial agreements to change their use pattern. (40n)

The AKRSP's experience in Gujarat leads to a similar conclusion. In villages where from the start an attempt was made to recruit women as members in the village development associations (which also undertake forest protection), both membership rates and attendance at meetings are high. In some of these villages, 50 percent of those attending are women (Chandran 1995). An AKRSP project officer told me that sometimes even the way the idea of membership in committees is introduced can make a difference. Earlier they had said, "There should be at least two women" (since that was the stipulated minimum for the working committee in the Gujarat JFM resolution). Now they say, "Anyone who wants to can become a member," leading to larger numbers of women joining.

Critical Mass

The presence of a critical mass of vocal women also appears necessary to give women an effective voice in mixed forums. Some women whom Britt (1993) interviewed in VP villages emphasized that "without a good majority of women present it is impossible to express opinions" (146) and that men would find it difficult to ignore larger numbers of women. The women I interviewed from the Gujarat villages where VIKSAT is working were clear that "more women should be involved; that will help."

Of course, new initiatives usually involve a process of learning, and what may begin as women-exclusive initiatives can become more women inclusive over time. Again, however, the presence of vocal women in large enough numbers appears important. Even in the Chipko movement, I believe, women's high participation has been an important factor, enabling them on occasion to take independent initiatives without the men's support, and at times even in opposition to the men.

Nevertheless, the question of whether or not women should organize separately by forming all-women groups remains a vexed one.

All-Women Groups?

In general it is noted that village women are more comfortable and vocal in all-women groups than in mixed ones. Women village leaders also argue that separate groups will enhance women's participation. For instance, in West Bengal's Midnapore district, the woman *panchayat pradhan* (council head) of Kesiary block said that women's participation in JFM would increase only if separate meetings were convened and women's special constraints dealt with. She recommended that there be an equal number of women and men in the FPCs and executive councils (Guhathakurta and Bhatia 1992). Similarly, in the village of Durgala (Sambalpur district, Orissa), Mohini Naik, the local woman activist who initiated a women's association in 1988 to replant and manage the VC land, argues that women can motivate women better and that there should be more female members in FPCs. She herself is a key member in the local FPC and feels her presence in the committee has enhanced her status in the community (Swedish International Development Authority 1991).

The experience of other NGOs is mixed. PEDO began by setting up all-women groups to regenerate wastelands in parts of Rajasthan, but found that this generated a great deal of hostility and suspicion among the village men. This led PEDO to change its policy and constitute groups of both men and women, but with ambiguous results. An evaluating team noted that:

> Joint meetings of men and women, while successful in reducing male hostility and securing their cooperation, tended to diminish free expression and articu-

lation by women. The need to create a separate forum for women, in which they could express their views and concerns uninhibited by the presence of men, was strongly felt. (Sarin and Sharma 1991, 20)

They found a way out by starting women's savings groups from the money earned through the sale of surplus grass from the protected land. This also provided a rationale for holding separate women's meetings without antagonizing the men.[25] In other words, the solution accommodated existing gender relations rather than challenging them, although over time it may well empower the women to do just that.

VIII: On Gendered Responses and Environmental Action

What does the experience of the emergent community institutions tell us about gender differences in responses to environmental degradation?

In an earlier paper I formulated the concept of feminist environmentalism (Agarwal 1992). I argued that people's responses to environmental degradation need to be understood *in the context of their material reality*, their everyday interactions with nature, and their dependence on it for survival. Gender-specific responses can typically be traced to a given gender division of labor, property, and power, rather than primarily or solely to the notion that women are closer to nature than men, as is suggested by the ecofeminist perspective.

The emergent initiatives described in this paper offer further support for the feminist environmental approach to understanding environmental action. It should be noted that rural men also have actively responded to severe deforestation and the degradation of VCs by seeking to contain and reverse these processes. This can be traced to the threat to their livelihood systems and dependence on common property resources for supplementary income, or for small timber for house repairs and agricultural tools, which are mainly men's responsibility. Women's responses are linked more to the availability of fuel, fodder, and nontimber products, for which they are more directly responsible, and the depletion of which has meant undertaking ever-lengthening journeys. In other words, there is clearly a link between the gender division of labor and the gendered nature of the responses.

The women I interviewed from some Gujarat villages where VIKSAT is working were unambiguous about this:

Q: On what issues do men and women differ in forest protection committee meetings?

A: Men can afford to wait for a while because their main concern is timber. But women need fuelwood daily.

However, whether these concerns get translated into *effective* action is dependent on whether women's rights in common property resources are explicitly recognized, on the influence women command in the community, and on their access to public decision-making forums. A case study of autonomous forest management initiatives in three districts of Orissa highlights both the gendered impulse for forest protection and the unequal distribution of power that has enabled men's interests to supersede those of women. Commenting on the factors that led to the formation of all-male initiatives in the region, the study notes:

> In most of the cases protection efforts started only when the forest had degraded and communities faced shortage of small timber for construction of houses and agricultural implements. Although there was a scarcity of fuelwood, it hardly served as an initiating factor. (Swedish International Development Authority 1993, 46)

Clearly women's concerns, even if pressing, do not automatically translate into environmental action by women themselves or by the community. For poor women to move from being the main victims of environmental degradation to being effective agents of environmental regeneration is not likely to be easy, although, as noted, a gender-progressive NGO or separate women's associations can make an important difference. Sarin and Sharma's (1991) observation of women's participation in the regeneration of VCs in Rajasthan also underscores this point:

> There is nothing "automatic" in the extent of women's active participation in the development of village common lands, no matter how acute their hardship of searching for fuel and fodder. Even in the villages where women took the initiative and played a leadership role, this was preceded by enabling them to interact with other women's groups through *melas*, visits, training programmes and awareness generation camps. Continuous interaction with PEDO's women staff has been another crucial input for facilitating women's genuine participation. (39)

The considerable regional and community variation in women's status is also likely to impinge on women's responses. In particular, there are significant differences across states and between social classes and communities (tribal/nontribal, Hindus/Muslims, upper-caste/lower-caste Hindus, hill dwellers/plains dwellers) in the emphasis on female seclusion and segregation, and hence in the constraints on women's mobility, freedom to participate in public meetings, ability to speak out in mixed gatherings of men and women, and so on. Therefore regional, community, and class differences in women's ability to organize collectively may be expected. Female seclusion practices among Hindus, for instance, are strongest in northern India and virtually nonexistent in south and northeast India; within northern India, they are strongest among the upper castes located in the plains and little practiced among upper castes in the hills or among lower-caste and tribal communities anywhere.[26] Seclusion practices among Muslims, although not identical to those for Hindus, show a similar regional and community variation. Manifested less in the practice of veiling (which is not widespread) and more in the gender segregation of public space (e.g., women being discouraged from spending time in spaces where men congregate, such as in the market place), such practices severely restrict women's free interaction in public forums.[27]

Of course, even in the absence of overt strictures, the social construction of appropriate female behavior (the emphasis on soft speech, deference to male elders, etc.) operates in some degree everywhere, including in the hill and tribal communities I have largely been discussing. But since women in such communities are not explicitly restricted and play a visible and substantial role in the economy in all parts of the country, this tends to reduce the importance of the regional dimension. The effect of this dimension on women's ability to undertake collective action is likely to be more significant for upper-caste Hindus and for Muslims. For instance, one would expect it to be much more difficult for upper-caste Hindu women in the northwestern plains of India to participate than those from south India, for the reasons noted. And it would be important to map these regional and community differences for understanding women's responses to the environmental crisis and the possibilities of their acting collectively, as an increasing number of communities that are not tribal or located in the hills get involved in forest and VC management.

IX. Summary Comments

The colonial and immediate post-Independence periods in India saw a notable shift in property rights over forests and VCs, from substantial community control and management to increasing State and individual control and management. This had particularly adverse consequences for poor rural households, and especially for women in such households, because of their greater and everyday dependence on these resources for basic necessities. More recently, however, there has been a small but important reversal toward a reestablishment of greater community control over those resources, through the emergence of numerous forest management groups, some initiated by the State, as under the JFM program, others initiated by villagers, and yet others by NGOs.

However, unlike the old systems of communal property management, in which all villagers, including women, had some form of use rights by virtue of being members of the village community, under the new formalized system of control of common property resources, rights depend more directly on formal membership in the emergent community institutions, from which women are often excluded.

I have argued here that this exclusion of women has serious negative consequences not just for gender equity, but also for the efficient functioning and long-term sustainability of these initiatives and for women's empowerment. Concerns of equity, efficiency, and empowerment therefore all point to the need to ensure greater participation of women.

A range of factors, however, constrains more gender-balanced participation, including the rules governing the new bodies; social barriers stemming from cultural constructions of gender roles; responsibilities and expected behavior; logistical barriers relating to the timing and length of organizational meetings; and male bias in the attitudes of those promoting these initiatives, including forest department personnel, village leaders, and sometimes even the intermediary NGOs.

At the same time, the fact that in several regions women have formed their own informal associations for forest protection, and in some cases their formal participation has also increased over time, suggests that these barriers are not insurmountable. Among factors that can enhance women's formal participation in the emergent community institutions, the most significant appears to be the presence of a gender-progressive NGO, and especially a women's association. Involving women at the very beginning of the initiative and the presence of a critical mass of women are also important for their effective and sustained participation in mixed forums.

Finally, the emergence of these varied institutional arrangements highlights the problematic nature of the ecofeminist argument that women, simply by virtue of being women, have a special relationship with nature and thus a particular stake in environmental protection that is seldom shared by men. As the examples described above show, both women and men whose livelihoods are threatened by the degradation of forests and VCs are interested in protection and regeneration, but from different concerns, related to their respective responsibilities and nature of dependence on these resources. Women are more concerned with fuel, fodder, and nontimber products, while men are more concerned with timber and cash benefits. Women are also more dependent on communal resources due to their limited private property access.

Moreover, to translate their concerns into practice, women need to overcome existing social and political barriers and to contend with the preexisting advantages that men as a gender (even if not all men as individuals) enjoy in terms of greater access to economic resources and public decision-making forums.

The benefits of overcoming these barriers would, however, flow not just to women, but to the entire household as well as to the larger community—the latter by enabling a more ecologically balanced and sustainable regeneration of forests and VCs.

ACKNOWLEDGMENTS

I thank Kenneth Keniston for his detailed comments on an earlier version of this paper and the participants of the Massachusetts Institute of Technology workshop, where I presented the paper in May 1995, for a stimulating discussion. I also thank Jeffrey Campbell for sharing with me his store of case studies, especially on JFM, and the staff of AKRSP and VIKSAT for facilitating my field visits in Gujarat in 1995. Somewhat different versions of this paper have appeared in *Transitions, Environments and Translations*, edited by J. Scott and C. Kaplan (Routledge, 1997), and *Development and Change*, 28 (1), 1997.

NOTES

1. See Kerala Forest Research Institute (1980); Fernandes and Menon (1987); Viegas and Menon (1991); and Sarin (1995).

2. In India, the term "state" relates to administrative divisions within the country and is not to be confused with "State," used throughout this paper in the political economy sense of the word.

3. This was supplemented by steps to reduce fuelwood use by promoting biogas plants and improved wood-burning stoves (Agarwal 1986a).

4. On communal management of forests and VCs, see Guha (1985); Gadgil (1985); and Moench (1988). On firewood-gathering practices, see Agarwal (1986a): firewood for domestic use in rural households was customarily collected in the form of twigs and fallen branches, which did not destroy the trees. Fifteen years ago, an estimated 75 percent (and 100 percent in some areas) of firewood used domestically in rural northern India was in this form (Agarwal 1987).

5. Viegas and Menon (1991); Gadgil and Guha (1992); personal observation in northeast India in 1989.

6. See Dasgupta and Maler (1990); Bromley and Cernea (1989); and Baland and Platteau (1994).

7. Sex ratios are particularly female adverse in the agriculturally prosperous northwestern states of Punjab and Haryana. On the causes of the regional variation, see Agarwal (1986b) and Miller (1981).

8. It is not clear whether the northeastern states were taken into account in Kaur's calculations.

9. In the early 1980s, in parts of Gujarat, even a four- to five-hour search yielded little apart from shrubs, weeds, and tree roots, which do not provide adequate cooking energy (Nagbrahman and Sambrani 1983).

10. In the village of Kumaon, 84 percent of the sample households reported purchasing part of their fodder now, but only 8 percent said they did so two decades ago. The number of large animals that rural households can afford to keep has also fallen in all the regions surveyed, with the decline in grazing lands and the hike in fodder prices.

11. Kumar and Hotchkiss (1988) report this for Nepal.

12. In 1988 there were an estimated 1,300 Forest Protection Committees (FPCs) in three districts: Midnapore, Bankura, and Purulia. Even today West Bengal has the largest JFM coverage: an estimated 350,000 hectares of forest area are being protected by some 2,350 FPCs (Society for the Promotion of Wastelands Development 1994).

13. That excludes the Kashmir part of Jammu and Kashmir.

14. These are also termed *mahila mangal dals* in some regions.

15. Personal communication, Chandi Prasad Bhatt, 1995.

16. Among specific case studies that highlight resource scarcity and relative socioeconomic homogeneity as conducive to successful group action for forest protection are those by Kant, Singh, and Singh (1991) for Orissa and Sarin and SARTHI (1994) for Gujarat.

17. The AKRSP works on a variety of village development issues, of which forest protection and wasteland regeneration are major ones. In villages under its ambit, therefore, AKRSP has encouraged the formation of Gram Vikas Mandals (village development associations), which cover a wider set of issues than just forest protection. In some villages, women's development associations have also been formed in AKRSP areas.

18. In the village of Gamtalao Khurd (east Surat district of Gujarat), villagers harvested twelve tons of firewood and fifty tons of fodder from cleaning operations after about a year of protection. They also report adding forty-seven milch animals to their herds (Arul and Poffenberger 1990). In Pingot (Gujarat), where AKRSP is working, in 1987–88, soon after protection began the villagers were able to earn Rs. 11,000 from the sale of grass (Arul and Poffenberger 1990). In the Dhenkanal district of Orissa, tribal women can now get supplementary employment from leaf-plate making for six months of the year (Kant, Singh, and Singh 1991). In 1989, in Tiring (West Bengal), thinning operations in the protected area yielded 344 cart loads of leaves and twigs. Of this, 25 percent was distributed to the villagers free of cost and the rest sold to them at a low price. Seasonal migration from the region has also fallen to half (Viegas and Menon 1991). For further information on benefits reaped from protection under JFM, see Raju, Vaghela, and Raju (1993) and Society for the

Promotion of Wastelands Development (SPWD) (1994). Similarly, in Nainital district, in some *van panchayat* villages studied by Mansingh (1991), grass yield had doubled after a year of protection, and thinning the dense shrub provided firewood.

19. For Orissa, see Pati, Panda, and Rai (1993), Kant, Singh, and Singh (1991), and Swedish International Development Authority (SIDA) (1993); for Gujarat, see Arul and Poffenberger (1990). As noted earlier, in some villages tribal women now have part-time employment from mat weaving and leaf-plate making through the raw material they collect from the protected forests (Kant, Singh, and Singh 1991).

20. Sarin (1995); Narain (1994); Singh and Kumar (1993); Shah and Shah (1995); personal observation in Gujarat.

21. Personal communication, AKRSP project officer, March 1995.

22. See Agarwal (1994); Venketeshwaran (1992); and Viegas and Menon (1991).

23. See Ahmed (1994) for Gujarat; for West Bengal, see Guhathakurta and Bhatia (1992); Mukerjee and Roy (1993); Roy et al. (1993); and Chatterjee (1992).

24. This gender divergence in choice of trees in tree-planting schemes is also noted in other parts of India (Brara 1989; Sarin and Sharma 1991) and can be attributed to gender differences in responsibilities and dependence on particular categories of trees and forest products.

25. The AKRSP similarly reports from its experience in Gujarat that starting a separate savings scheme for women increased women's attendance in village development association meetings. In this case, however, what the savings scheme did was to enhance women's motivation to attend.

26. The link between economic class and seclusion is more complex. On the one hand, women of richer households are often more restricted, but where they are more educated they may also be better able to bypass seclusion norms.

27. For a detailed discussion of regional variations in seclusion practices and more generally in women's participation in public activities, see Agarwal (1994).

REFERENCES

Adhikari, N., G. Ydav, S. B. Ray, and S. Kumar. 1991. Process Documentation of Women's Involvement in Forest Management at Mahespur, Ranchi. In *Managing the Village Commons, Proceedings of the National Workshop*, edited by R. Singh (Bhopal: Indian Institute of Forest Management).

Agarwal, A. 1983. The Cooking Energy Systems—Problems and Opportunities. Discussion paper, Center for Science and Environment, New Delhi.

Agarwal, B. 1984. Rural Women and the HYV Rice Technology in India. *Economic and Political Weekly* 19(13): A39–A52.

———. 1986a. *Cold Hearths and Barren Slopes: The Woodfuel Crisis in the Third World* (London: Zed Books; Delhi: Allied Publishers; Maryland: Riverdale).

———. 1986b. Women, Poverty and Agricultural Growth in India. *Journal of Peasant Studies* 13(4): 165–220.

———. 1987. Under the Cooking Pot: The Political Economy of the Domestic Fuel Crisis in Rural South Asia. *IDS (Institute for Development Studies, Sussex) Bulletin* 18(1): 11–22.

————. 1990. Social Security and the Family: Coping with Seasonality and Calamity in Rural India. *Journal of Peasant Studies* 17(3): 341–412.

————. 1992. The Gender and Environment Debate: Lessons from India. *Feminist Studies* 18(1): 119–156.

————. 1994. *A Field of One's Own: Gender and Land Rights in South Asia* (Cambridge: Cambridge University Press).

————. 1995. Gender, Environment and Poverty Interlinks in Rural India: Regional Variations and Temporal Shifts, 1971–1991. Discussion Paper No. 62, United Nations Research Institute for Social Development, Geneva.

Ahmed, S. 1994. The Rhetoric and Reality of Women's Participation in Joint Forest Management: The Case of an NGO in Western India. Paper presented at a conference, Women, Poverty and Demographic Change, 25–28 October, Oaxaca, Mexico.

Arul, N.J., and M. Poffenberger. 1990. FPC Case Studies. In Forest Protection Committees in Gujarat: Joint Management Initiative, edited by R. S. Pathan, N.J. Arul, and M. Poffenberger. Working Paper No. 7, Ford Foundation, New Delhi.

Baden-Powell, B. H. 1957. *The Indian Village Community* (New Haven, CT: HRAF Press).

Bahuguna, S. 1984. Women's Non-Violent Power in the Chipko Movement. In *In Search of Answers: Indian Women's Voices in Manushi*, edited by M. Kishwar and R. Vanita (London: Zed Books).

Baland, J. M., and J. P. Platteau. 1994. Should Common Property Resources Be Privatized? A Re-examination of the Tragedy of the Commons. Discussion Paper, Center for Research in Economic Development, Namur University, Belgium.

Ballabh, V., and K. Singh. 1988. Van (Forest) Panchayats in Uttar Pradesh Hills: A Critical Analysis. Research paper, Institute for Rural Management, Anand.

Bardhan, K. 1977. Rural Employment, Wages and Labour Markets in India—A Survey of Research. *Economic and Political Weekly* 12(26): A34–A48; 12(27): 1062–1074; 12(28): 1101–18.

Batliwala, S. 1983. Women and Cooking Energy. *Economic and Political Weekly* 18(52/53): 2227–2230.

Bhaduri, T., and V. Surin. 1980. Community Forestry and Women Headloaders. In Community Forestry and People's Participation, Seminar Report. Ranchi Consortium for Community Forestry, Ranchi, Bihar.

Blaikie, P. 1985. *The Political Economy of Soil Erosion in Developing Countries* (London and New York: Longman).

Brara, R. 1989. "'Commons' Policy as Process: The Case of Rjasthan, 1955–85. *Economic and Political Weekly* 24(40): 2247–2254.

Britt, C. 1993. Out of the Wood? Local Institutions and Community Forest Management in two Central Himalayan Villages. Draft monograph, Cornell University, Ithaca, NY.

Bromley, D. W., and M. M. Cernea. 1989. The Management of Common Property Natural Resources. Discussion Paper No. 57, World Bank, Washington, D.C.

Chand, M., and R. Bezboruah. 1980. Employment Opportunities for Women in Forestry. In Community Forestry and People's Participation, Seminar Report. Ranchi Consortium for Community Forestry, Ranchi, Bihar.

Chandran, A. 1995. Involvement of Women in the GVM and Forestry Activity:

Strategy Paper. Photocopied note circulated at Agha Khan Rural Support Group meeting, Baruch, Gujarat, March.

Chatterjee, M. 1992. Women in Joint Forest Management: A Case Study from West Bengal. Indian Institute of Bio-Social Research and Development, Calcutta. Mimeograph.

Dasgupta, P., and K. Maler. 1990. The Environment and Emerging Development Issues. Paper presented at conference, Environment and Development, World Institute for Development Economics Research, September, Helsinki.

Fernandes, W., and G. Menon. 1987. *Tribal Women and Forest Economy: Deforestation, Exploitation and Status Change* (Delhi: Indian Social Institute).

Folger, B., and M. Dewan. 1983. Kumaon Hills Reclamation: End of Year Site Visit. OXFAM America, New Delhi. Mimeograph.

Gadgil, M. 1985. Towards An Ecological History of India. *Economic and Political Weekly* 20(45–47): 1909–1918.

Gadgil, M., and R. Guha. 1992. *This Fissured Land: An Ecological History of India* (New Delhi: Oxford University Press).

Guha, R. 1983. Forestry in British and Post-British India: A Historical Analysis. *Economic and Political Weekly* 18(44): 1882–1896.

———. 1985. Scientific Forestry and Social Change in Uttarakhand. *Economic and Political Weekly* 20(45–47): 1909–1918.

Guhathakurta, P., and K. S. Bhatia. 1992. A Case Study on Gender and Forest Resources in West Bengal. World Bank, Delhi, June 16. Mimeograph.

Howes, M., and M. A. Jabbar. 1986. Rural Fuel Shortages in Bangladesh: The Evidence from Four Villages. Discussion Paper No. 213, Institute of Development Studies, Sussex.

Jodha, N. S. 1986. Common Property Resources and Rural Poor. *Economic and Political Weekly* 21(27): 1169–1181.

Kant, S., N. M. Singh, and K. K. Singh. 1991. *Community-based Forest Management Systems (Case Studies from Orissa)*. Swedish International Development Authority, New Delhi; Indian Institute of Forest Management, Bhopal; and ISO/Swedforest, New Delhi, April.

Kaur, R. 1991. Women in Forestry in India. Women in Development Working Paper WPS 714, World Bank, Washington, D.C.

Kerala Forest Research Institute. 1980. *Studies on the Changing Pattern of Man-Forest Interactions and its Implications for Ecology and Management* (Trivandrum, India: Kerala Forest Research Institute).

Kulkarni, S. 1983. Towards a Social Forestry Policy. *Economic and Political Weekly* 18(6): 191–196.

Kumar, S. K., and D. Hotchkiss. 1988. Consequences of Deforestation for Women's Time Allocation, Agricultural Production and Nutrition in Hill Areas of Nepal. Research Report 69, International Food Policy Research Institute, Washington, D.C.

Malhotra, K. C., et al. 1990. Joint Management of Forest Lands in West Bengal: A Case Study of Jamboni Range in Midnapore District. Technical Paper No. 2, Indian Institute of Bio-Social Research and Development, Calcutta.

Mansingh, O. 1991. Community Organization and Ecological Restoration: An Analysis of Strategic Options for NGOs in the Central Himalaya, with Particular Reference to the Community Forestry Programme of the NGO Chirag. Master's thesis, University of Sussex.

Mazumdar, V. 1989. Peasant Women Organize for Empowerment: The Bankura Experiment. Occasional Monograph, Center for Women's Development Studies, Delhi.

Mencher, J. 1989. Women's Work and Poverty: Women's Contribution to Household Maintenance in Two Regions of South India, in *A Home Divided: Women and Income Control in the Third World*, edited by D. Dwyer and J. Bruce (Stanford, CA: Stanford University Press).

Miller, B. 1981. *The Endangered Sex: Neglect of Female Children in Rural India* (Ithaca: Cornell University Press).

Ministry of Forest and Environment, Forest Survey of India. 1991. *The State of Forest Report 1991.* (New Delhi: Government of India).

Moench, M. 1988. Turf and Forest Management in a Garhwal Hill Village. In *Whose Trees? Proprietary Dimensions of Forestry*, edited by L. Fortmann & J. W. Bruce (Boulder, CO: Westview Press).

Mukerjee, R., and S. B. Roy. 1993. Influence of Social Institutions on Women's Participation in JFM: A Case Study from Sarugarh, North Bengal. Working Paper No. 17, Indian Institute of Bio-Social Research and Development, Calcutta.

Nagbrahman, D., and S. Sambrani. 1983. Women's Drudgery in Firewood Collection. *Economic and Political Weekly* 28(12): 33–38.

Narain, U. 1994. "Women's Involvement in Joint Forest Management: Analyzing the Issues," Draft unpublished paper, May 6.

Pati, S., R. Panda, and A. Rai. 1993. Comparative Assessment of Forest Protection by Communities. In Proceedings of the Workshop on Joint Forest Management, Bhubaneshwar, Orissa, May 28–29.

Poffenberger, M. 1990. Joint Forest Management in West Bengal: The Process of Agency Change. Working Paper No. 9, Ford Foundation, Delhi.

Raju, G., R. Vaghela, and M. S. Raju. 1993. *Development of People's Institutions for Management of Forests* (Ahemdabad: Vikram Sarabhai Center for Science and Technology).

Rao, C. H. H., S. Ray, and K. Subbarao. 1988. *Unstable Agriculture and Droughts* (Delhi: Vikas).

Roy, S. B., R. Mukerjee, and M. Chatterjee. 1992. Endogenous Development and Gender Roles in Participatory Forest Management. Indian Institute of Bio–Social Research and Development, Calcutta.

Roy, S. B., R. Mukerjee, D. S. Roy, P. Bhattacharya, and R. K. Bhadra. 1993. Profile of Forest Protection Committees at Sarugarh Range, North Bengal. Working Paper No. 16, Indian Institute of Bio–Social Research and Development, Calcutta.

Ryan, J. G., and R. D. Ghodake. 1980. Labour Market Behaviour in Rural Villages of South India: Effects of Season, Sex and Socio-Economic Status. Indian Crop Research Institute for Semi-Arid Tropics, Patancheru, Andhra Pradesh.

Sarin, M. 1995. Regenerating India's Forest: Reconciling Gender Equity and Joint Forest Management. *IDS Bulletin* 26(1): 83–91.

Sarin, M., and R. Khanna. 1993. Women Organize for Wasteland Development: A Case Study of SARTHI in Gujarat. In *Women and Wasteland Development in India*, edited by A. Singh and N. Burra (New Delhi: Sage).

Sarin, M., and SARTHI. 1994. The View from the Ground: Community Perspectives on Joint Forest Management in Gujarat, India. Paper presented at symposium, Community Based Sustainable Development, 4–8 July, Institute for Development Studies, Sussex.

Sarin, M., and C. Sharma. 1991. Women's Involvement in Rehabilitation of Common Lands in Bicchiwara Block of Dungarpur District, Rajasthan. Paper presented at workshop, Women and Wasteland Development, January, International Labour Organization, Delhi.

Sen, A. K. 1990. Gender and Cooperative Conflicts. In *Persistent Inequalities: Women and World Development*, edited by I. Tinker. (New York: Oxford University Press).

Shah, M. K., and P. Shah. 1995. Gender, Environment and Livelihood Security: An Alternative Viewpoint from India. *IDS Bulletin* 26(1): 75–82.

Sharma, A., and A. Sinha. 1993. A Study of the Common Property Resources in the Project Area of the Central Himalaya Rural Action Group. Mimeo, Indian Institute of Forest Management, Bhopal.

Sharma, K., K. Nautiyal, and B. Pandey. 1987. Women in Struggle: Role and Participation of Women in the Chipko Movement in Uttarakhand Region of Uttar Pradesh. Occasional Monograph, Center for Women's Development Studies, Delhi.

Singh, A., and N. Burra, eds. 1993. *Women and Wasteland Development in India* (New Delhi: Sage).

Singh, N. M. 1993. Regional Workshop of Forest Department on Joint Forest Management at Keonjhar. In Proceedings of the Workshop on Joint Forest Management, Bhubaneshwar, Orissa, May 28–29.

Singh, N., and K. Kumar. 1993. Community Initiatives to Protect and Manage Forests in Balangir and Sambalpur Districts. SIDA, New Delhi. Mimeograph.

Society for Promotion of Wastelands Development. 1994. *Joint Forest Management Update, 1993* SPWD, New Delhi.

Swaminathan, M. 1984. Eight Hours a Day for Fuel Collection. *Manushi*, March–April.

Swedish International Development Authority. 1991. *Social Forestry: Case Studies from Orissa*. SIDA, New Delhi.

————. 1993. *Forests, People and Protection: Case Studies of Voluntary Forest Protection by Communities in Orissa*. SIDA, New Delhi.

Venkateshwaran, S. 1992. *Living on the Edge: Women, Environment and Development* (New Delhi: Friedrich Ebert Stiftung).

Viegas, P., and G. Menon. 1991. "Forest Protection Committees of West Bengal: Role and Participation of Women," paper prepared for the ILO Workshop on "Women and Wasteland Development," International Labour Organization, New Delhi, 9–11 January.

Wasteland News. 1993–94. 9(2).

III. The Question
of Modernity

INTRODUCTION

"Environmentalism," we noted at the outset, does not
name a single doctrine. Environmental theorists have not
succeeded in defining either a comprehensive set of principles or a political agenda acceptable to most people who
are concerned about the degradation of the natural world.
Indeed, self-styled environmentalists hold myriad views of
the nature of the problem and the measures required for its
resolution. Their divergent views reflect differences in national origin, gender affiliation, social status, as well as religious, political, and philosophical beliefs.

Despite their differences, most versions of environmentalism share a certain ambivalence toward modernity itself.
They are haunted by the very real possibility that the form
of life characteristic of technologically "advanced" societies
is inherently—perhaps irremediably—damaging to the natural environment. They fear that limitless consumerism and
economic development are placing an insupportable burden
on global ecosystems. As evidence they point to explosive
population growth, the depletion of nonrenewable resources, unprecedented kinds (and degrees) of atmospheric
pollution and climate change, and an accelerating rate of
species extinction. The essays in this section examine a few
salient theoretical problems raised by the complex, often
ambiguous, impact of modernity upon the environment.

Among the distinctive features of modernity that have

influenced environmental thinking is a marked change in the status of women and the consequent emergence of feminism and feminist movements. To explore the interaction between feminism and environmentalism, Jill Ker Conway and Yaakov Garb discuss the implications of the differing social locations occupied by women in modern and modernizing societies. According to a received stereotype of Western culture, women are "closer to nature" than men. (As the catchphrase has it, "Women are to nature as men are to civilization.") This prevalent notion is congenial to patriarchal thinking: it serves as a rationale for the belief that women's role in society is to bear and raise children, maintain the home, and provide the emotional (nurturing) complement to the distinctively male capacity for abstract thought; physical, technical, and military prowess; and political governance. In the nineteenth century, the presumed proximity of women to nature also provided a justification for denying them the democratic political rights then being claimed and won by "the common man." After the Civil War, when just liberated male slaves were granted the vote, it continued to be denied to women.

As women slowly gained a degree of social, economic, and political equality—more access to education, the workplace, and the public arena—their alleged identification with nature was turned to quite different ideological purposes. It served, for example, to explain their seemingly greater responsiveness to, and superior skill in resolving, environmental problems. In a patriarchal society, it was thought, women instinctively become the self-appointed defenders of nature. Hence the emergence of "ecofeminism," a programmatically gendered environmental doctrine. A central theoretical issue raised by ecofeminism, and by the role of women in the environmental movement generally, is the validity of "essentialism," the idea that the apparent distinctiveness of women's relations with nature in some degree derives from the very essence of femininity: not merely its transitory, culturally constructed traits, but its intrinsic biological attributes.

In their essay, Conway and Garb examine the implications of gender for the defense of the environment. They are concerned with the import of gender in general, and especially with the validity of essentialist views of femininity. Accordingly, they analyze two contrasting historical episodes: (1) the "social feminism" of the 1890s in the United States, which involved the contrast between Alice Hamilton and her concern with "industrial medicine" on the one hand and Teddy Roosevelt's highly masculinized version of conservationism on the other, and (2) the ideologically complex politics of ecofeminism in the 1970s and 1980s. Today, they show, environmental thinking is still closely bound up with gender issues, but in the end they call into question the old habit of coupling ideas about nature with the gender of those who advance them.

In the second essay, Anton Struchkov examines the wide-ranging arguments about the impact of modernity on the environment in the former Soviet Union. The special political circumstances of the USSR—in a word, Stalinism—provide a unique vantage point for an assessment of some the "worst case" environmental consequences of modernity. Not only did Russian theorists have before them the appalling environmental record of their own totalitarian regime, but their geographical, cultural, and ideological distance from the West also provided them with an unusual perspective from which to appraise the environmental practices (and consequences) of market-driven Western economies. From their distant vantage they saw aspects of capitalist practice that were less obvious from within. They were able to compare Western practices with both the utopian promises and the dystopian realities associated with the command economy of the Soviet Union.

In addition, in thinking about the devastation of their nation's environment, Russian intellectuals drew upon the peculiarly sharp contrast between a quasi-feudal past and a programmatically technocratic version of modernity. Within living memory many Russians had experienced these two radically different kinds of society: a coercive, totalitarian program of industrialization based on a secular Leninist ideology and a traditional society with a landed aristocracy, an established Orthodox Church, and a predominantly peasant population, a large segment of which held on to the eternal values of blood and soil. By examining the ideological import of the clash between modernity and the environment, as manifested in the debates of the Russian intelligentsia, Struchkov brings the passionate conflict between modernity and the environment into sharp relief.

No aspect of modernity is more complex or ambiguous than the theoretical debate that has swirled around the idea itself. The terms "modernity" and "modern" are among the most contested in the serious discourse of this century, and since the environmental crisis is widely regarded as a peculiar by-product of modernity, it would be immensely helpful to clarify that relationship. In the third essay, Louis Menand undertakes just that task. Although mindful that these terms refer to conditions that are manifested on both socioeconomic and cultural (or ideological) planes, he takes as his starting point the late nineteenth-century effort of European artists, writers, and intellectuals to formulate a distinctive "discourse of modernity." Though modernity, as a social condition, was too multifaceted and contradictory to yield anything like a single, coherent viewpoint, it did provoke the emergence of a singularly vigorous and innovative cultural style called "modernism." The most striking achievements of modernism in the expressive arts embody a complex response—an unstable mix of celebration and repudiation—to the peculiarly "modern" features of contemporary societies, including, of course, their problematic impact on the nonhuman environment. A similar ambiguity marks the successor style of thought and expression that emerged in the 1970s (roughly coincident with the new wave of environmental activism) "postmodernism" and its theoretical corollaries, "poststructuralism" and "deconstructionism."

The salient feature of postmodernism, so far as it bears on environmental thought, is its unequivocally antifoundationalist tenor. That aspect of the doctrine is a contemporary version of philosophical skepticism, an antimetaphysical viewpoint much like that of the American pragmatists, William James, Charles Sanders Peirce, and John Dewey. It begins with the assumption that we cannot gain access to a single, context-free, or "objective" foundation for our knowledge. This viewpoint, which in effect relativizes all knowledge claims, obviously puts in question certain kinds of assertion, routinely made by environmentalists, about the degradation of the environment. To accept the limitations implied by this relativistic perspective, Menand argues, does not entail the complete abandonment of the optimistic Enlightenment outlook or a repudiation of science and technology. Although the inadequacy of the technological and instrumental view of life has become a truism, we have no choice but to meet the environmental critique of modernity with solutions made available by—and with resources created by—modernity itself. To that end, Menand persuasively argues, we should reorganize human knowledge in accordance with postdisciplinary, holistic principles.

In the final essay, I examine the ambiguity surrounding the role of science and technology, especially with respect to the defense of the environment. Today, scientists and engineers are simultaneously cast as complicit in the devastation of nature and as the most indispensable defenders of nature. To appreciate the reasons for this uncertainty, it is first necessary to recognize the special responsibility of the scholars in the humanities in getting at the cultural roots of the current environmental "crisis." To put it differently, the ambiguous role of science and technology has a revealing history. To understand today's conflicting American attitudes toward the use of science-based technological power in transforming—and often degrading—the environment, accordingly, it is essential that we grasp their history. It is, among other things, a history of rival belief systems, each one embodying a distinct version of the American myth of national origins.

In this essay, I discuss three contrasting interpretations of the myth—a progressive, primitivist, and pastoral variant—and the adaptation of each to a corresponding expression of today's heightened environmental consciousness. Each version of the myth has its counterpart in the discourse of contemporary environmentalism. The effect of all this, it should be said, has been to intensify the ambiguity attached to the social role of scientists and engineers. The perceived role of science and technology with respect to the environment depends on, among other things, the place they occupy within the worldview of the perceiver. As I suggest, however, they could become active agents in this process;

scientists and engineers could help to dispel some of the ambiguity that is currently attached to them and their social behavior. Such an outcome would entail a redefinition of their social roles, and their endorsement, as a chief criterion in the choice of research projects, their probable effect upon the life-enhancing capacities of the natural environment.

LM

JILL KER CONWAY AND YAAKOV GARB

Gender, Environment, and Nature

Two Episodes in Feminist Politics

Introduction

ENVIRONMENTAL ISSUES are shaped by gender in several overlapping ways. First, gender defines sociological locations that give differing shape to people's environmental interactions and knowledge, to their perception of and vulnerability to environmental hazards and degradation, and to their motivation and resources for doing something about such hazards. Second, the feelings that link people to nature and the ideologies that frame environmental concern and action have typically been gendered. Indeed, the uses and purview of the notion "environment" are in part dependent on gendered divisions of labor.

Third, in the late nineteenth century, and again in the 1970s, gender and environment became key concepts for both feminist reformers and males trying to adjust ideals of masculinity to changing American social conditions. Feminists in both periods needed to find persuasive new rationales for their claims to an enlarged set of political and social responsibilities, because older modes of argument were unsuccessful. Thus the maternalist and domestic science traditions of 1880 to 1950 achieved their protoenvironmental agendas through feminizing concern for urban and domestic health and safety; cultural feminists of the 1970s were able to use an essentialist conflation of woman and nature as a galvanizing analytic and rhetoric for the critique of consumer society and the technological excesses of modern industry.

Since the 1980s, poststructuralist philosophers such as Donna Haraway and Judith Butler have directed feminist lenses to the dismantling of the concept of "nature" itself, the linchpin of environmental thought. They and others have pointed to the ties between "naturalizing"—the ideological work of marking off the underlying, inherent, inborn, and enduring from their opposites—and the oppressive stabilization of gender categories and roles around those polarities.

Our aim in this essay is to keep all of these intersecting lines of thought in mind as we explore episodes in the history of environmental thought over the last two centuries. We begin with examples of ways in which gender—as social location—shapes contemporary environmental responses. Next we consider how feminists in critical periods in the nineteenth and twentieth centuries used an ideological coupling of women to nature as a political tool. We then consider the hazards of this coupling by contemporary development agencies and in political life, because the conflation can work both for and against dominant power structures. We conclude by asking whether contemporary environmentalist thought is as deeply gendered as in the historical instances cited and how this might affect future environmental action.

Gender as Social Location

To the degree that gender determines the social location of the sexes, which in turn shapes their environmental interactions, gender is a critical component of environmental analysis. In particular, we must be aware of the environmental consequences of gendered divisions of labor, power, property ownership, and daily experience. One of the first and most prominent analyses of how forms of gendered social and economic stratification impact on environmental concerns and activism was given by Indian economist Bina Agarwal in her work on how women and the poor experience the degradation of the commons and forests in rural India.[1] She shows that rural poor women in particular spend more time in direct contact with the environment as they gather fuel, fodder, drinking water, and food from common lands; that they are more vulnerable nutritionally to degradation, privatization, or state appropriation of these resources, since they have lower access to food within households; that women are more exposed because of the nature of their daily tasks to a range of health hazards (e.g., waterborne diseases and pesticides) while health care is less accessible to them; that they are more dependent on social support networks that are destroyed by large irrigation projects or deforestation; and that by virtue of their location in the division of labor, women often have a greater stock of indigenous ecological knowledge that is being devalued and rendered obsolete with disappearing natural resources. Women also have less access than men to private property resources and therefore face a greater dependence on common property resources. Thus a convincing materialist explanation can be given for the commitment of rural women to the preservation of common property such as village commons and forests. Agarwal offers this formulation (which she terms "feminist environmentalism") as an alternative to ecofeminist claims that this commitment is evidence of women's greater inherent proximity to nature and natural processes.

A linking of environmental behavior to gender as social location, rather than to the feminine as essential principal, means that women are not ahistorically linked to proenvironmental behavior. Such linkages, when they exist, can be a function of quite mundane and fragile circumstances. As Cecile Jackson demonstrates in her work on gender and development,[2] women's social location can encourage environmental-destructive behaviors and attitudes, and struggles for gender equality (e.g., for full participation in development) can be at odds with efforts for environmental protection. Women may prefer gathering dead wood to cutting living trees not because of their reverential attitudes to nature and commitment to future generations, as some might claim, but because dead wood is lighter and easier to carry (400). Under systems of patrilineal inheritance women will be more mobile than men, have less of a stake in a particular place or property, and may therefore be *less* motivated to adopt conservation practices than men who hold primary land rights (406). Under conditions of stress it may be men who are in closer contact with the land, common property resources, and their care (408–411) and who exhibit the "conserving" characteristics seen by eco-feminists as inherently associated with women and female roles.

This kind of analysis of the linkages between social location and environmental behavior is most often applied to Third World situations, but it is equally relevant to the highly industrialized nations. Accounts of the predominantly female engagement with community in organizing against toxics, for example, can be read not so much as an expression of feminine caring and environmental connectedness as a function of the gendered patterning of daily time and space trajectories (women notice different things), divisions of labor (women tend ailing children), and social resources (women are part of a network of neighborhood friend-ships and PTA affiliations).

These material or sociological aspects of the gendered differentiation of social location shade into gender differences in ideological location and subjectivity that also have environmental consequences. Men and women have available to them different repertoires of ideological claims they can mobilize to get their grievances addressed. Relatedly, environmental claims (e.g., for the safeguarding of household water supplies or the preservation of game species) are often linked to gender-patterned desires. In the nineteenth century, the convergence of these material, sociological, and ideological factors is well illustrated in the formation of two disconnected forms of what can retroactively be called "environmental" concern.

The Politics of Gender, Environment, and Nature at the Turn of the Century: Maternalism, Domestic Science, and the Origins of Preservationism

To describe the impact of maternalist thought on environmental issues in the Progressive Era, we need to understand how this historical example fits into the larger picture of Western thought about woman and her relationship to the nonhuman. That larger picture has been shaped by the political fact of the subjection of women, a situation that has resulted in the definition of the female as less than the male, whether in theology, social contract theory, psychoanalytic theory, or the mainstream of evolutionary thought. Gender boundaries have thus been drawn to define women as closer to nature than man. In those revolutionary moments when social hierarchies were dismantled, this differential proximity to the natural was used to bar women from participation in the new political rights granted to the "common" man.

Western feminists have naturally framed their arguments advocating raising the status of women in terms of the major intellectual movements of their day. Eighteenth-century women drew upon Enlightenment rationalism when they stressed the universality of human reason as a basis for educating women, whereas nineteenth-century feminists used the stress on the differences between male and female present in both Romanticism and evolutionary thought to argue for complementary but equally important roles for men and women. This focus on difference is the intellectual frame for maternalist thought, though its proponents are often unclear whether it is the experience of giving birth, the potential for doing so, the experience of nurturing the young, or the social training to prepare women for doing so that produced in women the special qualities claimed as necessary in the public sphere.

In the line of feminist thinkers from Mary Wollstonecraft to Carol Gilligan the claim of women's essential difference from men, whether in moral sense, patterns of reasoning, or motivation for service, has been used to put a positive value upon human qualities denigrated in patriarchal society, and the existence of those qualities has most frequently been explained through women's supposed closeness to nature. There are, of course, sound historical and social reasons for American women's interest in environmental hazards and for Third World women's concerns with land use and the preservation of common lands, but the traditional pattern in feminist thought has been not to seek such explanations but to essentialize the capacity to be concerned with the natural environment.

These essentialist claims, however, feed a conservative view of the female, because they classify women in terms used to suppress women under patriarchy. Ecofeminist thinkers have accepted the "closer to nature" claim in their assertion that women's role in agriculture and the domestication of animals in prehistoric times led to the development of peaceful and nonviolent matriarchal societies. Such arguments romanticize agriculture, which is in itself a disturbance of the biophysical environment, and the domestication of animals, which involves power relationships and not simply nurture. Nonetheless, the "woman closer to nature" theme has been the basis on which feminist agitation for change has been most successful in modern societies, because it often permits the entry of women into new gendered social and political territories without disturbing traditional male/female power relationships. Thus maternalist thought in feminism is double-edged: accepting qualities assigned to women under patriarchy, even while providing a strategy that may open up, at least for the short term, new areas of social and intellectual life for women within a changing social and economic system.

It was just such a set of short-term opportunities that appealed to American "social feminists" during the 1890s. Earlier nineteenth-century feminist thought had drawn on Enlightenment ideas about the universality of human reason and argued for votes for women because they shared the same intellectual capacities and civic concerns as men. But following the Civil War, when black males received the vote while women's postponed claims were denied, new strategies were needed. Feminist leaders divided over whether to continue to argue for women's rights on the basis of equality or on the basis of their capacity to bring special sex-linked qualities to the political process.

Social feminism drew upon evolutionary thought to claim for females a special set of essential qualities that were necessary to foster racial and social progress—qualities of nurturance, empathy, and altruism, sex-linked qualities essential to the female biological role. On this basis, social feminists argued with varying degrees of success for protective legislation limiting the exploitation of female and child labor; for pacifism as an international policy, as opposed to the militarism fostered by male aggressiveness; and for training for women in home economics and domestic science so that they could continue their role of caring for the health of their families within an urban and industrial society.

It was in the establishment of a special system of education promoting "domestic science" that the social feminists were most successful in influencing the curriculum of American schools and colleges. They created a new field, opened a new teaching role for women scientists, and made instruction in domestic science mandatory for girls in two-thirds of the U.S. public education system. In the United States, they also fostered the development of a new medical field, "industrial medicine," which initially had only female practitioners. Its major proponent, Alice Hamilton, developed her interest in industrial pollutants because she was shut out of more conventional forms of medical research. In a period when female empathy was seen as desirable in women, Hamilton could be concerned with the health of the working poor, even though a man with her scientific interests in the United States would have been seen as sentimental. "It seemed natural and right," she wrote, "that a woman should put the care of the producing workman ahead of the value of the thing he was producing. In a man it would have been seen as sentimentality or radicalism" (49–50).[3]

By contrast it was legitimate for Hamilton's contemporaries, Theodore Roosevelt, Gifford Pinchot, and other male conservationists, to express deep concerns for preserving wildlife and wilderness, because hunting and the outdoor life were identified with the regeneration of masculinity, seen as undermined by city life and the supposed closing of the frontier.[4] The gendered nature of our perception of what constitutes "environmentalism" is illustrated by the omission of domestic science and industrial medicine from the canonical story of American environmentalism of this period, which has been exclusively a narrative about the preservation of the wild as a transcendent good or about the battle for a rational conservation of nature.[5]

Masculine forms of concern with preservation came not from the motivation to nurture others, but from what has been called "the crisis of American masculinity," a crisis triggered by the shrinking ranks of individual farmers and small entrepreneurs and the rise of urban industrial capitalism, a process highlighted by the 1893 announcement by the U.S. Census Bureau that the "frontier" was closing.[6] Early wilderness preservation efforts must be seen against American men's preoccupation with the assertion of masculinity in the face of these changes.

Whereas a previous generation had encouraged young men to develop qualities of piety, thrift, and industry—the virtues of nineteenth-century evangelical Christianity—these idealized traits were replaced in the closing decades of the century by the affirmation of a maleness that was vigorous, forceful, and muscularly adventurous. American football (invented at West Point) took cultural precedence over gentler recreations and came to be played in giant stadia modeled on the gladiatorial combats of the late Roman Empire. The supposedly disappearing frontier fostered a new concern with the outdoors and organized leisure activities that developed the skills of the backwoodsman, such as the Boy Scouts. These changed relationships to the outdoors and recreation were matched by the effort to transform professional education in fields such as medicine, law, and business into strenuous agonistic activities that produced "tough-mindedness," a desired quality in a society that, in William James's terms, was searching for "moral equivalents to war."[7]

Theodore Roosevelt embodies the ambiguities of the relationships to nature and culture subsumed within the American response to the closing of the frontier and the recognition that urban society was the pattern of the future. His encounters with the natural world as soldier, cowboy, hunter for big game, and early explorer of Brazil were epic and widely publicized. The Boone and Crockett Club, which he founded in New York in 1887, was designed to foster "manly sport with a rifle," travel and exploration of unknown parts of the globe, and the preservation of "the large game of this country" by promoting conservation legislation. Admission to the club was through proof of killing "in fair chase" at least three of various kinds of American game. This club was the center of influence that lobbied for the passage of the Forest Reserve Act of 1891 and for the designation of the country's first national park, Yellowstone (est. 1872), as a wildlife refuge, and in this respect the sportsmen were the founders of the conservation movement.

Yet there was a gender contradiction in the promotion of sports hunting in a country becoming more aware of the need to conserve diminishing game populations. The need to sustain the potential and sites for masculine regeneration based on untrammeled adventure and killing ran counter to the prevalent stereotyping of preserving wild creatures and of opposition to killing as feminine.[8] Much of Theodore Roosevelt's thinking about conservation was an attempt to negotiate this dilemma. This he did through promoting the idea of a "fair chase" and decrying wanton killing as "unsportsmanlike," an attitude to game that resulted in his fabled rescue of the bear cub and the creation of the Teddy Bear. The proponent of "the strenuous life" and vigorous pursuit of game thus worked, as president, to create fifty-three wildlife reserves, sixteen national monuments, five other national parks, and numerous forest reserves.

In this era, both feminists and proponents of vigorous American masculinity essentialized important aspects of maleness and femaleness in American culture and linked their existence and future strength to a particular relation to the "environment," be it the natural one or the human and built environment of the modern city. The male linkage to the environment has been made central to the history of American conservation. The female linkage has been of central importance in the history of American feminism. These paired essentialisms are dependent on each other and vary dynamically within larger cultural changes, sometimes being fundamentally reworked in moments of cultural crisis within American society.[9]

The Politics of Gender, Environment, and Nature in the 1970s and 1980s: The Emergence of Ecofeminism

In the early 1960s "environmentalism" had come to denote not only traditional turn-of-the-century concerns with preservation of open space and wildlife but newer concerns about the health consequences of environmental degradation and pollution. The publication of Rachel Carson's (1962) *Silent Spring* was the key event in forging these two ideological concerns into a single movement.

Contrary to claims by some feminist historians, *Silent Spring* contains few explicitly feminine (much less feminist) motifs (Vera Norwood). Nor was Carson a feminist or the center of female networks (Hynes).[10] Yet the incorporation of the kinds of concerns voiced in *Silent Spring* into the mainstream of public debate *was*, at a deeper level, an outcome of the maternalist and domestic care traditions launched by earlier feminists. A sizeable group of American women in high schools, colleges, and universities had been, for some seven decades, taught that nutrition, sanitation, and hygiene were important female concerns and that one should use one's scientific knowledge to assess, with some skepticism, the advertising claims of all those suppliers of products and services that impacted on the home and the health of its inhabitants. When we realize the size of the female population whose schooling included home economics, we can see that a potential mass audience for environmental issues had been formed. The extent to which two generations of American women were taught to think that nutrition, sanitation, and hygiene were peculiarly female concerns is indicated, Minakshi Menon reminds us, by the journals that Carson and her agent considered suitable for publishing her work.[11] They approached *Reader's Digest*, the *Ladies' Home*

Journal, *Women's Home Companion*, and *Good Housekeeping*, all of whom refused. Carson's book implicitly linked women's practical attention to domestic health and environmental safety, already manifested in the anti–nuclear testing movement of the same period, with concerns about chemical pollution, both culturally defined as central to women's maternal functions.

These environmental concerns became focused in the 1960s at a time when other cultural forces were prompting a profound reevaluation of what it means to be feminine. Arguments for zero population growth prompted by demographers' expectations of a world population crisis made many question the centrality of reproduction in the female experience. Moreover, the availability of the Pill placed in women's hands an effective form of birth control that had the consequence of completing a centurylong Western trend toward separating female sexuality from reproduction. Thus, in the 1960s it became necessary to consider maternalism not as function of actual childbearing but as a psychosocial trait, since it seemed clear that responsible women would not spend more than a small segment of their adult life in childbearing. At the same time, research on female sexual response ended the myth of the vaginal orgasm and changed perceptions of the nature of female erotic life. Moreover, systematic feminist criticism of the Freudian model of human sexuality focused attention on female-to-female bonds in human development, making women's networks and social bonds assume a position in social theory previously ignored. In religious life, feminist theologians, frustrated in their desire to dismantle patriarchy in both Judeo-Christian doctrine and practice, turned in the late 1960s to asserting the importance of female symbols of transcendence and reviving earlier female cult practices. A maternalist interest in environmental protection and its linkage to a redefinition of the female occurred within the context of these larger cultural trends.

While the cultural imagery of femininity was being redrawn, the social and economic forces that limited women's aspirations remained unchanged. The Presidential Report on the Status of Women of 1964 revealed the magnitude of economic discrimination continuing from the Depression, discrimination that was attacked by women activists who had been politicized by their participation in the civil rights movement. When women civil rights workers found that their political commitments were not respected and their participation was limited to helping roles, they began to work on theoretical analyses of the roots of oppression based on sex.

The sustained attack on economic and social discrimination launched by women and blacks during the late 1960s bore fruit in the 1970s with improved access for women to high-status professional work in fields such as law, medicine, academic life, and business. But work patterns and rewards in these occupations remained based upon the assumption of an exclusively male workforce and allowed women no time for domestic nurturing roles or for childbearing. Thus, when the high-achieving women pioneers in the male professions reached their midthirties in the late 1970s, they displayed a new interest in maternalist thinking and in the stereotyping of males as aggressive, acquisitive, and incapable of nurturance. More radical feminists extended these ideas to include a definition of male domination as a unitary and all encompassing social phenomenon, the root model of other forms of domination in all cultures and societies.

At the same time that the feminist movement of the 1960s and 1970s reached an analysis that made male domination the *ur* form of ruling power and women's relation to nature an essential component of femininity, the concept of maleness in American culture was called into question by a set of social transformations as profound as those that prompted maternalist thought. Traditional male military virtues were challenged by broadly based opposition to the American military role in Vietnam. These challenges were accompanied by widespread and publicly avowed male pacifism and efforts to evade the draft. While many of these challenges were male led, they did not betoken these men's critique of the war as rooted in masculine values. The feminist proponents of the antinuclear movement, on the other hand, characterized as masculine the power-crazed and irrational mentality of America's military-industrial complex, which, if unopposed, could lead to Armageddon. Meanwhile, the subsequent surging growth of the Japanese economy challenged America's position as the number one economic power in the world, a position hitherto thought to be earned by American economic and technological strength. These shocks to the national self-image were reenforced by the rhetoric of anti-imperialism as former Western colonies threw off European domination and criticized the idealized qualities of the white European male as masking an underlying greed and insecurity behind ostensible commitments to the "white man's burden" and "civilizing mission."

The internal crisis within the United States posed by profound divisions over the country's role in Vietnam virtually silenced proponents of the old masculine values, while the ecological damage wrought in Vietnam by the nation's use of chemical warfare appeared to lend substance to the ecofeminist claims of an essential linkage between environmental damage and the masculinity of militarism.

It was at this moment that environmental thinkers began to question whether improved management of resources and technological problem solving could avert a seemingly irresistible progression toward environmental disaster. Lynn White's famous essay of 1967, "The Historical Roots of Our Ecologic Crisis," marks a fundamental redefinition of environmental problems, with White arguing that the origin of modern technological society's devastating environmental impacts lay within Western society's overarching cultural orientation and was not, as previously argued, the result of simply technical mishaps within a basically sound and progressive enterprise. Later writers such as Paul Shepard (1967), Edward Abbey (1968), and Arne Naess (1973) reenforced the message.[12]

Thus by the late 1970s there was a shift in environmental thought toward framing environmental problems as evidence of a systemic tendency to dominate and exploit nature, a tendency shaped by the fundamental assumptions of western European culture, even while feminist analysis emphasized the importance of women's capacity as social nurturer and traced environmental destructiveness to a patriarchal society not guided by feminine values. The convergence of these two points of view produced ecofeminist claims that environmental problems are universal symptoms of patriarchal domination of nature and that women, because of their essential maternalism, are uniquely equipped to speak out against environmental degradation.

Ecofeminism

From the mid-1970s, ecofeminists applied themes in radical feminist thought—the long-standing maternalist tradition, the critique of domination in critical theory, and feminist psychoanalytic perspectives—to explain the special bond of women to nature and its counterpart of male alienation from nature. Ecofeminist writing, ranging from popular to academic, located the feminine bond to nature within a spectrum of ideas that had as its central component the following claims: the uniqueness of the female body and its association with reproduction and nurturance; a heightened sense of continuity with the natural world arising from the experience of reproduction; the maternalist claim that women's maternal role engenders in them nurturant capacities that extend to a concern for the environment; and the assertion that women's long history of oppression has produced a viewpoint of "critical otherness" and a capacity to identify with victims of domination, including nature. The ability to perceive and speak for nature has been designated feminine because, according to ecofeminists, women are less prone to dualistic structures of thought, a superstructural correlate of patriarchal domination, which alienates people from natural processes, and because of the psychological dynamics of the early-childhood mother/daughter relation, in which women develop a personality structure different from men's, one with more easily permeable boundaries between self and other and thus less of a drive to dominate, control, and objectify.

Such ecofeminist claims were sometimes combined with attempts to describe the historical roots and key points of consolidation of a European worldview that was both hostile to nature and founded on the values of patriarchy. In particular, the archaeological writings of Marija Gimbutas and the work of science historian Carolyn Merchant were adapted, often uncritically, to support an account of early turning points in gender and environmental relations at which patriarchal antinature attitudes were strengthened. Gimbutas's work, an updating of Bachofen's theories of a primordial matriarchy, synthesized archeological findings from across Old Europe into an account of the conquest by warlike Kurgan hordes of a peaceful Neolithic matriarchy that, prior to the domination of both women and nature by men, had revered the Great Earth Mother and celebrated the sanctity of immanence rather than transcendence.[13] Merchant's *Death of Nature* examined shifts in gender relations and gendered imagery of natural process accompanying the rise of the seventeenth-century New Philosophy.[14] Her portrait of how an increasingly mechanical worldview went hand in hand with environmental arrogance and misogyny was used to support ecofeminist claims about the masculinist origin and assumptions of science and technology.

In the early 1980s, Western ecofeminists increasingly drew on the experience of Third World peasant women in their accounts of the relationship between patriarchy and environmental degradation. The women of the Chipko (tree-hugging) movement of northern India and the work of Wangari Mathai of Kenya became popular case studies of women's heightened awareness of and vulnerability to environmental degradation and their greater commitment to environmental conservation. These ecological sensibilities were often presented, whether explicitly or by implication, as intrinsic to women's nature rather than as a product of material and social circumstance of the women activists involved.[15] A series of Western women—Ellen Swallow Richards and other Progressive Era domestic scientists, Rachel Carson, the whistleblower Karen Silkwood, and Lois Gibbs and other antitoxics activists— were similarly framed as exemplars of ecofeminist values.

Over time, "ecofeminism" began to refer loosely to a range of studies and movements dealing with women and environment, not just those that explicitly shared early ecofeminist premises. At the same time, some of the more unabashedly essentialist ecofeminist claims that had been influential in academic and feminist reform circles and in development studies until the mid-1980s began to be qualified. Widespread criticism of essentialism by feminist theorists led to a move to a more complex (historical, cross-culturally sensitive, and philosophically nuanced) basis for ecofeminist claims. After the first flush of ecofeminist enthusiasm, more attention was now paid to the political usefulness and hazards of essentialist claims.

Today, therefore, essentialist couplings of gender to environmental behavior are only part of the heterogenous category referred to as "ecofeminism" within contemporary feminist thought. They remain influential, however, and have become embodied in some influential locations, such as the development policies of large foreign aid institutions (e.g., World Bank and USAID). The counterpart of this essentialist form of essentialism—interpreting stereotypical Western male qualities as leading to behaviors that endanger the natural environment—is also prevalent, and, not surprisingly, this view of maleness has been challenged.

Gender and Nature in the 1990s

Theodore Roosevelt's conservationism and 1890s-style social feminism were based upon complementary assumptions about maleness and femaleness and about men's and women's differing connection to nature as desirable expressions of gendered sensibilities. We can discern, through examining the relations between ideals of maleness and femaleness in the 1890s and the popular understanding of nature (i.e., the nonhuman setting for human activities) that gender roles and "nature" are systemically related, varying in response to political, economic, and demographic change. The important question for the student of gender and environmental thought and action in the 1990s is whether that systemic relationship still holds for contemporary America.

Before we can consider whether or not ecofeminism is part of a general remapping of gender categories and of our understanding of "essential" nature, it is important to reemphasize the mutability and variety within "masculinity" and "femininity" as categories. The ideologies of maleness and femaleness we have described have been held by white, heterosexual, middle-class males or white, educated, middle-class women. To the extent that these groups represent the "mainstream" of American culture they are relevant groups to study, although it may be argued that with the high value placed on cultural diversity in contemporary America, the mainstream is becoming more a varied pattern of channels than a single river.

Even with these qualifications, we can usefully speak about contemporary events that have unsettled hitherto accepted definitions of masculinity. Women's continued movement into the ranks of management and the professions has triggered hostility in men, who see decreasing opportunity for themselves in restructured corporate management or in efforts to regulate fee-for-service professions, such as medicine. The lightning rod for such feelings has been the role of women in the military and the currently popular attack on affirmative action.

In the 1980's, American economic supremacy, once unchallenged, was called in question by the growth and achievement of other industrial societies like Japan and Germany, so that the (male) American entrepreneur no longer seemed the model for the world. Middle-class white males, once the idealized type of the country, saw their cultural position challenged by immigrant, gay, foreign, and nonwhite leaders and celebrities.

Even though the objectives of feminist leaders of the 1970s and 1980s remain to be secured politically, and even though the U.S. economy of the 1990s is now preeminent, the concerns of feminists occupy considerable ideological ground across the political spectrum. The feminist critique of the stereotypical masculine character, whether stridently rejected or apologetically accepted, has achieved wide cultural currency. Secular liberals, the religious right, and left radicals must all deal with feminist claims, even if only to denounce them and inveigh against their dangers. Because of feminism's place in the ideological foreground, many of the overlapping idioms of masculinity now available to the middle-class male can be seen, at least in part, as responses to feminism. The liberal male apologetically knows he should be sharing child care and chores and be freer about voicing his feelings—even if he does none of these things. Even the hypermasculinity of Rambo and Terminator type heroes is shaped in almost deliberate opposition to the feminist criticism of masculine destructiveness.

Certainly, the men's movement is self-consciously a response to feminist ideas about males. Robert Bly's best-selling *Iron John* is a case in point. The book proudly celebrates masculine difference in terms that are reactive to maternalist feminist claims and are as profoundly essentialist. A major theme of the men's movement is that men are as deeply damaged by prevailing gender roles as women. Moreover, just as maternalist feminists lauded the qualities of the female, the men's movement has tried to rework the much denigrated imagery of masculinity to emphasize the positive aspects of the male role. The monarch was not simply a patriarchal tyrant but a model of noble action and the exercise of legitimate authority. The Old Testament patriarch was not simply indulging his ego but evinced deep concern for the well-being of his tribe. The warrior could display qualities of fierceness that were needed in the battle for social justice. While Bly was one of the early celebrants of the Great Earth Mother that was to become the central motif of some forms of ecofeminism, his interests mutated so that his early devotion to fertility goddesses was countered by the cult of the wild-man, the green man, and a pantheon of other male gods represented by the men's movement as embodying socially valuable qualities. The maternalist feminist emphasis on the psychological centrality of mother/daughter bonds found a counterpart in the quest for the lost father and renewed attention to the tribal bonds between males. Thus the feminist critique of male rationality was opposed by a deliberate quest for male myths and by cults designed to revitalize male emotional experiences of nature. The quest was eclectic in the extreme, drawing on Greek mythology, African and Native American symbolism and ritual (especially through sweat lodges, ritual drumming, and vision quests), as well as the more recent languages of addiction and recovery, therapy, and New Age religiosity.

The questions posed most vividly by the men's movement are whether, absent the feminist attack on accepted forms of masculinity in the 1970s and 1980s, maleness would have continued to be viewed implicitly as the neutral ground of the public sphere of life and whether this forced redefinition of masculinity affected the mapping of gender categories on the nonhuman "natural" environment.

Whatever the interactive social, economic, and cultural forces requiring the remapping of gender on nature in the 1980s and 1990s, we can generalize about some of the continuities and changes shaping current environmental thought. Clearly, new modes of being male or female are stabilized and made to seem natural or essential through reference to nature. In this process, nature is both a philosophical concept and a physical space for certain kinds of human activity. Today, as in the nineteenth century, the mutability or fixedness of many gendered phenomena (mathematical ability, aggression, and infidelity) is being charted through locating them in "nature"—now appearing in its newest dress in the human genetic material. And, of course, the outdoors is still a setting for inducting boys into manhood, whether via off-road vehicles and hunting or wilderness vision quests.

As in the 1890s, stands taken on wilderness preservation or the conservation of natural resources are still articulated in gendered terms. Hindsight makes it clear that Theodore Roosevelt was successful in offering his contemporaries a new mode of masculinity and an accompanying stance toward the natural environment, even though both were constructs that kept inherent contradictions in uneasy balance. We may need the perspective of another century to decipher the nature/gender matrix in the style of Reagan/Bush and Clinton/Gore, but we can already discern a strongly essentialist and strikingly convergent coupling of masculinity and nature in the macho gender balancing of both "wise use" antienvironmentalists and their Earth First! opponents.

At the turn of the century, what we would now call "environmental" concern, whether for the preservation of wilderness or for the protection of worker and family health in cities, was inseparable from gender. All aspects of both forms of concern—including the fact that they were separate enterprises—were shaped by gendered differences in the access to political power, the separate social settings inhabited by men and women, differences in the natural landscapes and environmental settings they encountered, and differences in the idioms of caring allowed to men and women. Is gender an equally pervasive and influential factor in contemporary environmentalism, or have less polarized gender roles made environmentalism, and other spheres, a more gender-neutral enterprise? Does gender, in other words, still shape environmental concern and action, and if so, how?

While real political advances have been made since the turn of the century, the level of representation of women in political life still lags significantly behind that of men. The glass ceiling that prevails in the business world means that women exert less influence than men on the activities of the multinational and public corporations responsible for the major part of contemporary environmental destruction. And while women are more represented than men in the membership of the ten largest American environmental groups, they are significantly underrepresented in their leadership. The hands at the helms of environmental destruction and of institutionalized resistance to it are male. Indeed, women's relative failure to achieve the sought-after political and economic success may be an important impetus behind the compensatory embrace of the more spiritually oriented forms of ecofeminism.

As in the nineteenth century, when male conservation concern drew upon the social bonds already formed in hunting and outdoors clubs; female conservation, from women's clubs; and the female urban "environmentalism" of the settlement houses from friendships formed in women's colleges, so today the milieu from which environmental movements emerge is often gender segregated. The predominantly female membership of grassroots antitoxics movements, for example, relies on networks of acquaintance established in neighborhood friendships, PTA committees, and homeowners associations. And while the importance of the social milieu of largely male hiking and sport activities for the formation of connections among the male leadership of environmental organizations is decreasing, it remains significant.

Considerable gender differentiation also remains in the experiences of nature and of environmental hazards from which proenvironmental behavior often emerges. The motivations for proenvironmental behavior often stem from sensibilities gained in gender-polarized activities such as hunting and gardening. While such polarities remain in some spheres, however, contemporary women increasingly engage in formerly male activities, such as hiking and other outdoor sports. To the extent that wilderness activity is an antidote to the pressures of fast-track careers, its embrace and defense are likely to remain, in part, the prerogative of the males likely to occupy those stressful (and rewarding) social niches. Strenuous outdoor activities, the protection of which remains a mainstay of environmental movement support, continue to be a setting for the proving of masculinity, even though women are increasingly using the outdoors in their own gender-specific ways.

On the urban front, women are far more likely to have the kinds of experiences and knowledge that have galvanized many grassroots antitoxics activists: caring for repeatedly ill children, noticing correlations between smells and ailments, and detecting the prevalence of particular maladies within a few neighborhood blocks. Gender-differentiated response to various kinds of pollution is likely to increase in the future as more is discovered about the sexual specificity of the most prevalent industrial chemicals (o,p'-DDT, kepone, methoxychlor, and other organochlorine pesticides or their degradation products). For these chemicals act as endocrine disrupters that mimic estrogens, thus disrupting one of the few sexually specific systems of the human body. By binding to estrogen receptors they act as estrogen agonists, an action that may cause increased male reproductive problems and breast cancer in women. The rising incidence of breast cancer has already prompted new environmental concerns on the part of women.

Finally, although considerable blurring has taken place in the gender stereotyping of emotional life, certain forms and expressions of caring are still regarded as more legitimate in women than men. A calculating managerial stance toward natural resources remains more seemly for men, and a concern for future generations and the young can be more readily expressed by a woman. In this respect, Al Gore is perhaps a transitional character between traditional and emerging modes of masculinity: he is part of a generation of postfeminist liberal men who know that expressions of tenderness and concern are called for, but who still lay their masculinity open to question in many quarters by doing so. It is probably no accident, therefore, that the most fervent and radical expressions of environmental concern that have emerged onto the contemporary scene—the actions of Earth First!—were associated with a swaggering machismo. Just as Theodore Roosevelt could endorse reform movement principles (and feminist planks in his Progressive party platform) precisely because his paradigmatic masculinity was so firmly established in the public eye, so the most far-reaching contemporary calls for wilderness preservation have emerged from the unambiguously male Edward Abbey and his followers. It is still easier for women to make appeals based on deformed infants and for men to talk of the need for the wild and to monkey-wrench construction site bulldozers.

History leads us to believe that as long as gendered divisions of power, property, and labor prevail, gender is likely to remain an important component of environmental discourse and actions, though the precise nature of gender-nature linkages is still contested. Ecofeminist claims for a female role as uniquely qualified protectress of the natural world don't occupy a stable position: they don't sit well alongside traditional male claims to nature as a key locus for the formation of masculinity on the one hand, and they are not easily accommodated to feminist wariness of essentialism on the other. Thus an easy formulation of a specifically feminine connection to nature is no longer available. The different attempts to shape a specifically masculine relationship to nature, such as those found in the men's movement and Earth First!, are likely to be similarly problematic. Perhaps as social relations become more equal, androgynous approaches to mobilizing feelings about nature will replace those based on divergent forms of essentialism.

At the same time, "nature" itself is mutating: less a dangerous force from which we must shelter ourselves, for example, and more a delicate remnant to be shielded from human activities; present in most people's lives more as territories imagined through genetics and molecular biology and less through daily contact with forests and rivers. It is unclear how these shifts will affect articulations of gender and nature and whether the tradition of basing political expressions of environmental concern on gender-polarized imagery will be maintained.

NOTES

1. See, for example, Bina Agarwal, "The Gender and Environment Debate: Lessons from India" *Feminist Studies* (1992): 119–158. The most prominent and critiqued example of the ecofeminist appropriation of the experience of third world peasant women, and especially of the women of the Chipko movement, is the writing of Vandana Shiva, Staying Alive: *Women, Ecology, and Development* (London: Zed, 1986).

2. Cecile Jackson, "Women/Nature or Gender/History? A Critique of Ecofeminist Development," *Journal of Peasant Studies.*

3. Alice Hamilton, cited in Robert Gottlieb, *Forcing the Spring: The Transformation of the American Environmental Movement* (Washington, D.C.: Island Press, 1993).

4. On wilderness as a space for the regeneration of masculinity, see T. Christie Jesperson, *Engendering the Frontier. Men, Women and the Adventure in the United States from 1880–1925.* (Ph.D. diss., Rutgers University, 1997).

5. See, however, Gottlieb's envisionist history, *Forcing the Spring.*

6. Frederick Jackson Turner "The Frontier in American History," (Melbourne, FL, Krieger Publishing Co., 1976).

7. William James, "The Moral Equivalent of War" in *Essays on Faith and Morals,* (Longmans Green, New York, 1949, 311–328.)

8. On this dilemma and for a gendered reading of Roosevelt, see Jesperson, "Engendering the Frontier."

9. By essentialism we mean the attribution of male and female characters and gender roles to natural rather than social origins. Essentialist thought conceives of gender attributes as inborn, enduring, and primary rather than acquired and of culturally constructed significance.

10. Vera L. Norwood, "The Nature of Knowing: Rachel Carson and the American Environment," SIGNS, 12 (4), 19–29; H. Patricia Hynes, *The Recurring Silent Spring,* Elmsford, NY: Pergamon, 1989.

11. Minakshi Manon called our attention to this choice of journals, described in Paul Brooks' biography of Carson, *The House of Life*, (Boston: Houghton-Mifflin, 1972.)

12. Lynn White, Jr. "Historical Roots of our Ecologic Crisis," *Science*, 155, 1967; Paul Shepard, *Man in the Landscape: A historic View of the Esthetics of Nature*, (New York: Knopf, distributed by Random House, 1967); Edward Abbey, *Desert Solitaire: A Season in the Wilderness*, (New York: McGraw-Hill, 1968); Arne Naess, "Deep Ecology for the 22nd Century," *Deep Ecology for the 21st Century*, Bill Devall and George Sessions eds. (Boston: Shambhala Publishers, 1994).

13. Marija Gimbutas, *The Goddess and Gods of Old Europe, 6500–3500 BC: Myths and Cult Images*, (Berkeley: University of California Press, 1982).

14. Carolyn Merchant, *The Death of Nature: Women, Ecology and the Scientific Revolution*, (New York: Harper Collins Publishers, 1989).

15. For an analysis of the global circulation of Chipko stories, and of how essentializing accounts of women's participation in Chipko undermine feminist goals, see Yaakov Garb, "Lost in Translations: Toward a Feminist Analysis of Chipko," edited by Joan Scott and Cora Kaplan, *Transitions, Environments, Translation: The Meanings of Feminism in Contemporary Politics*, (New York, Routledge, 1996).

ANTON STRUCHKOV

Modernity and the Environment as a Public Issue in Today's Russia

The Precipice of Modernity: Western Way Down, Russian Way Out?

A PROFESSIONAL mathematician known internationally in academic circles, Igor Shafarevich has arguably brought one of the most thorough-going critiques of modern civilization into the public discourse of today's Russia. In the following account, I do not discuss his long dissidence with the Soviet regime, but focus on his present dissidence with modern civilization as a whole, and in particular, on two recent articles (1989, 1993) in which Shafarevich dwells on the issue of "modernity and the environment."

Let us start with the following passage from the latter one, entitled "Russia and the World Catastrophe," which gives a condensed account of his view on modern civilization.[1] "Is it justified at all," Shafarevich asks at the outset, "to speak of the present-day world as a unity, despite the evident contrast between the democratic West and the totalitarian East, the wealthy North and the poverty-stricken South?" (101). To answer this question, he invites us to look at our world with the eyes of an imagined archeologist living, say, a thousand years from now. If all written sources have disappeared for some reason, this archeologist

> will conclude that it is precisely by an extraordinary *uniformity* of life that . . . our epoch differs from those preceding it. On the overwhelming part of the Earth he will see a civilization of a single type. This type of civilization manifests itself everywhere by such features as: the gathering of people in gigantic

---

This paper attempts to consider how the concept of modernity figures in contemporary responses by Russian authors to environmental issues. To prepare it, I have looked through the relevant articles which appeared in the country's various mainstream journals from 1988 on. (Using 1988 as the starting-point was due to my impression that it is since that year that one may observe a real abolishment of press restrictions in Russia.) As it has turned out, several authors have been addressing the subject of "modernity and the environment" in a more or less regular manner, and in what follows I shall try to give accounts of their views. In doing so, my intention is to present, not an inclusive grouping of Russian authors addressing this subject, but rather a representative panorama of views.

cities; living in the anthills of colossal houses; the dominant role of machines in the economy and daily life; the orientation of all life toward technology based upon science, almost entwined with it; and a severe ecological crisis. . . .

And if he can distinguish different archaeological layers and reconstruct the process in time, then he will see that the unheard-of tempo of change is peculiar to our epoch (in comparison with the preceding ones). And he will discover that this civilization, having emerged in the small region of Western Europe, in the course of only a few centuries subjected the whole world to itself. He will discern a civilization incomparably more aggressive and intoler- ant than all preceding it. As distinct from all others, this does not tolerate [the existence of] other modes of life by its side, does not allow other . . . ways of development for humanity. (101)

The essence of this "universal civilization," holds Shafarevich, "consists in that human beings are becoming more and more subordinated to the demands of production and technology. . . . And, even more broadly, technological civilization ousts everything living, everything naturally grown, from the world, and supplants it with the artificial" (101).

Shafarevich emphasizes that modern technological civilization is of western European origin and has reached its greatest expression in west- ern Europe and North America. But he argues that the development of technological civilization should not be viewed simply within the frame- work of "the habitual antithesis of 'capitalism vs. socialism.'" "The very spirit of technological civilization," he writes, "is close to the spirit of the socialist utopia" as well (106). Indeed, the central idea of his earlier ar- ticle was to show that the Soviet "command system" of socialism and the Western "liberal-progressive system" of capitalism "are but two ways of realizing the . . . utopia of 'organization' of nature and society on the principle of 'megamachine,' with the maximum elimination of the hu- man and, in general, of the living element" (158); hence the title of that article: "Two Roads Toward One Precipice."[2]

> The Western way . . . is more gentle, is based more on manipulation than on straight violence. The way of the command system is bound up with violence on an enormous scale. This difference in methods makes the two trends look like irreconcilable antagonists, but actually they are moved by the same spirit, and their ideal goals coincide in principle. (159)

Despite this conclusion, in that earlier article Shafarevich found it necessary to ask how it was possible to combine the humaneness and respect for human personality inherent in the Western liberalism with the downright antihumaneness of the Soviet command system. His an- swer was that the humaneness of the former pertains only to those societies that have accepted the principles of "liberal-progressive" ideology and does not apply in any way to the rest of humanity. And this, he argued, is bound up with an important feature of this ideology—namely, its "hypnotic conviction that it opens the only path of development for humanity." It is this conviction that "engenders the notions of 'advanced' and 'backward,' of 'developed' and 'developing' countries." From this standpoint, civilizations other than the Western liberal-progressive one, appear, as it were, as "obstacles in humanity's path toward progress. That is why the West is capable of combining lofty humaneness within [its own type of civilization] with extreme cruelty to everything that is outside" (162).

Such comments suggest a total depreciation of modern technological civilization. But, as is evident from the following passage, in 1989 Shafarevich did exhibit a certain ambivalence:

> Technological civilization arose to take the place of a civilization that had been basically peasant, with the overwhelming part of the population living amongst nature, in constant contact with animals. Labor had been directly linked with its result; its meaning had been clear. The plan of work had been made up, every important decision had been taken, by the peasants themselves—their labor had been creative. But on the other side of the scale had been [such things as]: hard and exhausting physical work, lack of confidence in the future, frequent starvation, and enormous mortality, especially among infants. Almost every adult person had had to suffer the death of his or her child. Technological civilization managed to obviate all these misfortunes to a considerable extent, but, as it turned out, on its own conditions. Humans had to give up their human claims to life, and to submit to the logic of technology. On these conditions, this civilization proved extremely productive—and not only in the production of nuclear weapons, but also in the production of energy or in the ability to feed an enormous population. (159)

In the 1993 article, no trace of this ambivalence remains. On the contrary, the positive effects of modernization are now called in question by the claim that these effects are paid for by what Shafarevich calls "the extended reproduction of environmental troubles": "For example, the concentration of people in large cities raises the question of supplying them with foodstuffs; this problem is being removed by the mechanization of agriculture; mechanization destroys soil and decreases its fertility; this is being surmounted by the use of pesticides and artificial fertilizers; and this leads to the poisoning of soils, food products, waters, and, finally, of humanity itself" (105–106).

Furthermore, in this article Shafarevich incessantly emphasizes the role of the West in begetting and spreading "technological civilization," dealing harshly with Protestantism as its "spiritual basis" and pointing to the hopeful "otherness" of Russia's traditional "peasant civilization" and Orthodox religion. Indeed, he argues, "the attitude of the Russian Orthodox religion comprises the awareness of kinship of all creation, as well as the ideal of self-restraint expressed in the commandment of patience . . . that is to say, precisely those basic spiritual principles" upon which our "salvation" (as he puts it) from the imminent disaster of technological civilization might depend (123–126).[3]

At the start of the twentieth century, holds Shafarevich, Russia— deeply permeated with such religion, remaining an overwhelmingly peasant country—was becoming one of the most industrially developed countries. Yet, "it had by no means become another standard component of technological civilization." On the contrary, according to him, the Russian path of development was an "alternative to technological civilization." For its part, the West "perceived Russia as a foreign body" and developed "the concept of 'Russia as an obstacle in the path toward progress.'" This perception, Shafarevich argues, allows us to "comprehend the phenomenon that would otherwise be mysterious: [that] the socialist revolution in Russia was financed by the capitalists!" (120–121).[4]

According to Shafarevich, "If we look at the revolution as a means of turning Russia from its original path and including it as an element into the system of technological civilization, the participation of Western capital in this operation becomes comprehensible, justified, and paid for itself." He refers to the subsequent seventy years as "the epoch of destruction of precisely those structures which had comprised the basis of Russian society," namely, the peasantry, the Orthodox church, and "the very belief in the existence of Russia as having a certain historic individuality." This was "the negative part of the work." The next stage of "restruction" [*perestroika*] was meant "to accomplish the inclusion" of Russia into technological civilization: "The country's resources and economics are being placed under the control of the West. To facilitate this operation, [the country's] army, intelligence service, [and] military industry are being destroyed, and the whole country is being smashed to pieces. And the people's type of consciousness is being systematically destroyed, remade in imitation of the 'spirit of capitalism'" (122).

Returning to his "Two Roads Toward One Precipice" with this picture in mind, one may wonder if Shafarevich would now want to revise that earlier article in the light of his later conclusions. Depicting Russia's path of development after the revolution as just an inroad made upon the country by Western technological civilization, would he still speak of the Soviet "commanding system" of socialism and the Western "liberal-progressive" system of capitalism as *two* roads toward the same precipice? Whether or not Shafarevich would like now to revise it, however, it is from his earlier article that I want to pick out his "modest proposal" for Russia. (In "Russia and the World Catastrophe" I have found no specific suggestions to deal with.)

As he sees it, the Russian way out of "the precipice of modernity" is not to be found by repeating the Western path:

> In order to acquire a full-fledged copy of the Western pattern of life (even if with all its shortcomings and dangers), it is necessary to have, as the starting-point, its Middle Ages, and to live through its subsequent path. . . . Provided we copy only some results of this development, we shall obtain, most probably, something like Latin America rather than the USA and Western Europe. That is, colossal debt to advanced countries (which is not small already), the ruination of nature, crying inequality in respect to property, terrorism and totalitarianism.

Searching for models to overcome the current social-ecological crisis, he argues,

> For us, the most near and comprehensible is that peasant civilization amongst which the life of our ancestors flowed only so recently. Coming back to it is in no way possible—no return at all is possible in history. But it may become for us the most valuable model of an organically developed pattern of life, from which a great deal may be learned, and, most importantly, of cosmocentrism—living in a state of stable social, economic, and ecological equilibrium. (164)

This model, he holds, is offered to us by the so-called "village literature." For Shafarevich, the model is neither a matter of merely ethnographic interest nor a "requiem . . . for a beautiful lost civilization." Instead, village literature

> resurrects peasant civilization—if not in life, then in our emotional experience. We come into contact with it, . . . [and] there comes to us an example of organic, stable social structure based on the deep unity of humans and the cosmos: a model of that way of life, the quest for which is the main problem of today's humanity. (165)

What motivates Shafarevich to draw the sharp dichotomy between Russia and the West? To answer this question, consider an American historian's account of modernization in the USSR. Connecting the eventual failure of the USSR to sustain its drive toward modernization with the vanishing of its citizens' faith that this drive would lead to a bountiful and humane society, Loren Graham writes:

> The erosion of faith accelerated as the citizens of the Soviet Union became increasingly aware that although their country had become a great industrial power, their standard of living matched that of third-world countries. By the seventies the Soviet Union was the largest producer in the world of steel, lead, asbestos, oil, cement, and several other basic industrial goods. But the cost in human and environmental terms of a blind fixation on output was perilously high. Food and consumer goods were often unavailable because the political bosses insisted on producing steel for heavy industry and the armed forces. Life expectancy declined until the Soviet Union ranked thirty-second in the world. Infant mortality rose until the Soviet Union ranked fiftieth in the world, after Mauritius and Barbados. The environment was a disaster, especially around industrial cities like Magnitogorsk and in areas requiring irrigation, such as Central Asia. (101)[5]

This passage underscores the negative effects of modernization in the Soviet Union in terms of human welfare, not to mention environmental well-being. It is these effects, I believe, that specifically color Shafarevich's viewpoint. In other words, my impression is that the sharp Russia/West dichotomy he draws may come from some deep personal need to find an escape from the sadness of Russia's modern reality—an escape achieved by focusing on the unattractive features of Western modernity and simultaneously erecting an ideal realm of Russia's "peasant civilization." It is an "ideal realm" because the "peasant civilization" Shafarevich speaks of is not the kind that actually existed in the Russian historical past, but rather the kind that ought to have existed according to his ideals. In fact, Shafarevich ignores the features of Russia's "peasant civilization" that do not correspond to the "deep unity between humans and the cosmos" he attributes to it. Do we find "the awareness of kinship of all creation, as well as the ideal of self-restraint," for example, in the operations of the Peasant Bank in the black earth region of Russia, which, following the 1861 emancipation of the serfs, was entrusted with ten thousand square kilometers of the region's state forests and managed to destroy them almost to the last stump by 1911?[6]

A Warning against "Green Plague"

In the work of another writer we will later see the dangers involved if such turning away from the actual as exhibited by Shafarevich is adopted as a deliberate strategy and carried to an extreme. The writings of Sergei

Kara-Murza reveal how the Russia/West dichotomy can be pushed so far as to lead him not only to roundly condemn Western industrial modernity in all its features, but also to postulate the fundamental continuity of Russia's development, in no essential way broken by the 1917 revolution. This postulate—and the corresponding assertion that Russia has not yet fallen into a "grave crisis of identity" (which *is* underway in the West)—becomes the cornerstone of a grimly totalitarian vision.

Before turning to these writings, it may be helpful to listen first to an author who is anxious about totalitarian visions in general, and the perspective of "green totalitarianism" in particular. This author, the notable journalist Maxim Sokolov, shows a decided antagonism to the patterns of thinking represented by Shafarevich and Kara-Murza. According to him, "It would be a frivolity to think that in an epoch of sharp societal stratification, inflation, Westernized mass culture, crisis of family, and economic devastation, there will be no one in our country who might become tempted with a fundamentalist utopia" (23).[7]

These words suggest that Russian society is suffering a grave internal crisis, and further imply a warning against precisely the kind of "Russian way out" proposed by Shafarevich. For, as Sokolov sees it, because of the grave reality of this crisis, today's Russia is protected "neither from the Great National Revival, colloquially called the brown plague, nor from the Great Green Utopia—the *green plague*." Arguing that "ecological crisis offers rather good prospects for the green plague," Sokolov opposes the perspective of "green totalitarianism." His argument proceeds from the juxtaposition of "exploitation of humans by humans" with "exploitation of nature by humans," which he thinks "have engendered somewhat similar crises" (24).

As for "the crisis caused by the rapacious exploitation of humans by humans," maintains Sokolov, it has turned out that overcoming it "was possible both by way of the real *struggle against rapacity* (through trade unions, public charity, the state regulation of labor-selling conditions, social insurance) and by way of the *struggle against humans* themselves—the way which comrades Stalin, Mao Zedong and the like chose to lead us." Today's environmentalism faces the same sort of "fork": "Either judiciousness will prevail (which means the maintenance of the already existing norms and their subsequent strengthening; strict sanctions for damaging the environment; economic stimulation of clean industries), or the next supergoal—the green utopia—will gain victory" (24).

Assuming that the "judicious" way of overcoming ecological crisis is similar to overcoming exploitation of humans by means of "struggle against rapacity," one may still question the equation of "green utopia" and the "struggle against humans themselves." Sokolov draws our attention, first, to "the character of the goal set before the builders of a new society" by "comrades Stalin, Mao Zedong, and the like":

> It was not the development of democratic self-government, not the development of the country's economy, not social policy, not the calm development of science and culture. Properly speaking, there was no goal, but a *supergoal:* the creation of a novel, unprecedented society; the unity of working people to create Paradise on earth."

This cannot truly be called a goal, he argues, because "what is expressed here is an aspiration, not for something concrete and tangible, but for the supervaluable, for going outside of history. For an absolute" (22).

Turning to the concept of green utopia, Sokolov maintains that "absolutizing survival, [it] aims at renouncing effluent-producing creative activity, and the reverse dissolution of humans in nature." He sees two possible ways of realizing this aim: "Either civilization is abandoned as such," which entails "not just the suppression of personality, but its simple absence," or civilization takes the form of a "new green order" (25).

Focusing upon the latter, Sokolov adds, "But, the division of labor, characteristic of civilization, can be based either on ecologically impure and dumb instruments, that is to say on machines, or on ecologically pure talking instruments, that is to say on human beings." Hence he is sure that the transition to this "green order"—where a few "writers-ecologists will inhale clean air and scribble their [pastorals], while talking instruments will plough the land or sit in the slave ergochairs"—"will require such omnipotence of the state and such a repressive apparatus . . . that comrade Stalin will look like a lamb" (25).

Sokolov is trying to expose what he calls "the lie of a false good—another humanity-saving supergoal" (25). For him, the false good of the green utopia is "the idea of survival of humanity as the absolute value." If we are to "survive *at whatever cost*," he asks,

> what is the preserved life itself for? From the creation of the world until today, all human culture was based on [the idea] that there are values *higher than life*. If the mode of thinking which renounces such values becomes universal, this in fact would mean the appearance of a new race of living things, having [nothing in common] with humanity. (24)

Hence, concludes Sokolov, "If we don't want to live in [such a] hell, let us learn to live in history" (25).

The "Ungreening" of a Totalitarian Vision

I now turn to the analysis of scientist-cum-journalist Sergei Kara-Murza, and trace the development of his vision through a series of articles he published in 1990 through 1993.

In "Science and the Crisis of Civilization," which appeared in 1990, Kara-Murza undertakes to expose the disastrous consequences for civilization of its development based upon modern science.[8] From the outset, he emphasizes the dominant role of science in modern culture: "It is modern European science upon which the methodology of thinking, the system of education, the views on the world, humanity, and society are based. Based on it are technology and the life-style it shapes, which are offered to the whole world as the standard" (3).

Viewing modern science not as a mode of cognition but as an ideology that lies at the root of industrial society, he goes on to deconstruct its keynote categories, namely, "freedom" and "progress." With regard to freedom, he argues that "the views of the Medieval European on man and society were based, first and foremost, on the categories of justice, faith, honor, and fidelity." The modern category of "freedom," he stresses, "could not emerge until the world-picture had become dominated by: atomistic views, the belief in the reversibility of basic processes, and the idea of infinity": "Only man-atom is genuinely free. And not simply an atom, but an atom as a mechanical body, devoid of properties (affinity, valency). An atom that enters into reversible processes of collision . . . An atom-man has an atom-voice in a democratic society" (3–4).

According to Kara-Murza, this idea of freedom could come into the foreground only when coupled with another aspect of modern science-ideology, namely, the conception of the world as a stage of predominantly reversible processes. In a world where errors are normally viewed as remediable, maximum freedom of action can be conceived of and only those "anomalies" that lead to irreversible consequences (such as murder) are subject to strong control. "The world of the market is free, because everything is reversible: money—goods—money." As a result of the extreme reductionism of this conception, maintains Kara-Murza (borrowing a phrase from Konrad Lorenz), modern humanity "knows the price of all things and doesn't know the value of anything," and this fact largely determines "the 'non-moral' character of freedom in industrial civilization" (5–6). Furthermore, the idea of freedom dominates

only when there is no awareness of . . . *limits* . . . The world-picture of people of industrial civilization was shaped under the influence of geographic discoveries, the opening up of American spaces, the colonization of lands with inexhaustible resources. Later came the belief that terrestrial limits are unessential: if necessary, we shall go out into space, or make use of thermo-nuclear synthesis, etc. The idea of freedom presupposes the possibility of continuous expansion. (6)

Bound up with continuous expansion is another key feature of modern science-ideology: the concept of "progress," which, according to Kara-Murza, undergirds "the entire ideology of industrial society, the capitalist system of extended production and the life-style associated with it." In relation to "progress," he writes, "the idea of freedom acts as a means, allowing [us] to ignore limits and irreversibility" (7).

Finally, in order for the ideas of freedom and limitless progress to prevail,

it was necessary that in the world-picture man should be placed beyond the bounds of nature, that he should be opposed to it, that he should conquer it, get to know it, and extract from it the resources he needs. If man is the crown of nature, he is the crown independent from it. This sensation rouses the melancholy of loneliness, but also makes the experience of freedom as full as possible. (7–8)

Kara-Murza asks, "What are the results of humanity's development under the dominion of the culture of industrial society and European science as its key element?" He answers that they fall into three main headings:

1. "The loss by humanity of the instinct of self-preservation," the main symptom of which is the destructive effect that "the life-style based on scientific technology" exerts upon terrestrial environment. "This life-style, characteristic of consumer society," maintains Kara-Murza, "is practiced today by about 13% of the Earth's population. They absorb about 70% of nonrenewable resources, and emit about the same percent of polluting substances." Furthermore,

The progress of the technosphere on which the 'first world's' free market economy is based is governed by the optimization criteria that ignore natural limitations. The prices at which resources are extracted from the bowels of the 'third world' countries have nothing in common with the real value of these resources for humanity from a reasonable perspective . . . These prices are determined by the expenditures for bribing the elite of developing countries; and if this elite becomes too greedy, it sometimes proves cheaper to resort to military force. (9–11)

2. "The unlimited growth of artificially created needs and the corresponding expansion of industry [which] have caused the real threat of destroying not only the environment, . . . but also humanity itself as a system." According to Kara-Murza, it has become obvious that the "first world" simply cannot allow the development of the "third world": "The natural limitations of our planet do not permit in principle to extend the type of consumption established in the 'first world' to all humanity . . . The underdevelopment of 70% of humanity is a sad necessity; without it, there cannot be consumer society for 13%." He writes that this "destruction of humanity will sooner or later lead to sharp social conflicts." The South American drug industry is one example, which he views as "not just a business, but a way of taking vengeance upon the 'first world'" (12–13).

3. "Yet, perhaps the deepest crisis," he concludes, "is the crisis of identity of the very 'atoms' of industrial civilization": "However we may avoid speaking of this, it must be seen that the maintenance of [its] lifestyle and the old trajectory of progress is only possible by breaking off altogether with that very system of norms of Christian morality on which our civilization was founded." As evidence of the "transformation that has already happened," Kara-Murza offers the following picture:

> An average contemporary American, if he sees a child in danger of death, will rush to rescue that child at the risk of his own life. In this case he is a Christian. But millions of children in the 'third world' currently die from hepatitis-B. The vaccine has by now been made, and inoculation costs one dollar. Yet the corporations don't want to produce it, because there is no one to pay. At the same time, however, an average American spends $267 per year on alcoholic drinks, and doesn't wish to spend less. (14)

In a later article, "The Fraud," which appeared in 1991, Kara-Murza's focus shifts to the free market economy as the moving force of Western industrial society.[9] What he calls the "fraud" is the claim that the free market economy is ecologically more benign than the planned one. He blames the free market economy for all the aspects of the aforementioned threefold crisis.

(1) As regards environmental problems, he argues, it is only superficially that the free market economy seems to solve them; in reality, it tends to "solve" them by removing them from its "show-case" to the developing countries (15). Further, Kara-Murza considers it fraudulent to compare the ecological situations in the USSR and the First World countries (which are, as he puts it, "at an altogether different stage of technological revolution"). He emphasizes that

the First World, too, has gone through the stage of cheap dirty technologies destroying the environment; it is precisely at this stage that those means have been accumulated which are now invested in nature protection . . . but this is being done after this very [free market] economy has caused almost irremediable damage to the world ecosystem! (16)

(2) The free market economy is "extremely antiecological in regard to humanity as a system." From the very beginning, he points out, it was "ruthlessly destroying cultures, ethnicities, social groups alien to itself" (17). This economy is sure to continue destroying "the 'ecosystem' of humanity," since "it has become obvious that the combination of progress and freedom [this economy provides] for 13% of humanity is possible only on the condition of suppressing progress and freedom of the rest" (16).

(3) According to Kara-Murza, the contemporary internal crisis of Western society (i.e., "the crisis of identity of the very 'atoms' of industrial civilization") is also due to the free market economy, to the extent that it promoted industrialization through the atomization of society and demolition of traditional structures.

In contrast to the West, Kara-Murza contends, Russian society did not suffer anything like this, either in the process of industrialization or even in consequence of the 1917 revolution. About Russia, Kara-Murza says:

> Even after having undergone the stage of initial industrialization, we remained a traditional society. The Revolution and the Civil War destroyed a lot of things, but *did not shake the structures of thought and the human outlook of agrarian civilization* . . . Even having ruined the churches, we remained a people possessed of the 'religious organ': some kept professing Orthodox religion or Islam; others replaced icons by Stalin's portraits, transferred [their] religious feeling to the mausoleum and the Kremlin. But there was neither desacralization of the basic conceptions, nor destruction of the instinct of collectivism and solidarity. (18)

This last theme is developed in three "Meditations on Economics and the People," published in 1992.[10] Kara-Murza argues that the Russian revolution took place in a type of traditional society where

> man felt himself included into more or less large collectives . . . A large family, a village commune, a household, a church parish, an underground organization, or a gang of robbers—everywhere man felt himself part of a group united by relationships of solidarity and mutual responsibility. (154–155)

This type of traditional society, according to him, was not destroyed by the October Revolution: traditional values ("the ideals of equality, compassion, and social justice") did not vanish, but only "assumed socialist color or at least phraseology," and continued to exist under the "ritual shell" of the new regime.

The pattern of industrialization in the West, he maintains, was altogether different. The system of free enterprise, its moving force, "was based upon the 'atoms' which moved freely from one place to another and entered into commercial relations with each other. In order to make people into such atoms, it was necessary to uproot them from collective structures and provide them with a psychology that denied these structures." Instead of traditional values, continues Kara-Murza, "the principle of individual freedom was raised to the top of the scale of values; then, the entire system of human rights and the modern conception of democracy were developed on its basis" (155).

Kara-Murza undertakes to unmask the "fraudulent" character of freedom in Western industrial society. Pointing to the lack of freedom for the "weak ones" who happen to "lose in the conditions of the market economy," he quotes Palme (the late prime minister of Sweden):

> Today the overwhelming majority of people think that freedom from poverty and starvation is more desired than many other rights. Freedom presupposes the sense of safety. Fear of the future, of pressing economic problems, of diseases and unemployment turns freedom into a meaningless abstraction. (163)[11]

As for the wealthy, adds Kara-Murza, their "freedom is also rather limited," because they "constantly feel threatened" by the marginalized and poor strata of society. Hence, contemporary Western society contains "a large potential of mutual intolerance" which "is driving society . . . to the restriction of freedom and totalitarianism" (164).

In "The 'Liberalization' of Russia: The Way to Civilization or to the Common Grave?," published in 1992,[12] Kara-Murza argues that the real goal pursued by the present-day reformers in Russia is "not the building of liberal market economy, but the total destruction of this hateful, irregular country" and the infusion of its resources into world civilization (112). To prevent the reformers from reaching this goal, he writes, an urgent consolidation of "the state-patriotic movement" is necessary. He lists a number of obstacles to this consolidation, one of the most serious of which he sees in the fact that "the state-patriotic movement" has not yet worked out a single credo, perhaps because it was "initially split into . . . the *reds* and the *whites*." To consolidate, he concludes, it must "overcome" this basic split (120).

With these words in mind, Kara-Murza's apparently fraudulent picture of the Russian past seems to make sense as his contribution to the task of working out a credo that will overcome the basic split within the state-patriotic movement. Two articles Kara-Murza published in 1993 allow us to envisage the contours of this credo. In "The Annihilation of Russia,"[13] he juxtaposes the Soviet regime with that of today's Russia:

> Under the old regime, it was *hammered into everyone's head* that the peoples of the USSR are one family, that it is necessary to respect each other and to help each other . . . The reality was not unclouded, but the important point is *what* are those dogmas which are being hammered into one's head. The new regime has proposed *the law of market* as the basic principle of life, and is hammering into heads the corresponding dogmas (competition instead of solidarity, the personal against the common). (133)

Kara-Murza finds nothing wrong with the *procedure* of hammering dogmas into people's heads; his only concern is the *content* of those dogmas that state power hammers into us. (Even that concern becomes a question of minor importance in his next article.)

The next passage gives us his understanding of what the state power is for:

> It is said that the former regime 'suppressed contradictions,' and that this was very bad. Indeed, it did suppress them—so that nobody admitted even the thought of establishing an organization for killing people because of their nationality . . . But doesn't [state] power exist for the purpose of *suppressing* the destructive impulses of a frantic minority of instigators which are found in any nation? Isn't [state] power obliged to keep the peace and rights of the citizens? And the [former] regime did not fulfill this paramount function half badly. (133)

Kara-Murza here forgets the millions of peaceful citizens murdered by the Soviet regime in the process of "suppressing the destructive impulses" of a nonexistent "frantic minority of instigators." Such reminders, however, only allude to our human ideals, and ideals, Kara-Murza tells us in his most recent article,[14] "are useless to argue about; they are irrational." From the viewpoint of his credo,

> The rational thing to do is to discuss and debate [not ideals, but] those results which will follow from the purposeful instillation of these or those ideals (and the corresponding social structures) into a concrete cultural milieu. And [to discuss and debate] what will happen if these ideals are instilled in this or that

way, with the application of this or that social technology. These questions
yield to rational analysis. To our deep misfortune, in the meanwhile we see an
altogether different picture: opposing groups of people are arguing with one
another precisely about ideals with mounting passion, and nobody wants to
turn to those pragmatic problems of 'technology' which yield both to analysis
and control. (153)

Kara-Murza is progressing in working out these "pragmatic problems
of technology"—for example, he writes that those who are currently
dragging Russia into the Western industrial civilization "force us to re-
nounce the *ethic of religious brotherhood*." (Recall that two years ago he
called it "the instinct of collectivism and solidarity.") It is exclusively this
ethic, he adds, "that ensured the possibility of a thousand-year *strikingly
peaceful* coexistence of peoples" (155). (Earlier, you'll recall, he wrote that
the Soviet regime "*hammered into everyone's head* the dogma that the
peoples of the USSR are one family.")

Where does "the environment" enter into these latest writings of
Kara-Murza? It does not! Discussion of environmental issues has altogether dropped out from his six articles published in 1992 and 1993
(157).[15] Does this mean that Sokolov's warning is no more than pointless
invective? It may seem so to the extent that the credo of Kara-Murza
nowhere refers to such "supergoals" as survival or environmental well-
being. These are what Kara-Murza would call "ideals," and the rational
thing, he told us, is not to discuss them but to turn to the pragmatic
problems of their "purposeful instillation" into our heads. But Sokolov's
warning need not lie dormant. For the crux of his argument is that we
should worry not about survival—which is meaningless—but about a
society worthy to survive.

Looking beyond "Modernity" and "Environment"

In Shafarevich's work, we have seen an increasingly negative attitude toward modern "technological civilization," together with praise for Russia's traditional "peasant civilization," portrayed as a "stable social, economic, and ecological equilibrium." However, his emphasis on the
Western origin of "technological civilization" and its subsequent disastrous assault upon Russia fails to take seriously the extent to which the
very attributes of modernity he condemns had developed in the prerevolutionary Russia as well, irrespective of any corrupting Occidental influence. Thus, his way of preparing the medicine, so to speak, to cure

the diseases of modernity ignores the fact that this medicine contains ingredients that contribute to the continuation of those very diseases.

The work of Kara-Murza reveals the dangers if this ignorance takes the character of a deliberate strategy. In comparison with Shafarevich, whose vision of Russia's "peasant civilization," though historically inaccurate, at least displays a certain grandeur of yearning to restore our human bond with the cosmos, the stance of Kara-Murza is far more disturbing. His insistence on the organic continuity of Russia's development, which he believes in no essential way to have been broken by the 1917 revolution, is not just blatantly incompatible with historical reality. It also reveals his strategy of working out a unifying ideology of resistance to Western industrial civilization, which he sees threatening to destroy Russia. Ironically, as his opposition to Western modernity, initially motivated by concern for environmental well-being, has developed into this ideology, environmental concerns have vanished without a trace. Moreover, his grim ideological scheme exhibits so little regard of the integrity of human personality that perhaps only a few most foolhardy modern technocrats might be content with it.

At the other extreme is the viewpoint of Sokolov, who insists that we must learn to compromise with an imperfect modern reality rather than submit to another grand-sounding "supergoal." The endorsement of freedom, toleration, and self-determination vibrates through his polemic against the yearning for a new "green order." Significantly, his commitment to the modern liberal tradition may also be seen in his failure to recognize that it contains the germs for its own transcendence. Thus, the liberal strategy of "struggle against rapacity" within human society, which Sokolov advocates against the utopian impulse to recast society totally for some "supergoal," stems from the acknowledgment of the intrinsic worth of society's exploited and oppressed members and endorses their freedom of self-determination. In arguing for this strategy against "green totalitarianism" in our dealings with nature, however, Sokolov fails to consider it in terms of liberating nature for its own sake. He thus seems unwilling to explore expanding the emancipatory impulse of modernity to a new way of relating to the natural world, in which our genuine self-respect is tied together with genuine respect for the rest of life.

At one important point, however, the tendency to accept modern civilization, represented by Sokolov, meets the tendency to reject it, represented by Shafarevich and Kara-Murza. That point is that both positions exhibit hardly any real signs of transcending the framework of modernity. It is symptomatic that Shafarevich speaks of the "resurrection" of his "peasant civilization," which, moreover, may occur in our "emotional experience" rather than in real life. Sokolov, on the other hand, seems to choose not to expand the emancipatory impulse of modern liberal tradition so as to transcend the categorical opposition between "civilization" and "nature." When he urges that "we learn to live in history," it is the framework of modern history that he has in mind.

In the final group of three authors on the subject of modernity and the environment, however, there are real signs of stretching beyond the framework of modernity. These authors are the novelist and publicist Zalygin; the philosopher Kutyrev; and Archbishop Kirill of the Russian Orthodox Church. By treating them here as a single group I do not wish to claim point-by-point similarities between the theocentric Christian outlook of Archbishop Kirill and the nonreligious viewpoints of Zalygin and Kutyrev. Nor do I give detailed accounts of their differing viewpoints. Rather, I indicate a common sensibility that vibrates through their meditations on the meaning of "nature" in modernity.

To begin with, they all have a keen awareness of the effects of modern civilization on the global environment, and none have illusions about the extent to which those attitudes toward nature that produced this effect have long been diffused through Russia's own cultural milieu. Thus, Zalygin, a towering figure in the country's nature protection movement, was professionally trained in the field of land reclamation in the 1930s, schooled (as he puts it) "in the actual spirit of the nature-transformer, believing that . . . it is precisely [human] intelligence and enthusiasm which nature lacks" (10).[16] Kutyrev maintains that the utopian quest for the triumph of reason on a cosmic scale to transform nature and construct a harmonious universe is by no means particular to the West, but is a long-standing element of Russian culture as well.[17] Likewise, Archbishop Kirill, connecting the crises that are now threatening "the integrity of creation, the life on our planet" with the character of European modernity, speaks of them as "the sins of single Europe, from the Atlantic to the Urals" (120).[18] In contrast to the accounts of Shafarevich and Kara-Murza, which distinguish Russia from the West, these authors stress what Archbishop Kirill calls "common sins": "The East is more guilty of some of them, the West of others, but in the end these are common sins. Confession of these common sins entails acknowledgment of common responsibility, and this is already a real step on the path to common actions" (120). Seeing modern environmental crises in terms of the East/West dichotomy will lead us nowhere.[19]

Connecting the "spiritual essence" of modern crises in general, and the current ecological crisis in particular, to "the alienation of human beings from God, from one another, from nature, and to the destruction of the integrity of the human person," Archbishop Kirill traces the roots of these crises to the disappearance of the medieval concept of the world as *ens creatum* (created being), from the Renaissance on.[20]

Kirill's point can be brought out by referring to the work of a Canadian author, Neil Evernden.[21] He holds that a particular transformation of the humanity-nature relationship is indicated by Leonardo da Vinci's *Mona Lisa*, which reveals two things: the emergence of the individual and the emergence of landscape. "The famous enigmatic smile," writes Evernden, "reveals a realm of privacy which we can glimpse but never know or possess, and the true individual is born. But the individual is created by pulling significance inward, and nature retreats outward as the thing we know as landscape." He cites J. H. van den Berg:

> The landscape behind [Mona Lisa] is justly famous; it is the first landscape painted as a landscape, just because it was a landscape. A pure landscape, not just a backdrop for human actions; nature, nature as the middle ages did not know it, an exterior nature closed within itself and self-sufficient, an exterior from which the human element has, in principle, been removed entirely. It is things-in-their-farewell, and therefore is as moving as a farewell of our dearest. It is the strangest landscape ever beheld by human eyes. (126)

This passage lays out what Archbishop Kirill points to as the mainstream tendency of subsequent societal development: that is, humans came to think of themselves as autonomous subjects, and the world became a secular aggregate of things, an object of study and exploitation.

Zalygin strikes a similar note, depicting a sort of vicious circle within the "modern enterprise."

> Our potentialities arise from our requirements, requirements from potentialities, and the circle becomes closed. It isolates itself from all the rest of the world, from those conditions in which the Earth and its nature exist, from nature's principles and arrangements. Man the producer and man the consumer; there is no longer anything else that he . . . represents . . . Producing the infinite series of articles for his consumption, he himself joins this series as number one. But the fact that article number one is the article-constructor, while all the other articles are constructions, does not change the essence of the matter. The series is the same, and being within it, the constructor inevitably acquires the properties of his own constructions, while he reduces all the properties of the ambient world to the same parameters as well. It is no longer in the ambient world that man gets to know himself, but in the world of his own requirements, even if [these are] unreal. (4)[22]

Where does this tendency finally take us, in terms of our relationship with nature? This point is captured by Kutyrev:

> From the life of humans, earth is vanishing as the natural soil on which they used to walk. Asphalt and footwear have eventually divorced humans from it. We see it now in the form of mud (in the city).
>
> From the life of humans, rivers are vanishing. Brooks are already being forgotten; only the last remnants are left of the rivulets; and large 'rivers' are 'regulated' into depositories of water.
>
> From the life of townspeople (and they make up the majority of population), nature on the whole is vanishing. There is, not snow, but 'precipitation'; not air, but 'oxygen' . . . ; not fog, but 'smog.'
>
> Nature, in the exact sense of the word, is not going to perish. It has already perished. Lawns, parks, canals, correlated with the needs of humans, and appearing in the capacity of nature, have no significance and being of their own. Nature has turned into 'environment.' (32)[23]

This passage evokes the earlier comment on the *Mona Lisa:* it is a picture of things-in-their-farewell. Only today the "things" are different. In Leonardo's time, it was the orderly scheme of creation sustained and governed by God that was vanishing; nature was coming into being as an autonomous entity. What is happening in our days is the disappearance of "nature" as the enveloping presence with a being of its own, and the emergence of "environment" as the problem-place of human activity.

Zalygin has expressed this point forcefully:

> The fact that nature is the world surrounding us is known to people without books and education; but why don't we realize that we surround nature too, only much more tightly and brutally than it surrounds us?! That we fix it up in the GULAG system, the system which . . . neither the convicts nor the escort—no one—will outlive? [Why don't we realize] that we are the environment of nature? (102)[24]

Many have raised the question of whether modernity is compatible with long-term environmental well-being.[25] But the foregoing calls that question itself into question. For if it is correct to suggest that the emergence of "nature" as the enveloping presence with being of its own lies at the beginning of modernity, then the emergence of "environment" as the problem-place of human activity indicates its ending.

I do not wish to claim that the modern concept of nature fostered a uniform attitude. Nor do I wish to panegyrize some sort of postmodernism. But as long as we think in terms of our relationships with the "environment," we are *not* moving into a new way of relating to the world, but rather concealing from ourselves a crucial fact about our present condition—that *we have become the environment of nature*. While at the outset of modernity human civilization was like a group of loosely connected islands surrounded by the immense ocean of self-sufficient nature, now the opposite is true. Accordingly, the real question is not Is modernity compatible with environmental well-being? but rather, What kind of environment may we moderns become for the rest of life?

The work of Zalygin hints at the two polar possibilities, epitomized by the figures of the GULAG and home. At the heart of the first option lies the impulse totally to manage and control the environment, so that it could serve human ends better than nature left untouched by human intervention. While its adherents often refer to this scenario as the "humanization of nature" to promote human welfare, I think "GULAG" is an essentially correct description of it. The wholesale application of the "ultimate humanization" project to human society in the case of my country resulted in an enormous network of prison camps and a rude disregard of the dignity and worth of the human person both inside and outside those camps. The writings of Kutyrev suggest that the "terrestrial GULAG" will be no better. It is his particular merit to have revealed the inherent irony of the "humanization of nature" scenario, namely, its tendency to dispense with the human person.[26]

This said, let me turn to the opposite concept of the "environment we may become for the rest of life," which Zalygin refers to as "home." Convinced that "neither the convicts nor the escort" will outlive the "terrestrial GULAG," he argues that nothing else remains for humans "but to shelter nature in this house of theirs, and shelter it not as a poor relation at all, but on the condition that it is nature who will define the regime and the order of life of a new home" (17).[27]

Thus, what this figure of home stands for is not domestication, but rather *domesticity*—our learning to dwell intimately with our earth-born fellow creatures. This home may be viewed in a more or less anthropocentric way, but it need not be anthropocentric at all, as everyone knows who has read, for example, Henry Beston's *The Outermost House*. And my favorite example comes from another American author, William Barrett, who brings out this figure of home beautifully in the following passage.[28] Addressing the mounting alienation of modern humanity from the natural world, Barrett writes:

> For some time now we have been told that man has become cut off from nature; and sometimes this is put in more imposing terms, that he suffers from

the loss of 'cosmic consciousness.' These descriptions are nonetheless true for being banal by this time. But the way out, or the way back, may require from us a kind of discipline and patience that we had not suspected. If you sit down and brood on large abstractions like 'cosmic consciousness,' you are not likely to remedy your situation. Better to turn your attention to things nearer at hand. In that simple and unself-conscious relation you have established with your dog or cat you have already affirmed a bond with the great unconscious life of nature. You have already taken a step beyond the prison of a narrow and excessive humanism. Boxed in a city apartment, you may begin to discover a curious kinship growing between you and the plant you water daily. You and it, after all, are partners in the same pilgrimage: you share the one life together on this earth. Dwell in that bond and let your thinking start there. (319–320)

The common element in the three Russian authors I have just discussed is this sensibility of our human partnership in the same pilgrimage with the rest of life. Archbishop Kirill conceives of this pilgrimage as the journey toward God, toward which all creation is predestined. Kutyrev finds the pilgrimage itself to be good, albeit not directed toward any highest point of development. Zalygin remains agnostic. Despite these obvious and important differences, they all share the same sensibility of partnership, of our being at home in the world with its other inhabitants.

The quietly adventurous figure of home, of our togetherness in joy and pain with the rest of life, blurs many of the categorical distinctions that haunt modern thinking—"nature" versus "civilization," the "sacred" versus the "profane," the "spiritual" versus the "commonplace," the "wild" versus the "tame"—and thereby stretches beyond the framework of modernity. Yet this stretching out is not a simple rejection. The notion of home implies a realm of privacy, the inviolability of the personal, thus fostering what modernity aspired to but failed to sustain. And the personal, of course, is to be thought of differently: not as my or your property, but as my or your way of relating to the world.[29] Thus, as Kutyrev has put it, "the phenomenon of interrelationship, coming to the foreground, appears as something possessed of its own integrity and value" (8–10).[30]

What will the world become when we have moved out of today's mazes of "environment," to be at home in it? Surely it will no more be modernity's "nature closed within itself and self-sufficient." But if it is bound to be anything, then it is our current task to pave the way for what it is to be. While there are many difficult things to be done for the world to become something, let me conclude by pointing to one thing that I believe is wholly in our power to do. It is simply this: to quit speaking and thinking of the world as the "environment."

NOTES

1. Igor Shafarevich, "Rossiia i mirovaia katastrofa" [Russia and the world catastrophe], *Nash Sovremennik*, no. 1 (1993): 100–129.

2. Igor Shafarevich, "Dve dorogi—k odnomu obryvu" [Two roads toward one precipice], *Novyi Mir*, no. 7 (1989): 147–165.

3. To emphasize the profound difference between these principles and the foundations of Western technological civilization, Shafarevich (Ibid.) refers to Max Weber, pointing to Protestantism, whose ethic "helped to break down the 'traditional' mode of thinking (which oriented labor and social activities toward attaining real human ends), in order to sanction the ideal of economic activity as . . . an end in itself." Shafarevich also refers to Lynn White's well-known thesis connecting Christian values to the present-day environmental crisis; however, he does so only to emphasize that "the features bound up with the ideology of technological civilization are considered by White to be more characteristic of the Western Christianity (Catholicism and Protestantism) than of the Eastern (Orthodoxy)." If this is so, muses Shafarevich, then, perhaps, "the riddle of the birth of technological civilization lies in the peculiarities of the Western European peoples." (110) To solve this riddle, he proposes to view the acceptance of Christianity as "reproducing the incarnation": "Christianity becomes 'incarnated' in a certain nation by dressing in the 'flesh' of its preceding history, culture, national tradition" (112). Applying this thesis to the issue of technological civilization, he writes:

> Incarnated in the world of the German peoples—whose central myth contained the prophesy of the death of the Earth, the revelation about the forces of chaos, temporarily fixed by the gods of order, about the Wolf and . . . the Serpent who will free themselves and destroy the Cosmos—Christianity has produced, for example, the theology of Meister Eckhart, with his teaching of the 'shapeless and deserted abyss of Deity,' where the soul 'loses all desires, image, comprehension, form, and dies its highest death.' It is from the same source that Protestantism emerged. In the end of this line of development we see the birth of technological civilization: the material creation of the 'shapeless and deserted abyss' of the technological Nothingness, in which the human individuality, the soul 'dies its highest death.' (112)

4. In support of this thesis, he refers to the facts presented in Anthony Sutton's *Wall Street and the Bolshevik Revolution* (New Rochelle, N.Y.: 1974).

5. Loren Graham, *The Ghost of the Executed Engineer: Technology and the Fall of the Soviet Union* (Cambridge, Mass.: Harvard University Press, 1993), 101.

6. I have taken this example from the writings of Andrei P. Semenov-Tian-Shanskii, one of the most distinguished figures in the Russian conservation movement in the first decades of the twentieth century (see my article "Nature Protection as Moral Duty: The Ethical Trend in the Russian Conservation Movement," *Journal of the History of Biology* 25 (1992): 413–428). What I have found in my study of this

author (as well as other early Russian conservationists) points to a strikingly different picture of Russia's "peasant civilization" than that postulated by Shafarevich. Thus, long before the socialist revolution allegedly came to turn Russia from its "original way" (praised by Shafarevich for having been ecologically benign), Semenov-Tian-Shanskii, commenting upon such examples as the one just mentioned, had been particularly troubled by the evidence that "this evil arose not as much from necessity" as from "human greed" and "savage destructive inclinations" (quoted in my "Nature Protection," 419). His prerevolutionary articles devoted to conservation presented an abundance of pictures such as "the barbarous cutting of an age-old alley of linden trees along the road, which did not bother anything or anyone. Ask any passing peasant: 'Don't you feel sad that these trees are dying needlessly?' And he will answer: 'What's to be sad about? We have lots of them.' This is the usual philosophy of our aborigines" (419).

If, indeed, Shafarevich was ever to descend from the Olympian heights of his totalizing vision of Russia's "peasant civilization" to those "savage destructive inclinations" of its paradigmatic representatives, which had moved such people as Semenov-Tian-Shanskii to raise their concern for nature in the first place, he would have found himself, to say the least, steering in very subtle water.

7. Maxim Sokolov, "Vozmozhen li 'zelionyi' totalitarizm?" [Is there a possibility for "green" totalitarianism?], *Vek XX i mir*, no. 10 (1989): 22–25.

8. Sergei Kara-Murza, "Nauka i krizis tsivilizatsii" [Science and the crisis of civilization], *Voprosy filosofii*, no. 9 (1990): 3–15.

9. Sergei Kara-Murza, "Obman" [The fraud], *Priroda i chelovek*, no. 4 (1991): 14–18.

10. Sergei Kara-Murza, "Razrushenie kul'tury—neobkhodimyi etap perekhoda k ekonomike svobodnogo predprinimatel'stva" [The destruction of culture is a necessary stage for the transition to the economy of free enterprise], in Idem, "Razmyshleniia ob ekonomike i narode" [Meditations on economics and the people], *Nasn sovremennik*, no. 1 (1992): 154–155.

11. U. Palme, as quoted by Kara-Murza in "Kakuiu svobodu daiet ekonomika kapitalisticheskogo rynka?" [What Freedom Is Being Given by the Capitalist Market Economy?], in Idem., "Razmyshleniia ob ekonomike i narode" [Meditations on economics and the people], *Nash sovremennik*, no 1 (1992): 163.

12. Sergei Kara-Murza, "'Liberalizatsiia Rossii—put'k tsivilizatsii ili k bratskoi mogile?" [The 'liberalization' of Russia: The way to civilization or to common grave?], *Nash Sovremennik*, no. 5 (1992): 112–120.

13. Sergei Kara-Murza, "Unichtozhenie Rossii" [The annihilation of Russia], *Nash Sovremennik*, no. 1 (1993): 131–135.

14. Sergei Kara-Murza, "Tsivilizatsionnyi slom" [Civilizational breakdown], *Nash Sovremennik*, no. 8 (1993): 152–160.

15. A single, purely rhetorical reference to environmental issues occurs only in the last article, when Kara-Murza claims that the current endeavor to force Russia into the Western industrial system will quite certainly "lead to the catastrophe incompatible with the very existence of the biosphere" (*Ibid.*, 157).

16. Sergei Zalygin, "Literatura i priroda" [Literature and nature], *Novyi Mir*, no. 1 (1991): 3–17.

17. In fact, his polemical writings are directed primarily against a certain trend

in Russian thought, known as "noospheric" (or, more generally, as "Russian cosmism"). See, for example, Vladimir Kutyrev, "Utopicheskoe i real'noe v uchenii o noosfere" [The utopian and the real in the concept of noosphere], *Priroda*, no. 11 (1990): 3–10.

18. Archbishop Kirill (Gundiaev), "K ekologii dukha" [Toward the ecology of spirit], *Simvol* 22 (December 1989): 117–143.

19. To continue this comparison a little bit, it is worth mentioning Archbishop Kirill's succinct view on Russia's recent history:

> The revolution of 1917 inscribed on its banners [the summons for] the establishment of a new society in the name of the human person. The indubitable truth of those ideas consisted in that human personality—and especially that of the worker and the peasant—was to become the subject possessed of intrinsic rights, instead of having been the object, the thing that might have been legally exploited. But the fatal course of events, whose meaning yet remains to be understood, resulted in that the enthusiasm for building the future outshined the present, the collective rights and the striving for equality [eclipsed] the individual rights and liberties . . . [T]he present generation came to be viewed as a means for building happiness of the future one. Not only the individual rights and freedom of the living, but even their very lives were being sacrificed in the name of the future. In this historical context, one could hardly expect the emergence of any attitude to nature, other than that which was expressed in a popular slogan of the 1930s: 'We cannot wait for favors from nature; our task is to take them from it.' (*Ibid.*, 121)

Obviously, this view differs not only from the praising of Russia's prerevolutionary "peasant civilization" and the outright anathematizing of the 1917 revolution we have seen in the case of Shafarevich, but also from the undiscriminating picture of Russian history we have seen in the case of Kara-Murza.

20. It is in that epoch, Archbishop Kirill writes, that the mainstream tendency of subsequent societal development was made manifest, consisting in that

> politics, economics, science, and technology definitively became autonomous spheres, disavowing [the vault of] the moral law and the spiritual realm over themselves. Egoistic interests and the craving for power in politics, the thirst for enrichment in economics, nationalism in the lives of peoples, the power of technology over human beings—all this is the outcome of autonomous development, cut off from and insubordinate to the spiritual source. Since the Enlightenment, there also emerged a novel attitude of humans to nature . . . Secularization embraced not only . . . human society, but nature as well; in the human consciousness of nature, it came to exist by itself, without any connection to God. Alienating God from nature and secularizing it, humanity changed the conditions of its own relationship to nature as well. From the subject [it had been], nature became the object of study and exploitation. ("K ekologii dukha," 124–125)

21. Neil Evernden, *The Natural Alien: Humankind and Environment* (Toronto: University of Toronto Press, 1985).

22. Sergei Zalygin, "K voprosu o bessmertii. Iz zametok minuvshego goda"

[Concerning the question of immortality. From the notes of the past year], *Novyi Mir*, no. 1 (1989): 3–48.

23. Vladimir Kutyrev, "Algebra, ubivaiushchaia garmoniiu" [Algebra killing harmony], *Chelovek*, no. 3 (1991): 31–36.

24. Sergei Zalygin, "Ekologicheskii roman" [An ecological novel], *Novyi Mir*, no. 12 (1993): 3–106.

25. As Leo Marx has put it in a 1994 letter to me, "The question that lies back of this series is whether modernity—understood as comprising, or as being an intellectualized epiphenomenon of, the more or less universal attributes of 'advanced' (non-traditional) cultures and societies—is compatible with long-term environmental well-being."

26. Arguing against such viewpoints as Vladimir Vernadsky's belief in our right, responsibility, and ability to undertake "the task of the reconstruction of the biosphere in the interest of freely thinking humanity as a single totality," Kutyrev emphasizes that the actual embodiment of this "noosphere" is threatening both "the existence of nature as a self-standing entirety" and "the very existence of humankind" ("Utopicheskoe i real'noe," 4–5). In "Universal'nyi evoliutsionizm ili koevoliutsiia?" [Universal evolutionism or coevolution?], *Priroda*, no. 8 (1988): 3–10), he explains that according to the "noospheric" ideology,

> we humans—first of all, in our natural hypostasis—are . . . also but a moment [in the ascending evolution], which, following this approach, is to be . . . absorbed by some new form, the higher 'posthuman' reason . . . In contemporary contexts, . . . what is usually meant by the . . . highest point of development is the maximally unfolded earthly or extraterrestrial Reason (in the technocratic version, artificial intelligence). Thereby we are suggested to give up calmly, in a scientific manner, the view of ourselves as an intrinsically valuable form. (8–9)

To underscore the contrast between the real-life humans and the dreamed-of posthuman reason, he offers the following comparison:

> If one amputates the human being to the head, and spirituality to information, the difference with the computer will be only quantitive. On the other hand, if one endows the [artificial intelligence systems] with feelings, emotional experience, corporeality, the difference [in kind] will evanesce as well . . . Their distinction from humans is the absence of the soul, of spiritual sensations that arise in the interaction of thought with nature . . . Reason is the dagger stuck into the human body. The wound emerges, which is the human soul. Then external nature becomes spiritualized as well. Bereave human beings of the consciousness of mortality, of affection and suffering, of the feeling of love and the sense of beauty, and the cherished hope of scientism will be realized. (7–9)

27. Zalygin, "Literatura i priroda," 17.

28. William Barrett, *The Illusion of Technique: A Search for Meaning in a Technological Civilization* (New York: Anchor Press/Doubleday, 1979), 319–320.

29. This sense of the inseparability of the self and the circumstance rings through Zalygin's account of what for him is "the main, philosophically most general, loss" in our alienation from the ambient world at the end of modernity. Speaking of this

loss as the vanishing of nature as "the only intermediary between us humans and Eternity, communion with which, experience of which by each of us is the meaning of our existence, its cause and purpose, its energy," he points to the fact that formerly each of us

> could again and again find oneself amongst the woods, mountains, and plains of the Earth, on the bank of some river, on the shore of some sea. And, having found oneself, each could again and again commune with . . . Eternity . . . [I]t seemed that it could not be otherwise. But now we already see, not just one, but several visages of nature, changeable and fickle, crippled by humans, passing before one's eyes in the course of so brief a human life. Now, where and in what, then, is Eternity contained for us? Only in the pointless distances of outer space, perhaps? This is the novelty that has happened in these days, before our eyes: Eternity has sunk into oblivion. ("K voprosu o bessmertii," 47–48)

30. See Kutyrev, "Universal'nyi evoliutsionizm," 8–10.

LOUIS MENAND

Modernity and Literary Theory

1.

THE PURPOSE of this paper is to see whether an analysis of the concept "modernity" has anything useful to tell us about the historical and intellectual forces that have shaped Western practices and attitudes concerning the natural environment. My contention is that "modernity," like most cognates of "modern," such as "modernism" and "postmodernism," is too contradictory and many-sided a concept to serve as the name for a single worldview and that efforts to tie productive environmental policy to an overthrow of the "discourse of modernity" are therefore not only ineffectual but unnecessary. It might be more promising (not to say more plausible) to look within the contradictions of modernity for the philosophical and conceptual tools needed to address the question of our relation to the environment. In particular, developments within the organization of knowledge as it is reflected in the structure of the modern university seem to me to hold some potential for reaching a better understanding of that relation.

Many of the examples I use to demonstrate the instability of terms like "modernist" and "postmodernist" are drawn from literary studies and literary theory. Of course, this vocabulary belongs to a theoretical discourse that has come to be shared by scholars in all the humanistic disciplines, and it should be easy to see parallels to my literary examples in art, music, philosophy, and the history of twentieth-century culture generally. Implicit in my argument is the belief that, as Simon Schama has recently undertaken to demonstrate in his prodigious *Landscape and Memory*, humans can know only a human world.[1] There is no environment apart from our imagination, our memory, and our representation of it. "The trail of the human serpent is . . . over everything," as William James once put it; we can deal with what is not human only in human terms.[2] Those terms are always more than biological in the global sense, because large-order cultural pattern making is part of the specific biology of our species. We behave, but we also, incessantly, conceptualize about our behavior. Our behavior doesn't always change as a consequence of our thinking through our conceptions a little more critically; on the other hand, where else do we have to look?

2.

"Modernity" is a term used to name our present historical condition. It usually—and in discussions of ecology and the environment, it almost always—carries the suggestion that there is, ultimately, something exploitative, soul-denying, and unnatural about that condition, and that its tendencies are therefore, as they approach these ultimate points, to be resisted as hazardous to the environment. That most of the people who use the term in this sense could not themselves be more modern is not a shallow irony. It is the essence of the problem the term "modernity" is meant to solve.

Merely as a matter of chronology, "modernity" has the design of a portable telescope with none of the convenience: it extends and collapses to almost any length, but it can rarely be made to stay put. "Modernity" is sometimes said to name a period that begins with Darwin, sometimes a period that begins with Rousseau, and so on, backwards in history. The onset of modernity has been identified with Descartes, with Bacon, and with Luther. In some accounts, modernity is a capitalist project, and its key idea is the division of labor; in others, it is an Enlightenment project, and its key idea is rationality; in still others, it is a Renaissance project, and its key idea is humanism.

As the name for a worldview, rather than simply a length of historical time, "modernity" is similarly promiscuous. One set of ideas that arise when the term is mentioned includes technology, dualism, positivism and the scientific method, bureaucracy and social control, instrumental rationality, the colonization of the life-world, and gesellschaft. Another set of terms includes romanticism, individualism, transcendentalism, liberalism, democracy, the unconscious, and indeterminacy. Each set of terms seems to belong to a different weltanschauung, and yet both would have to be included in any definition we made of "modernity." In short, all concepts are fuzzy, but "modernity" abuses the privilege.

The adjective "modern" causes a similar aggravation. When used as an attribute of a work of art or literature, "modern" can mean, as Paul de Man defined it in "Literary History and Literary Modernity," simply that the primary or paramount impulse the writer feels in putting pen to paper is to do something new, to be of the moment—as opposed, for example, to validating or updating a traditional literary form. In this sense Sterne is modern because he turned the novel inside out, but the novel itself is modern because it treats each story as an original rather than an allegory. Even Dante is a modern in this sense, by virtue of his choice to write in the vernacular. And yet literature can never be completely modern, since if a piece of writing were radically new, we would not recognize it as literature. Talk about writing as "modern" runs, in other words, into paradox; in thinking through the relation between modernity and literature, de Man found himself coming to the conclusion that "the modernity of a literary period [is] the manner in which it discovers the impossibility of being modern" (144).[3]

The term "modern" does have, in academic humanities departments, a narrower, shop definition, which is the period in art and literature that begins around 1880 and ends around 1945. This particular pie, 1880 to 1945, has been sliced in too many ways to summarize, but a rough distinction has usually been observed between two kinds of modern art and writing: "high modernist" and "low modernist." In literature written in English, the high modernists are writers such as Pound, Eliot, Yeats, Joyce, Woolf, Stein, Williams, Faulkner, and Beckett; the low modernists are writers such as Kipling, Bennett, Shaw, Forster, Wharton, Crane, Robinson, Frost, and Fitzgerald. This distinction has always been predicated on an assumption about the relation between the particular writer and the historical condition of modernity. But as the perception of this relation has gradually changed, the whole distinction has now been flipped on its back.

The old assumption was that high modernism is essentially a reaction against whatever we take to mean by modernity as a social and political phenomenon. This definition of high modernism dates from Horace Kallen's article on the term "modern" in the 1933 edition of *The Encyclopedia of the Social Sciences*, in which he suggests, in a discussion of modernist painting, that modernist art (i.e., post-Impressionist art) is a reaction against modern art (i.e., nineteenth-century realist art),[4] and this contrast is central to the way the high modernist writers were thought about for much of this century. It was commonly said, for example, that the kind of modernism associated with Eliot or Lawrence made a self-conscious break with the literary and philosophical traditions of the nineteenth century. This was the view of nearly all the New Critics (Yvor Winters was virtually the only exception), who were the academic champions of literary modernism in the 1940s and 1950s. The New Critical view was eventually challenged, preeminently by Frank Kermode, who argued in *Romantic Image* that modernism is actually an *extension* of romanticism and the aesthetic tradition from Keats through the nine-teenth century to Eliot is continuous rather than discontinuous.[5] This claim split the critical consensus about high modernist writing in two, but neither the New Critical view nor Kermode's alternative account questioned the idea that the values of modernist literature were formu-lated in reaction against the values of modernity.

That idea is being questioned today. High modernist writing is now suspected of being not largely reactive against, but largely complicit in, the modern worldview. Features of high modernist writing that were re-garded as salutary by one generation of critics dissatisfied with moder-nity are being regarded as pernicious by a new generation of critics dis-satisfied with modernity. A generation ago, it was thought that high modernist writing emphasizes the autonomy and isolation of the subject and privileges the private, or epiphanic, moment against the deindividu-ating forces of mass society; that modernist writing proposes art as an ordering principle in what Eliot famously referred to as the anarchy and futility of contemporary history; that modernist writers pulled together the shards of a fragmented tradition to construct, in Frost's phrase, a momentary stay against confusion; and so forth.

To the current generation of literary scholars, these features don't read as antimodern at all. They read as consistent with the project of moder-nity. In this new view, the modernist emphasis on a radically subjectivist epistemology repeats the modern ideology of individualism; modernist formalism detaches aesthetic experience from political life in a manner appropriate to the dehistoricizing tendencies of mass consumer society; the modernist rage for order echoes the top-down systems of social con-trol characteristic of both the modern totalitarian and the modern liberal state; modernism's appropriation of the artifacts of global culture and its fascination with the primitive complement capitalism's colonialist proj-ect. The result of this reordering is the argument (articulated in Perry Meisel's *The Myth of the Modern*, for example)[6] that it is not the high modernists but the low modernists, writers like Forster and Lytton Stra-chey, who are truly counterhegemonic—who are the *real* modernist anti-moderns.

This reordering of assumptions about the relations between modern writers and modernity has coincided with the emergence of a fresh con-ceptual troublemaker, "postmodernism." "Postmodernism" recapitulates many of the ambiguities that surround the terms "modernism" and "mo-dernity." Like most words beginning in "post," in fact, "postmodernism" announces its own ambiguity in its name. Does the term mean that we have entered a new historical phase characterized by a break with modernity? This is more or less the meaning given to it by Jean-François Lyotard in *The Postmodern Condition: A Report on Knowledge* (although he acknowledges a sense in which postmodernism is an impulse within modernity).[7] Or does "postmodernism" mean simply that the modern project is now complete—that we are fully modernized, with no going back? This is the meaning emphasized by Fredric Jameson in *Postmodernism; or, The Cultural Logic of Late Capitalism*.[8] "Postmodernism" has, in other words, the same problematic relation to modernity that "modernism" has traditionally had: it's not clear whether it represents a fresh historical departure or just more of the same thing.

If we take "postmodernism" in Lyotard's sense, as the name for a conceptual break with modernity, we quickly find ourselves in difficulties. How does this meaning of the term align, for example, with Weber's famous definition of modernity as the progressive "disenchantment" of the world?[9] Becoming more modern, Weber thought, means having fewer illusions; it means entering into an increasingly rational, empirical, and demystified relation to life. Now, one literary form generally associated with postmodernism is metafiction. A work of metafiction reflects self-consciously on its own status as a work of fiction—as, in the most material sense, a text. Metafiction is fiction that breaks the frame: it calls attention to the fact that we are reading a book. But what is this except the disenchantment of art? What could be more quintessentially modern, in the Weberian sense, than this kind of postmodernism?

The alternative view, the view that we're postmodern because, in our time, the triumph of modernity is complete, encounters similar analytic embarrassments. The modern project, Jameson says, was to transform nature into culture, and today, he thinks, we can agree that the transformation is complete. There is no nature left, no unmediated or prerepresentational access to the real. Everything is a simulacrum. The armature of signs has closed in around us.[10] This account of postmodernism has usually included a critique of high art's claim, in the modern period, to be a holdout against modernity. If modernist writers were trying to preserve the transcendental status of the work of art in a social formation characterized precisely by its hostility to transcendental agents, then postmodernism must represent the final demystification of that aesthetic ideology. Postmodernism, in this view, isn't reacting against modernity (of which it is merely the consummation); it's reacting against modernism—against the kind of art that endeavored to protect art against modernity's disenchantments.

Is this a plausible account of high modernism? No work is more central to the conventional understanding of Anglo-American high modernism than Eliot's *The Waste Land*. For years after its appearance, in 1922, the poem was regarded as a lament for a world worm-eaten by modernity ("A crowd flowed over London Bridge, so many / I had not thought death had undone so many"). Against this vision of entropic disorder was believed to stand the order of the poem itself, which pulls together the bits and pieces of a fractured culture and shapes them into a structure of symbol and allusion ("These fragments I have shored against my ruin").[11]

But this reading of the poem leaves something out. Eliot wrote the notes to *The Waste Land* on the occasion of its book publication, and they have appeared in every subsequent reprinting. The notes are almost always treated as only a gloss on the poem proper, a student's aid for some of the references the poem makes. When the poem is reprinted in the *Norton Anthology*, for example, the Norton editor's notes are mixed right in with Eliot's notes at the bottom of the page. But on what authority? The notes are not a separate commentary on *The Waste Land*; they are part of the text of the poem. No matter how Eliot intended them, there are no legitimate hermeneutical grounds for considering the notes as anything other than a literary text, subject to all the same interpretive difficulties a poem is subject to. *The Waste Land* is a poem that includes its own interpretation *as part of the poem*, and that interpretation (like any interpretation) can be regarded only as an unreliable and partial guess at what the poem "really" means. Since one of the subjects of the poem proper is the reinterpretation of primitive ritual in Christianity, this highlighting of the difficulty of interpretation by the notes must be itself a warning not to regard the authority of the poem as final, a warning that the poem, too, is only an interpretation. The reader is thus confronted at every juncture with the problem of poetic meaning. What use, in this instance, is the distinction between a naive and reactionary modernism (to which a work like *The Waste Land* is imagined to belong) and a skeptical and thoroughly semiotized postmodernism?

The principal theoretical component of postmodernism is poststructuralism. A little of the ambiguity surrounding the meaning of "post" that makes the term "postmodernism" confusing attaches to poststructuralism, as well. We can say that we're poststructuralist simply in the sense that we can never understand the world in a nonstructuralist way again. But most poststructuralists have, understandably, a better opinion of themselves than that, and they regard poststructuralism as signaling a rupture with the theoretical assumptions of structuralism. This is certainly the way deconstructionists regard themselves.

The idea of structure is a fundamental principle of modern forms of knowledge in the human sciences. Structuralism understands meaning, in any social formation, as a function of the relations among representations. It conceives of consciousness as a structure of pictures, or "texts," of the world, and it proposes that these pictures have value and significance by virtue of their place in that structure, rather than by virtue of being representations of something outside consciousness. There *is* nothing outside consciousness, or outside signs. Deconstruction, beginning with Derrida's famous Johns Hopkins paper in 1966,[12] accepts this model but destabilizes it by rejecting the idea that the structure can have a center, since that would require us to posit the existence of one part in the sign system that, unlike every other part, is not a function of relation. Deconstruction yanks the gyroscope out of the structuralist system and plunges us into a world of indeterminacy, of inverted hierarchies of value, of what deconstructionists call "the free play of signifiers."

The critique deconstruction made of structuralism has been extremely influential, and many features of contemporary critical theory can be said to follow from it. What deconstruction suggested was that although cultures take the form of structures—languages, kinship systems, gender roles, social and economic hierarchies, sexual norms, belief systems—there is nothing *natural* about these structures, since there is no transcendental point around which they are organized and no extra-representational reality to which they refer. There is, in Derridean language, no transcendental signified, nothing that stands outside the textual system and stops the proliferation of meanings within it. We are, in Wallace Stevens's image, in Tennessee, but there is no jar.

Why is it that we don't feel as though we're living in a deconstructed universe? Why do we have the sense that things more or less hang together in a way that feels more or less reliable? The poststructuralist answer is that what holds systems of relations together is power, and the most important set of relations in the relational web in which all meaning inheres is therefore the set of power relations. This is where Derrida links hands with Foucault. Power reifies the cultural systems that enforce it by making those systems seem natural. (Or vice versa: the systems reify existing power relations by making *them* seem natural.) In Derrida's writing, structures are usually analyzed as a series of dichotomies in which one term is privileged over—or, in effect, defines—the other: male/female, speech/writing, self/other, nature/culture, subject/object, signified/signifier, cause/effect, and so forth. These hierarchies pervade the sign system and thus make the relations of power they describe seem, to socialized consciousnesses, inevitable and right.

Deconstructing these hierarchies means demonstrating that it is, in fact, always the second-order term in the dichotomy that defines and makes possible the first. Ethnocentrism is made possible only by our group's invention of the other—those-who-are-not-ourselves—and, having established this entity, by keeping it stable and in place. The center could not exist without the margin. A political poststructuralism attacks ethnocentrism, patriarchy, racism, heterosexism, and other features of contemporary power relations by attempting to show that they rest on hierarchies of value that are culturally constructed and are therefore (on a good day, anyway) susceptible to subversion and transformation. The Derridean critique applies to Western metaphysics in its entirety. Platonism and Christianity are paradigmatic cases of what Derrida calls logocentrism, but it is clear that once all that has been accepted as *given* is regarded as *made*, a number of dominoes start to fall, among them, of course, many of the dominoes of modernity. Rationality, objectivity, the mind-body split, the fact-value distinction—these are all chickens awaiting the poststructuralist's axe.

Three terms generated by the poststructuralist critique have become widespread: "indeterminacy," "antifoundationalism," and "intersubjectivity." Indeterminacy is what happens when a poststructuralist reads a text. From a poststructuralist viewpoint, there can be no single correct interpretation of a piece of writing—not because, or not simply because, different readers will interpret from different positions within the cultural system, but because language doesn't stop referring. Words don't point to things, in some determinate way; words point only to other words. And since there is no fixed point within or outside the universe of these signs, meanings simply proliferate, contradicting and overcoming each other, and the meaning of the text as a whole becomes indeterminate. Meaning keeps on being generated, in the deconstructionist phrase, right into the abyss.

Antifoundationalism is the consequence of the assertion that meaning and value are not grounded in a set of a priori categories. Values don't have foundations and things don't have essences, since the existence of values and things is always and only a function of the sign system, the language used to talk about them. Thus truth can be regarded only as the consensus position on a particular subject. It's the point at which we have agreed, for the moment, to stop talking. But it has no greater claim to a privileged status than that; we may decide, at any moment, to start talking again, and thus to shift our consensus. This is where poststructuralism recapitulates various theories of knowledge that have become familiar without necessarily being labeled poststructuralist, such as Thomas Kuhn's *The Structure of Scientific Revolutions* and Richard Rorty's *Philosophy and the Mirror of Nature.*[13]

Intersubjectivity follows from the notion that human beings are socially constructed all the way down, that there is no essential human nature, only culture on top of culture. The individual, the great invention of Western humanism, is, in this view, an abstraction. Our identity, our subjectness, like everything else, is a function of relation. Individuals cannot be autonomous, since they are the loci of social and cultural determinants and take their consciousness of existing as discrete entities only from their relations to others.

What, then, is the relation between poststructuralism and modernity? Poststructuralism is often discussed as a force undermining the assumptions and principles upon which modern Western culture has been constructed. Some people talk about poststructuralism this way with alarm; some people talk about it this way with glee. This is how Lyotard describes the dispensation postmodernism and poststructuralism are supposed to have overcome: "I . . . use the term modern to designate any science that legitimates itself with reference to a metadiscourse . . . making an explicit appeal to some grand narrative, such as the dialectics of Spirit, the hermeneutics of meaning, the emancipation of the rational or working subject, or the creation of wealth . . . I define postmodern as incredulity toward metanarratives."[14] Is this a fair account of modern science and philosophy?

Historically, it is not the case that indeterminacy, antifoundationalism, and intersubjectivity are ideas that required a postmodern moment before they could emerge. They are all present at the heart of modernity, in the philosophy and science of the turn of the century. In America alone, William James and John Dewey were outspoken philosophical antifoundationalists. So was Oliver Wendell Holmes, Jr. Far from being a challenge to liberal political theory, antifoundationalism is vital to liberal thought and practice. The notion of truth as consensus is the idea behind liberal principles such as freedom of expression, which Holmes did so much to establish. That notion also enables one of the techniques of liberal polity most frequently condemned by critics of modernity, instrumentalism. Instrumentalism was, in fact, one of the names Dewey gave to his philosophy.

The idea of intersubjectivity is present in Hegel and in Marx; it is implicit in Darwin; it forms the basis for the social psychology of George Herbert Mead and for nearly everything in the thought of Mead's great friend John Dewey. In Dewey's view, an intersubjectivist notion of consciousness is what underwrites democracy, which is the great modern political concept. And indeterminacy, far from constituting a rebuke of the pretensions of science, was itself, in a certain respect, a scientific discovery. We owe our idea of the importance of indeterminacy in understanding the way the universe is constructed to Planck, Einstein, Heisenberg, and Bohr, and before them, to Maxwell, Boltzman, and Peirce. It was the recognition of indeterminacy that enabled scientists, through the development of quantum physics, to conquer the atom, and there is surely no story more emblematic of everything that is usually evoked by the term "modernity" than the story of the making of the atomic bomb.

Theoretically, poststructuralism presents another kind of ambiguity. Bluntly, is it going to help us out or not? The helping-out view follows from the belief that once it has been shown that everything is potentially up for grabs, that present conditions are not reflective of some preordained and universal order but are the consequence of political choices and coercions in the here and now, we can remake the systems in which we have our being. Antifoundationalism, in this view, is liberating. The other view, the view that poststructuralism just describes our dilemma, is the kind of attitude you find articulated by writers such as Stanley Fish, who insists that acknowledging the provisional and political nature of our judgments doesn't open the door to social transformation, since there is no other way for us to form judgments except by pretending that they are "right" or "just" or "true."[15] We can knock down any system that asserts a claim to foundational legitimacy, but it will be replaced by some other system making the same kinds of claims. A great deal of applied poststructuralism is devoted to this dispute between the hopeful and the resigned, and there is nothing within the philosophical suppositions of poststructuralism that decides the outcome.

3.

Let's see if there is a way of getting the toothpaste back in the tube. One way to describe the historical process that began in the West in the late eighteenth century and whose impact on the environment, in particular, has come to seem pernicious is to call it the liberation of self-interest.

At that point in Western history, it became not only possible but socially and culturally encouraged for people to act as agents for themselves. The terms that describe this event are "democracy," which is designed to allow people to vote their own political interests; "laissez-faire," which is designed to accommodate the pursuit of individual economic self-interest; and "romanticism," which introduces self-expression as a primary value into art and literature. These concepts are ideal types; they do not manifest themselves in a pure form at any particular moment in the history of the last two centuries, and it is, in a sense, part of their internal nature to provoke reactionary as well as revolutionary responses. Nevertheless, it is possible to see their march since, say, 1776 as a generally forward one.

There is, within this development, a contradiction or, better put, a tension. The tension is between the values of science and reason that make the economic and political systems run and the values of nature and spirit that sustain individual life within this system—between rationalism and romanticism. The crucial fact is that these tendencies run together. The progress of technocracy is accompanied at every step by its cultural critique. It is the business of liberalism to insist that romanticism is necessary, and it is the business of romanticism to point out that liberalism is not enough. Modernity has given us the Uniform Commercial Code, and it has also given us the *White Album*.

This is why it seems pointless to argue (except for the purposes of professional advancement) about whether romantic literature or modern literature or postmodern literature is complicit with or subversive of the forces of nineteenth-and twentieth-century life. You don't have to read Wordsworth to know that we murder to dissect; you can read the same message in a greeting card. We know that the *White Album* is more valuable than the Uniform Commercial Code, that love is better than money, that a sunset is more beautiful than a car, that organisms are better than machines, that the heart has reasons that reason can never know, that small communities are more humane than metropolises, that general stores are nicer than malls, that modernization has (as Dickens says in *Dombey and Son*) made it natural to be unnatural. We know these things because nearly all of Western art and literature and philosophy for the last two hundred years has insisted that the technological and instrumentalist view of life is inadequate.

This is the inconvenient fact that makes large-scale indictments of nineteenth-and twentieth-century Western civilization so peculiar. The reader of Alistair MacIntyre's *After Virtue* awaits impatiently the name of the figure whose worldview epitomizes the modern consciousness.[16] At last, the moment arrives, and the figure revealed is Benjamin Franklin. Modernity is always being represented in books like MacIntyre's by Franklin and Bethman and the Mills: you would think there were statues of these exemplars on every street corner in the Western world and that no one else had ever written. The same distortion appears in Christopher Lasch's *The True and Only Heaven*, an attack on the liberal ideology of progress.[17] Lasch tries to resurrect what he calls a populist tradition, dedicated to the way of life modernity is supposed to be sweeping away, but apart from some genuine eccentrics, currency reformers, and anti-Semites it does not seem a shame to have neglected, his populists turn out to be people like Emerson, and the values they promote can be found, on a much grander scale, everywhere in Western culture.

It therefore seems to be the case that the environmental critique of modernity has to come, and can only come, from within modernity itself. It is romanticism that teaches us that science and rationality are impoverishing, and it is science and rationality that are most likely to produce the technology required to overcome the depredations technology has wrought. The West—increasingly, the globe—operates an economic system that is predicated on the necessity of growth. We are reaching a point where growth is killing us. We don't need to step into a whole new worldview in order to address the problem.

We may, though, need a way of organizing knowledge that is less consistent with the assumptions of positivism and more consistent with the assumptions of romanticism. Our present way of organizing knowledge is embodied, of course, in the research university, an institution that, in this country, dates from the late nineteenth century. The progenitor of the modern university was not Daniel Coit Gilman or Charles William Eliot, but Charles Darwin. Darwin deconstructed, in effect, the transcendental signified that had held together the university's traditional conception of knowledge and thereby loosed the disciplines upon the world. After Darwin, knowledge could no longer be pursued by the light of some unifying theory—that is, theology. The response to this withdrawal was to pursue knowledge according to the methods and data appropriate to the particular subject matter. Transdisciplinary theory (which is what theology was) dropped out and was replaced by procedural principles, such as the principles of peer review, professional autonomy, and academic freedom.

The problem the Darwinian revolution caused for knowledge is a version of the central problem of modernity, which is the problem of legitimacy. If legitimation for a practice doesn't come from outside the system, where does it come from? Some efforts were made in the nineteenth century to replace theology, as the fixed center of the university's knowledge system, with philosophy. The argument was made in England by Sir William Hamilton, and it was tried in this country by various thinkers: Charles Sanders Peirce, for example, insisted that logic must be the metadiscipline, the prerequisite to every other branch of learning. Around the time of World War I, as part of a reaction against the positivistic model of knowledge that had become instantiated in the university by the establishment of graduate studies and the free elective system for undergraduates, it was proposed, notably at Harvard and Columbia, that the liberal arts, or the Great Books, might constitute the substantive core around which other disciplines could be imagined to arrange themselves. Later, at the University of Chicago in the era of Robert M. Hutchins, moral and political philosophy was proposed as the core discipline. This notion of a hub to which all the disciplinary spokes are connected is still a powerful one in the modern university, but in practice it takes the form of distribution requirements, a policy that reflects not a belief that there is a core to knowledge, but its opposite: an official insistence on treating every discipline as separate and equal and naming no discipline as queen. This tension between the organization of knowledge by institutionally efficient "departments," on the one hand, and the yearning for a common set of core values or principles, on the other, is, of course, emblematic of the Janus-faced nature of modernity itself. Nothing is more "modern" than bureaucratic systematization followed closely by appeals to organicism and wholeness.

The disciplines are so obviously factitious intellectually that there is no need to dwell on the matter. Most were created at the end of the nineteenth century through the breakup of larger scholarly organizations. The American Social Science Association, for example, founded in 1865 as a group for amateur students of a broadly defined range of social science subjects, split up into the Modern Language Association, which broke off in 1883, the American Historical Association (1884), and the professional associations of economists (1885), church historians (1888), folklorists (1888), and political scientists (1889). A similar process of professional specialization overtook the natural sciences around the same time. Historians, anthropologists, sociologists, political scientists, social psychologists, and economists all study the same thing, which is the behavior of people in groups. Yet they operate within the university as autonomous departments; they develop their own reigning theories and methodologies; and progress within them is driven by their own internal debates.

It's not as though professors haven't been aware of the artificiality of these boundaries; the usual way to overcome it has been to practice interdisciplinarity. In practice, though, interdisciplinarity is usually just a method of authority borrowing. Professor X in political science needs a view of human nature in order to resolve a dispute within the discipline of political science, so he goes downstairs to Professor Y in the anthropology department to borrow one. Professor Y obliges with a theory of human nature that is, in fact, highly contested within the discipline of anthropology, but which will sound impressive to people in political science. Professor X can then write into his argument a sentence that begins, "As anthropologists have discovered," and an interdisciplinary act has been consummated. The university is redeemed.

The inadequacy of this system of overcoming disciplinary boundaries has become apparent since the 1960s. One reason for this new awareness is the addition of subject matter that does not fit into—or that administrators have been reluctant to fit into—the traditional departmental schema. These are subjects like African American Studies, Third World or Postcolonial Studies, Women's Studies, Cultural Studies, Science Studies, and Gay and Lesbian Studies. Such subjects are interdisciplinary by definition, and therefore conventional interdisciplinary practice, which simply puts specialists from different disciplines into the same classroom, isn't enough. Knowledge in these areas needs to be worked up from more holistic premises.

This is not easy to do, particularly since studies centers are likely to have a marginal institutional status. The people who work in them must be credentialed, after all, in a particular discipline. The response to the need for a common language, intelligible across departmental boundaries, has been the rise, in all the humanistic disciplines, of theory. There is now little difference between what an art historian will have to say, as a theoretical matter, about "modernity" and what an English professor will have to say about it. The vocabulary of contemporary theory is metadisciplinary: it allows people to talk across the disciplines. The environment is a perfect instance of a subject that cannot be roped off in the way the research university has traditionally roped off subjects. The only way to think intelligently about the environment is holistically; it is, arguably, in part the fault of the bureaucratic impediments to holistic thinking that we have arrived at the ecological crisis that appears to face us. Useful study of the environment cannot be disciplinary, and it cannot be interdisciplinary. It has to be postdisciplinary. Postdisciplinarity is coming; but it comes with a daunting array of intellectual problems, since it will seem to many people simply a license for speculation and improvisation. Maybe it is. Thinking holistically while eschewing systematization is possibly something that can only be done badly. But we won't be certain until we try it.

NOTES

1. Simon Schama, *Landscape and Memory* (New York: Knopf, 1995).
2. William James, *Pragmatism* (1907; Cambridge, Mass.: Harvard University Press, 1975), 37.
3. Paul de Man, "Literary History and Literary Modernity," in *Blindness and Insight: Essays in the Rhetoric of Contemporary Criticism*, 2d ed. (1969; Minneapolis: University of Minnesota Press, 1983), 144.
4. Horace Kallen, "Modernism," *Encyclopedia of the Social Sciences* (New York: Macmillan, 1933).
5. Frank Kermode, *Romantic Image* (London: Routledge and Kegan Paul, 1957).
6. Perry Meisel, *the Myth of the Modern* (New Haven: Yale University Press, 1987).
7. Jean-François Lyotard, *The Postmodern Condition: A Report on Knowledge*, trans. Geoff Bennington and Brian Massumi (Minneapolis: University of Minnesota Press, 1984).
8. Fredric Jameson, *Postmodernism; or, The Cultural Logic of Late Capitalism* (Durham: Duke University Press, 1991).
9. Max Weber, "Science as a Vocation," in *From Max Weber: essays in Sociology*, ed. H. H. Gerth and C. Wright Mills (1919; New York, Oxford University Press, 1946), 139.
10. Jameson, *Postmodernism*, 302–313.
11. T. S. Eliot, *The Waste Land*, in *The Complete Poems and Plays, 1909–1950* (1922; New York: Harcourt, Brace & World, 1952), 39, 50.
12. Jacques Derrida, "Structure, Sign, and Play in the Discourse of the Human Sciences," (1966), in *The Structuralist Controversy: The Languages of Criticism and the Sciences of Man* (1966; Baltimore: The Johns Hopkins University Press, 1972), 247–265.
13. Thomas Kuhn, *The Structure of Scientific Revolutions* (Chicago: University of Chicago Press, 1962); Richard Rorty, *Philosophy and the Mirror of Nature* (Princeton: Princeton University Press, 1979).
14. Lyotard, *The Postmodern Condition*, xxiii–xxiv.
15. Stanley Fish, *Doing What Comes Naturally: Change, Rhetoric, and the Practice of Theory in Literary and Legal Studies* (Durham: Duke University Press, 1989).
16. Alistair MacIntyre, *After Virtue* (Notre Dame: University of Notre Dame Press, 1981).
17. Christopher Lasch, *The True and Only Heaven* (New York: W. W. Norton, 1991).

LEO MARX

Environmental Degradation and the Ambiguous Social Role of Science and Technology

SINCE 1970 concern about the degradation of the global environment has mounted rapidly throughout the world. Both the media and the public in the United States seem to have accepted the not implausible idea that a dire global "crisis" of the environment is upon us. One striking if little noticed side effect of the presumed crisis has been an intensification of the ambiguity surrounding the social role of scientists and engineers. Nowadays these experts often are treated with the highest respect as society's most knowledgeable, reliable, and effective protectors of the environment; yet, at the same time, they almost as often seem to be charged with complicity in—even responsibility for—environmental degradation. How shall we account for such contradictory perceptions of the ways that science and engineering affect the environment?

Concern about the global environment unquestionably has increased the prominence of scientists. It has led governments to grant them new, often unprecedented, social responsibility and decision-making power. The reasons are obvious. Because natural scientists are presumed to know most of what is known about the biophysical environment, it ("nature") is assumed to be their special domain, and they its delegated custodians.[1] It was alert scientists, after all, who first sounded the alarm about the irreversible damage that humanity may be causing by depleting the ozone layer and spewing more and more greenhouse gases into the atmosphere. It is scientists and engineers, moreover, on whom we necessarily must rely to identify, monitor, and analyze the results of human interventions in the biosphere, and on whom we also must rely to devise more benign alternatives to our present modes of intervention. The judgments that scientists and engineers now are being asked to make have potential consequences of extraordinary gravity, and in coming decades both the frequency of the requests and the gravity of the consequences may be expected to increase. As the social responsibility, power, and influence of organized science expand, so does society's dependence upon it. And as society becomes more aware, and more fearful, of the threat that its technological power poses to the life-sustaining capacities of global ecosystems, the greater the respect and the higher the status society may be expected to accord to the scientific and engineering professions.

Yet the rise of new environmentalisms also has coincided with the wider dissemination of a powerful critique of science and technology. This critique is by no means wholly new, but many of the inherited counter-Enlightenment arguments have been reformulated with environmental degradation in view, and the revised version that is emerging promises to be more radical, more comprehensive, and perhaps more widely diffused and effective than its precursors. Today, powerful antiscience and antitechnology attitudes, expressed with varying degrees of rationality, explicitness, and sophistication, seem to be gaining influence at all levels of contemporary American society.[2] At one end of this continuum of attitudes may be found the irresponsible, anti-intellectual, dismissive views of science held by believers in various populist, quasi-mystical, evangelical, antirational religious creeds and adherents of pseudosciences such as astrology, scientology, and creationism. Toward the center we find more reasoned views advanced by several countercultural movements, including those in favor of alternative, ostensibly less damaging, technologies (e.g., solar, wind, and tidal sources of energy) and those opposed to particular technologies (e.g., nuclear, chemical, and biological weapons systems and nuclear power generation). At the other end of the spectrum, there is a qualitatively different kind of criticism, based on serious, rational, intellectually responsible arguments put forward by reputable historians, literary scholars, and philosophers (including advocates of various "postmodernist" viewpoints)—arguments that reassess "modernity" itself, including its central faith in the conceptual foundations of modern science as embodied in the rationalist mainstream of Western philosophy from Plato to the Enlightenment and culminating in the epistemological realism of the logical positivists.[3]

Public receptivity to these variegated antiscience attitudes has been heightened, over the last half-century, by a frightening series of major disasters associated in one way or another with science and technology. Many of these calamities, which resulted in high death tolls and considerable damage to the environment, involved highly innovative, science-based technologies. Some are attributable to the malfunctioning of those technologies (e.g., Three Mile Island, Chernobyl, Bhopal, the Challenger explosion, the Exxon oil spill); some to their deliberate misuse (e.g., by militaristic or genocidal leaders in wartime: the Coventry and Dresden bombings, Hiroshima, the Holocaust); and some to the unforeseen costs or consequences of technological innovation (e.g., the extinction of species resulting from massive deforestation or the use of powerful new insecticides such as DDT or the water pollution caused by the use of chemical fertilizers). Whatever their presumed causes, these highly publicized catastrophes undoubtedly have enhanced the credibility of the various critiques of science and technology. So, no doubt, does the apparent enthusiasm, the lack of moral restraint or discrimination, with which scientists and engineers have been willing to serve leaders, organizations, or regimes intent upon the ruthless exercise of military or political domination. All of this helps to explain the sharp decline, in recent decades, of the extremely high respect, close to veneration, in which the public once had held scientists and engineers.

Taken together, then, the simultaneous heightening of these incompatible viewpoints indicates the existence of a widespread ambivalence about the social role of science and technology. Whereas one segment of the population holds scientists and engineers in some measure to blame for the late-twentieth-century degradation of the biosphere, another segment looks upon those same experts as the chief agents for its defense. (Still another segment doubtless is undecided, is inclined to entertain either view, or alternates between them, each individual thus exhibiting a personal ambivalence toward science and technology.) But here many questions arise that require closer examination: Is this supposed ambivalence of Americans attributable to an actual—as distinct from a merely perceived—ambiguity in the social role of scientists and engineers?[4] Even more important, can we confirm the apparent connection between the American people's ambivalence about science and technology and their growing concern about environmental degradation? Did that relationship in fact originate in recent decades, or does it have another, older, conceptually more fundamental origin? What is its history, and what are its larger, longer-term social and political implications?

In this essay, I sketch a partial answer to these questions. This obviously is not a task for scientists or engineers, and a secondary purpose of this paper, incidentally, is to exemplify one kind of contribution to environmental studies that we might expect from scholars in the humanities and humanistic social sciences. Let me begin, then, by suggesting why it is imperative to enlist them in our effort to understand certain salient problems that modern societies face in coping with environmental degradation.[5]

The Role of the Humanities in the Defense of the Environment

In spite of the intense concern aroused in recent years by the ecological crisis, it has barely begun to elicit a truly significant research effort from scholars working in the humanities and social sciences. Indeed, the idea that scholars other than scientists and engineers might make significant contributions to our grasp of environmental problems evidently surprises most people, in and out of the academy. Their doubts on this score are repeatedly reinforced by our habit of identifying environmental problems with—in the literal sense of naming them according to—their biophysical manifestations. Take, for example, Lester Brown's more or less typical list, in his 1990 *State of the World* report, (like the one cited earlier in this volume) of the conditions that constitute the global ecological crisis: "eroding soils, shrinking forests, deteriorating rangelands, expanding deserts, acid rain, stratospheric ozone depletion, the buildup of greenhouse gases, air pollution, and the loss of biological diversity."[6] Every one of the conditions named here points to an essentially biophysical problem. What is needed to begin correcting them, presumably, is a generous application of technical expertise.

This way of defining environmental problems is one reason why it is so difficult to imagine how the work of, say, philosophers or historians might contribute to the defense of the environment. How could a philosopher possibly help society reduce the level of carbon dioxide in the atmosphere? But the very absurdity of the question suggests that our present way of defining environmental problems may be dangerously narrow and misleading. Although the most conspicuous presenting symptoms of a set of problems like atmospheric pollution are likely to be biophysical, its primary causes and consequences, and many of the measures required for its amelioration, belong to the very different sphere of human behavior, institutions, and beliefs. This is the case, I believe, with every one of the dire conditions named by Brown. The point is that no adequate grasp of most forms of environmental degradation is possible until we understand the socioeconomic and cultural contexts from which they derive and those in which we will have to deal with them.

To put it differently, the participation of humanists and social scientists is needed to cope with environmental degradation because its origins lie deep in our history. Once the biophysical parameters and possible remedies of most environmental problems have been established, society's capacity to deal with them will in large measure depend on less tangible, largely unquantifiable political, institutional, and cultural factors. Scholars in the humanities are particularly well equipped to identify, interpret, and assess the cultural determinants of our relations with the environment. Their essential method is historically informed interpretation. Among its merits, this method helps to enlarge the temporal dimension within which we analyze problems that might otherwise be approached in misleadingly ahistorical or presentist terms. The "presentist" approach, to which scientists and engineers tend to restrict themselves, focuses chiefly, if not entirely, on manifestations of problems that exist in the present. The initial account, above, of the presumed link between today's environmental consciousness and the ambiguous role of science and technology exhibits that very shortcoming—which I will try to correct.

It is said, wrote Ralph Waldo Emerson, "that the views of nature held by any people determine all their institutions" (II:46).[7] The determining power of such fundamental beliefs and, by extension, of the bearing of cultural history upon environmental issues, occasionally is brought home to a wider public. In 1967, for example, the historian of science Lynn White Jr. set off something of a national controversy with his provocative essay, "The Historic Roots of Our Environmental Crisis."[8] There he contends that the aggressive, often destructive treatment of the environment that characterizes Western culture is traceable to attitudes deeply embedded in Judeo-Christian religion. Of all the great world religions, he claims, Christianity is unique in the extent to which it separates humanity and nature, and thus in fostering the belief that the natural world exists chiefly, if not exclusively, to serve humankind. Quite apart from the merits of that particular argument, which provoked many rejoinders, what is relevant here is White's premise as a historian, namely, that a people's treatment of the environment is mediated by its deepest, shared assumptions about the character of its relationship with the realm of the nonhuman, or nature—assumptions that underlie its most enduring religious, social, and political beliefs and practices.

As the controversy set off by White's essay indicates, we need to know a great deal more about the environmental implications of religious ideas and practices. Another inadequately understood aspect of prevailing attitudes toward the environment to which the humanities provide our sole

conduit is aesthetic. The widespread revulsion generated by news accounts of environmental degradation—such as vivid photographic reports of the damage inflicted on wildlife and the landscape by an oil
spill—is in large part aesthetic. That revulsion is the reverse side of the
human capacity to derive pleasure from the beauty of nature. Aesthetic
pleasure, in turn, is a vital element in the strong emotional attachments
that people form to particular places—to their native regions or "home-
lands." As any student of Romanticism recognizes, there are good reasons for thinking that in a secularizing era like our own, some part of that
ostensibly aesthetic attachment to "nature," or to natural phenomena, is
an expression of repressed, sublimated, or somehow redirected religious
feelings. Unfortunately, humanists have failed to give much scholarly attention to the potential efficacy of aesthetic motives in the defense of
the environment.

That failure, like other failures of humanistic scholarship in dealing
with environmental issues, is bound up with the anxious emulation, by
humanists, of the physical sciences and their exact, rigorous, objective,
or, to use the invidious word, "hard" forms of knowledge. An unfortunate offshoot of the prestige of science and technology has been a continuing faith in the long-term promise of the science-based technological fix. Scholars in the humanities and social sciences consequently have
been reluctant to credit the social efficacy of behavior perceived as emotional, subjective, and resistant to quantitative analysis—or, in short, as
easily sentimentalized—as the responsiveness of human beings to the
"beauty" of nature. It is in fact testimony to the intimidating primacy of
the "hard," scientific form of knowledge that the word "beauty" has all
but disappeared in scholarly discussions of environmental degradation or,
for that matter, any other subject. But avoiding the aesthetic, religious,
or metaphysical motives behind the deep attachment that many people
have to various aspects of the natural world is a serious mistake. Those
motives, and the larger belief systems to which they belong, do not figure
in the work of scientists or engineers, but they are among the most potent ideological or conceptual resources available to society in mobilizing broad popular support for environmental action. To that end, as I
hope to illustrate, the attention of scholars in the humanities is required.

The Cultural Roots of the Problem

To return, then, to the problem raised at the outset: how shall we understand the increasingly ambiguous social role of science and technology

in the context of mounting public concern about environmental degradation? The lacunae in my initial account of the problem underscore the need to approach it from the more inclusive, historically informed, viewpoint of the humanities. We cannot hope to clarify the relations in question without taking account of (1) the history of the specific attitudes involved and (2) the mediation of attitudes toward the social role of science and technology by (as a result of their inclusion in) larger belief systems.

A common presupposition of anthropologists and other students of culture nowadays is the centrality, within a people's collective mentality, of its shared assumptions about nature.[9] These assumptions seldom are made explicit or fully articulated. Rather, they are embedded in the common background knowledge that adherents of the culture acquire simply by virtue of being born into it. They include preconceptions about both the presumed nature of nature and the essential character of the interrelations among individuals, social groups, and nature. Although attitudes toward nature occupy a central place in all cultures, significant differences obviously exist in the degree of influence they exercise. Thus a strong case can be made for the particular salience of ideas about the interplay between society and nature in the dominant American (i.e., U.S.) worldview. This collective preoccupation is largely attributable to the special circumstances surrounding the conquest and settlement of North America by white Europeans.

Chief among those circumstances was the rare and relatively "late" (in the chronology of European development) coming together of self-selected groups from the most advanced Western cultures and a vast, rich, underdeveloped, seemingly uninhabited, unclaimed, wilderness. (The colonists' cultures of origin were "advanced" in the sense of possessing sophisticated knowledge, especially a highly innovative, rapidly developing science and technology.) The distinctiveness of this event lies in the conjunction of cognitive sophistication and socioeconomic (or geopolitical) "underdevelopment." Leaders of the colonizing Protestant sects saw the availability of this unimaginably large, rich land mass as a providential corroboration of the sacred character of their "errand into the wilderness." They believed that North America was made available at that time so that the reformed Christian churches and sects could fulfill the divinely ordained mission of humankind, as expounded in Genesis, to dominate the earth and all of the creatures upon it. "This state of things," as Tocqueville later recast this providential view of the uniqueness of the Great Migration in secular language, "is without a parallel in the history of the world" (II:36).[10]

American studies has yielded a large body of work on the cultural history of relations between nature and society in North America.[11] What requires emphasis here is that from the beginning, the arriving Europeans perceived the underdeveloped environment of the "New World" in the context of their collective power to transform it. In other words, they thought of the resources—the latent wealth—of North American nature as coupled, in a complex reciprocal relationship, with their own power; each of the polar concepts, the colonizer's transformative powers and New World nature, in effect derived a large measure of its meaning and value from its perceived relation to the other. Thus it is essential, if we are to understand today's conflicting American attitudes toward the use of science-based technological power in transforming and, often, degrading the environment, to trace their history at least as far back as the beginnings of European settlement. Early in that history, the American myth of national origins, whose narrative core invariably was a version of the trans-Atlantic migration, gave rise to at least three sharply divergent interpretations. They may be called, for expository convenience, progressivism, primitivism, and pastoralism.

Progressivism

According to the dominant interpretation of the myth, the journey of Europeans to North America represents the imposition of a righteous, enlightened, powerful civilization upon a disorderly, barbarous, unredeemed wilderness. The presence of Native Americans was not felt to be inconsistent with this belief, since Europeans tended to think of them, like other nonwhite peoples around the world, as savages, closer to the status of animals than of people and hence part of wild rather than human nature.[12] In the seventeenth century, this version of American beginnings lent expression to the New England Puritans' Calvinistic view of the natural world as "fallen," a "hideous wilderness" or "howling desert," requiring for its right ordering the redemptive intervention of the elect: the Christian "saints" of Protestant churches. In its unimproved state, they believed, nature was hostile, death-dealing, infertile, and ugly—in short, Satan's territory. As Max Weber was to emphasize, Protestant doctrine made a sacred duty of their capitalist zeal in converting natural resources into wealth.[13]

This religious version of the myth, which comported so nicely with the economic motives of the settlers, later (in the late-eighteenth and early-nineteenth centuries) was reformulated in the secular language of the Enlightenment ideology of progress. The fulcrum of that ideology, whose dominance in the United States was not to be seriously challenged until the middle of the twentieth century, is the quasi-mythic idea of history, or at least modern history, as progress. (In the modern era, secular conceptions of history have in large measure supplanted the older, supernatural myths and serve many purposes formerly served by them; these quasi-mythic ideas of history provide central, ordering principles for the belief systems of modern peoples and thus lend them a degree of continuity and coherence.) The secular myth of progress turns on the conviction that history is a record of the continuous, cumulative, steady, and (for the truest believers) preordained expansion of human knowledge of, and power over, nature. It would be difficult to conceive of an outlook more flattering to human beings and their cherished sense of themselves as the deliberate agents of their own destiny.

What, we may well ask, did the initiators of the idea of history as progress have to say about the danger of environmental degradation? The revealing fact is that that danger does not seem to have concerned either the Enlightenment thinkers who initially formulated the progressive view of history (e.g., Condorcet and Turgot, Paine and Priestley, Franklin and Jefferson) or their precursors among the founders of scientific rationalism (e.g., Galileo, Bacon, Newton, Descartes, Locke). It simply did not occur to these individuals, only two or three centuries ago, that the exercise of science-based technological power might cause serious damage to the environment. This striking fact underscores the distinctive, unprecedented character of the realization, during the last two decades, that our advanced technological societies already may be inflicting irreversible, irreparable damage on the atmospheric envelope of the biosphere.

We are reminded that whatever humans' professed beliefs, their practical behavior during most of their history can only be understood, finally, as the expression of a predominantly utilitarian idea of nature. They perforce have been chiefly concerned to use the resources of the natural world to feed, shelter, and clothe themselves and to adapt themselves to nature's evident inhospitality—the rigors of climate; the scarcity of food; the rapacity of other animals; and the devastation of fire, flood, and epidemic disease. We are likely to forget that human behavior rarely has been guided by solicitude for the environment (or for other creatures) and that technological innovation seldom has been considered anything but an unquestioned good. The special situation of European settlers in North America served, for obvious reasons, to accentuate the idea of an unending struggle, with heroic overtones, of technologically sophisticated human beings against a resistant, often antagonistic nature. Then in the mid-nineteenth century, Darwinism, as popularly interpreted by Herbert Spencer and others, provided a new, secular, "scientific" extension and reinforcement of the belief in progress: now all of human development could be seen as the story, like a heroic folktale, of the species' rise from humble origins to higher and higher states of civilization. Social Darwinism metaphysicalized progress.

The history of the idea of progress is too complex to recount in detail, but here it is necessary to note two related nineteenth-century developments that effected basic changes in its prevailing meaning. The first was the gradual abandonment of the Enlightenment's "means-ends" approach to the idea of progress, whereby knowledge—especially scientific and technological knowledge—was seen as a primary means of achieving the sociopolitical ends of republicanism. Most of the progressive thinkers named above were republican revolutionists for whom scientific rationalism and the practices it made possible were liberatory agents of change. Those practices were technical means to political ends. With the development of industrial capitalism and the elevation of the idea of progress to the status of quasi-official doctrine, that political commitment atrophied and the advance of science and technology increasingly was endorsed not merely as a means of arriving at political ends, but as an end sufficient unto itself. According to this new, technocratic concept of progress, if a steady increase in the power and efficiency of science-based technology can be assured, a steady improvement in most other aspects of life may be expected to follow.[14]

The appearance of a technocratic idea of progress was closely bound up with a change in the dominant character of technology itself. When the idea was first advanced, its exponents had invoked specific artifacts (e.g., the steam engine, the locomotive, and the telegraph) as evidence that history indeed was advancing. Technology was conceived in artifactual terms, as exhibited by the invention of new devices or machines. During the nineteenth century the characteristic form of technology changed, and the individual artifact was supplanted by complex, large, geographically far-flung technological systems such as the railroad system and the electric power generation and distribution system. These new systems required for their operation large capital investments, a corpus of technical knowledge, specially trained experts on duty day and night, a professional management with new skills, etc., and as a result the family-sized firm was replaced by the modern, impersonal corporation as the characteristic institutional form in which technologies operated. Mass production, the assembly line (Fordism), and the practice of scientific management (Taylorism) were characteristic innovations of the new era of large, bureaucratized technological systems. Their productive power lent credibility to the technocratic idea of progress.[15]

Although the concept of progress changed over time and the degree of the American public's commitment to it has fluctuated, its essential core—the idea of steady improvement in the overall conditions of life stemming, above all, from the expansion of scientific and technological knowledge—has remained a major, if no longer unchallenged, article of belief in the United States. Today, the official measures of national progress are indices of economic growth and consumption, notably the gross domestic product (GDP) and the average family income. Thus the most frequently invoked argument against costly measures for mitigating (or averting) environmental degradation is that they will diminish the annual rise in the GDP and, by inference, impede the nation's progress.

Primitivism and Pastoralism

During the first century of European colonization, the dominance of the utilitarian (later to be recast as progressive) view of nature was rarely challenged. Many early descriptions of North America consisted of mere lists or inventories of potentially useful resources. The exercise of maximum available technological power over the environment—in clearing the land; building shelters, harbors, and roads; hunting; fishing; planting; lumbering; and mining—was assumed to be an urgent necessity, initially for survival and then for national development. By the mid-eighteenth century, however, two opposed conceptions of the relation between technological power and environment came into view.

The primitivist viewpoint is the conceptual opposite of progressivism. Whereas the geographic locus of meaning and value in the utilitarian-progressive view of national origins lay to the east, in European civilization, as represented in the typical American mental map by the high culture of London, Paris, and Rome, the chief locus of the primitivist viewpoint lay to the west, in the raw, untrammeled, unspoiled wilderness of the frontier.[16] To adherents of the primitivist view, the trans-Atlantic migration represented a desirable retreat from the oppression, excessive constraints, and deprivation associated with the complex hierarchical societies of Europe epitomized by the tyranny of *l'ancien regime;* thus the mythic journey was seen as liberatory, holding forth the promise of re-covering the simplicity, spontaneity, freedom, and harmony associated with the "natural." To adherents of this view, of course, the fewer inter-ventions of science-based technology in the natural order, the better.

In the eighteenth century the most conspicuous expressions of the primitivist viewpoint were embodied in natural theology, the aesthetic philosophy of Rousseau and other romantic writers, and nature cults like that of the noble savage. Later, in the era of high Romanticism, the prim-itivist outlook informed the expression, and occasionally the way of life, of a gifted minority of writers, artists, intellectuals, and utopian commu-nards. Although primitivist ideas have yet to be adopted by a politically significant social group in the United States (their closest political ana-logues probably are versions of anarchism), they nonetheless have con-tributed, by virtue of the clear, extreme, radical character of their com-mitment to "the natural," to reinforce other, less extreme, more prudential critiques of science, technology, and the ideology of progress. It is only recently, as we shall see, that the urgency of the environmental crisis has had the effect, for radical environmentalists, of investing primi-tivist ideas with new coherence, energy, and significance.

Although the third, or pastoral, interpretation of the American myth of national origins occasionally appeared in the literature of exploration and early colonization, it was first given significant expression by Thomas Jefferson and his contemporaries. On this view, European mi-grants to the New World enacted a modern version of the ancient pasto-ral impulse: the centrifugal urge, similar to the initiating impulse of primitivism, to move away from organized society and cultural sophisti-cation (often represented by a city or royal court) in search of a simpler way of life "closer to nature." Unlike exponents of primitivism, however, advocates of pastoralism do not prefer unspoiled nature to the overdevel-oped urban landscape. Hence they invariably check their retreat in order to achieve a condition of harmony, or equilibrium, in a partially devel-oped state of nature. The pastoral ideal, as envisaged by Jefferson, for example, was to be a republic of the "middle landscape," midway between the overcivilization of Versailles and the savagery of the American fron-tier. This was, in fact, an updated version of a view of life whose origins at least three millennia earlier had been closely bound up with the ideal-ization of the herdsmen of the ancient Near East. The pastor was a "limi-nal," or threshold, figure who occupied the center of one of the most enduring and affecting of human fantasies: the achievement of a harmo-nious *via media* between humanity and raw nature.[17]

In 1785, when Jefferson argued against introducing the new European system of manufactures to the American republic, he anticipated the outlook of many late-twentieth-century environmentalists. He was not opposed to technology in the narrow, artifactual sense. On the contrary, like his European counterparts, the philosophes, he was enchanted by steam engines and other mechanical inventions. However, he rejected the new technology as it was employed in the early capitalist factory system and, even more important, as its social function was conceived within the emergent progressive worldview. "Let our workshops remain in Europe," he wrote.

> It is better to carry provisions and materials to workmen there, than bring them to the provisions and materials, and with them their manners and principles. The loss by the transportation of commodities across the Atlantic will be made up in happiness and permanence of government. The mobs of great cities add just so much to the support of pure government, as sores do to the strength of the human body. It is the manners and spirit of a people which preserve a republic in vigor. A degeneracy in these is a canker which soon eats to the heart of its laws and constitution. (125)[18]

Most striking, apart from the notion (advanced earlier by Adam Smith) that industrial production leads to the moral decay of working people, is Jefferson's anticipation of several fundamental principles of today's environmentalism. In his endorsement of the ancient pastoral ideal of economic sufficiency, he in effect repudiates the overarching goals of the program of industrial development (set forth in Alexander Hamilton's 1791 "Report on Manufactures"),[19] namely, to maximize the nation's production, consumption, and overall economic growth. In stating this preference, Jefferson, like the environmentalists of our time, also rejects the primacy of economic criteria generally in framing social policies. He is willing, indeed, to accept a lower national GDP, or average "standard of living," in order to achieve other, less tangible, qualitative social goods. In place of economic criteria, indeed, he would substitute the kind of nonmaterial, political, or "quality of life" standards implied by his phrase "happiness and permanence of government."

Jefferson's initial rejection of the manufacturing system and, by implication, the goals of urban industrial capitalism, embodies the ideological germ of what was to become a minority, dissident, or adversary culture in the United States. Much of the subsequent development of that culture, which embraced an increasingly explicit defense of the environment and a corresponding critique of the ideology of progress, involved efforts to reformulate and to realize the principles set forth by Jefferson. Another essay would be needed merely to summarize the high points in the history of that effort, including, for example, the ideas of George Perkins Marsh and the beginnings of scientific ecology; the work of Henry Thoreau and other classic American writers; the great landscape painting of the nineteenth century; the projects of Frederick Law Olmsted and other urban planners and landscape architects; the establishment of the National Park Service and Forest Service; the rise of the nineteenth-century conservation movement; and, of course, the emergence of the new environmentalism out of the political upheaval of the 1960s.

And yet, having said all that, it is necessary to add that Jefferson also recognized the relative weakness of this minority viewpoint. Soon after adopting it, he reluctantly changed his mind and, in effect, acquiesced in the development of American manufactures. What he thereby acknowledged, to oversimplify greatly, was the popularity and forthcoming dominance of the progressive worldview, a system of value, meaning, and purpose that was to be energized, and seemingly corroborated, by the well-nigh irresistible impetus of needs stemming from rapidly accelerating population growth, the pressure of infinite material wants, the dynamism generated by endless technological innovation, and the growing productive power of an expansionary capitalist economy.

The New Environmental Consciousness in Historical Perspective

By now all three ways of conceiving the interaction between society and nature—progressivism, pastoralism, primitivism—have been adapted to the expression of today's heightened environmental awareness. Indeed, each has its rough counterpart in the spectrum of late-twentieth-century views that extends from the prudent exercise of resource management at one pole, to a middle ground occupied by the ideas of the preservationists or ecocentrists, to the exponents of deep ecology, or biocentrism, at the other pole.

The assumptions of the dominant culture in the United States, as revealed by the official environmental policies and practices of the concerned government agencies (e.g., the Environmental Protection Agency, U.S. Army Corps of Engineers, Department of Energy, Bureau of Land Management, National Forest Service, National Park Service), continue to derive from the progressive worldview. They also inform the policies and practices of large business corporations. Many of the men and women who make policy in these institutions, and in such large private environmental organizations as the Sierra Club, consider themselves to be "environmentalists." (So, according to recent opinion polls, do some 70 percent of the American people.) They are committed to coping with many aspects of environmental degradation. But they tend to regard these problems as matters of "conservation" or "resource management." They may favor relatively stringent regulation of economic and government activities in the interests of reducing pollution, conserving resources, or protecting endangered species, but they characteristically do not question the compatibility of the market economy, the continuing acceleration in the rate of science-based technological innovation, and unlimited economic growth with the protection of the environment. On the contrary, many of the measures they advocate for dealing with these problems employ the method of cost-benefit analysis, a procedure that rests on "rational choice" and market principles. On balance, then, the attitudes toward the environment sanctioned by the official culture in the United States continue to reflect the view, central to the belief in progress, that nature exists chiefly to satisfy human needs.

Yet the fact remains that the new environmental consciousness is steadily undermining the public's belief that contemporary history entails a general improvement in the human condition. Much of the erosion of that optimistic viewpoint derives from the continuing revelation, month after month, week after week, of new and glaring instances of environmental degradation. The characteristic conflict that these revelations sets off in the United States is between spokespersons for environmental protection, who favor more stringent regulation of economic and governmental activities, and representatives of business or government, who argue that the threatened constraints will impede social "national security" or "progress." In the long term, the chronic conflict between environmental protection and economic profitability probably does more to erode the old faith in progress than the series of dramatic disasters I mentioned at the beginning of this paper. For it slowly insinuates the disturbing thought that the effective cause of the late-twentieth-century environmental crisis, epitomized by ozone depletion and global warming, may reside in the fundamental economic, social, and political arrangements—the very structure—of advanced industrial societies.

The legacy of pastoralism is manifest in the views of today's preservationists, or ecocentrists. These people, many of whom are active in environmental organizations (in other countries they would be members of green parties), favor the imposition of far more severe constraints on human interventions in the environment than those espoused by the resource managers. Like Jefferson, they reject the primacy of economic criteria in determining social policies. The chief idea they inherit from pastoralism is the need to bring human institutions into some kind of "harmony," or accommodation, with environmental imperatives. The use of science-based technologies should, in their view, be guided by (and if necessary curbed), the need to meet relatively severe standards of environmental protection. Unlike exponents of the progressive ideology, they regard human beings not as separated from and acting upon the environment but rather as an integral part of it. Like Jefferson, again, they would expect the rich (both individuals within the "advanced" nations and the richer nations) to sacrifice material superfluities in the interests of protecting the biosphere, including what remains of the wilderness. They would, in other words, replace the goal of maximizing production and consumption with the pastoral ideal of economic sufficiency.[20]

At the radical extreme of today's environmental ideologies is "deep ecology," a viewpoint that bears certain obvious affinities with Rousseauian ideas and primitivism generally. The distinguishing characteristic of this outlook is a refusal to grant the human species a privileged status vis-à-vis other creatures. All life, in the view of the small but articulate minority of deep ecologists, has intrinsic value. Human beings have a right to satisfy their vital needs, but no legitimate claim to other special privileges. To accept this viewpoint is to accept a personal obligation to effect sociocultural change and help maintain the greatest possible biological diversity. Taken seriously, the deep ecologists' viewpoint would entail the elimination of many technologies in use today, or their replacement with such benign alternatives as natural (solar, tidal, and wind) sources of energy, and a global campaign to curb population growth.

Conclusion

Recent anxieties about the deterioration of the global environment have had the effect of intensifying the ambiguity that surrounds the social roles of scientists and engineers. This has happened not merely, as suggested at the outset, because the environmental crisis has made their roles more conspicuous. Nor is it merely because recent disasters have alerted us to new, or hitherto unrecognized, social consequences of using the latest science-based technologies. What also requires recognition is that ideas about the social role of modern science and engineering are embedded in and hence mediated by larger views of the world. Within such American worldviews, moreover, the status of science and engineering is closely bound up with their perceived effect upon the environment.

In the dominant culture, accordingly, the respect given to scientists and engineers is in large measure dependent on their ability to play the central role assigned to them in the historical narrative about progress. As the ostensible heroes of that popular story, they are expected to lead society's way in realizing the promise of prosperity and general well-being. The environmental crisis surely has diminished the credibility of that story, thereby causing the social role of science and engineering to seem more dubious—more ambiguous. To be sure, the crisis also may have the effect, for very different reasons, of increasing the power and the responsibility of organized science. But the late-twentieth-century task of damage control cannot possibly elicit anything like the respect accorded to organized science by the earlier belief in progress.

It also is important to recall, finally, that the narrative of progress itself has undergone a disillusioning transformation. The early Enlightenment version of the story depicted scientists and engineers working in the service of a social and political ideal that all people could share. But the later technocratic concept of progress, with its sterile instrumentalist notion of advancing the power of science-based technology as an end in itself, is far less likely to inspire trust. Its patent inadequacies have had the effect of enhancing the appeal, if only by contrast, of the seemingly "antiscience" ideologies of pastoralism and primitivism. All of which might be taken to suggest that if the scientific and engineering professions want to recover some of the respect and status they once had, they would be advised to join with sympathetic humanists and social scientists in recuperating some of the idealism that the project of modern science formerly derived from its place within the ideology of progress. That might require them to sacrifice their technocratic posture of moral neutrality, dissociate themselves from people and institutions responsible for environmental degradation, and help in formulating a new concept—which is to say, new criteria—of progress to which they might commit themselves. A primary test of any proposed social policy under this new dispensation surely would be whether it would improve, or at a minimum protect, the life-enhancing capacities of the global ecosystem.

NOTES

1. In spite of the notorious confusions inherent in the word "nature," I use it in the now popular sense to refer to the biophysical environment, either as it is presumed to have existed prior to human intervention or that part of it that remains distinguishable from the works of humanity.

2. This was apparent at the joint U.S.-U.S.S.R. workshop entitled "Anti-Science and Anti-Technology Movements in the U.S. and U.S.S.R." held at the Massachusetts Institute of Technology, May 2–3, 1991. I am particularly indebted to Gerald Holton's paper, "How to Think about the 'Anti-Science' Phenomenon," and to Kenneth Keniston's comments on it.

3. Here I would associate myself with Keniston's insistence, expressed in his comments on Holton's paper (see note 2), on the importance of distinguishing between the various popular irrational, pseudoscientific attacks on science and these intellectually responsible critiques.

4. Similarly conflicting attitudes no doubt are discernible in other societies, but for present purposes I focus on the society and culture I know most about, the United States.

5. In arguing this last point, I draw on a report prepared for the Rockefeller Foundation by a group of Massachusetts Institute of Technology graduate students and fellows working under my supervision in 1990. See Leo Marx, "The Humanities and the Defense of the Environment," Working Paper No. 15, Program in Science, Technology, and Society, Massachusetts Institute of Technology, Cambridge, 1991.

6. Lester R. Brown, ed., *State of the World, 1990: A Worldwatch Institute Report on Progress Toward a Sustainable Society* (New York: W. W. Norton, 1990), 10.

7. Ralph Waldo Emerson, in *Works* [*English Traits*] (Philadelphia: John D. Morris, 1906), II:46. I would amend the motto to read "the views of their relations with nature held by any people."

8. Lynn White Jr., "The Historic Roots of Our Environmental Crisis," *Science* 155 (1967): 1203. White's argument accords with that of many historians of religion. See, for example, Mircea Eliade, *The Myth of the Eternal Return* (Princeton, N.J.: Princeton University Press, 1954). Eliade distinguishes the cyclical concept of time (history) characteristic of most world religions from the Judeo-Christian linear concept of time (and historical development) as a continuum of nonrecurring events and hence peculiarly receptive to the belief in history as a record of continuous, cumulative "progress."

9. There is no generally accepted terminology for reference to collective mentalities. Among the terms in general use today, the most prominent are: *myth, ideology, ethos, utopia, belief system,* and *worldview.* Here I am referring to a body of ideas like that which Clifford Geertz calls a "worldview": a people's "picture of the way things in sheer actuality are, their concept of nature, of self, of society . . . [containing] their most comprehensive ideas of order." *The Interpretation of Cultures* (New York: Basic Books, 1973), 127.

10. Alexis de Tocqueville, *Democracy in America*, ed. Phillips Bradley (New York: Vintage Books, 1945), II:36.

11. See, for example, Sacvan Bercovitch, *American Jeremiad* (Madison: University of Wisconsin Press, 1978); Leo Marx, *The Machine in the Garden: The Pastoral Ideal in America* (New York: Oxford University Press, 1964); Richard Slotkin, *Regeneration through Violence: The Mythology of the American Frontier, 1600–1860* (Middletown, Conn.: Wesleyan University Press, 1973); and Henry Nash Smith, *Virgin Land: The American West as Symbol and Myth* (Cambridge, Mass.: Harvard University Press, 1950).

12. According to the best current estimates, there were some two and a half million "Indians" living in North America when colonization began.

13. See, especially, Max Weber, *The Protestant Ethic and the Spirit of Capitalism*, trans. Talcott Parsons (New York: Charles Scribner's Sons, 1958).

14. For a more detailed account of the transition to a technocratic idea of progress, see Leo Marx, "Does Improved Technology Mean Progress?" *Technology Review* 90, no. 1 (January 1987): 32–41. Nothing demonstrates the hegemonic scope of the idea of progress, incidentally, better than its simultaneous embrace by leading apologists for capitalism and by that system's most systematic and influential socialist critics, the Marxists. The latter's version of the idea is rooted in Marx's conviction that the key to long-term human development and social change is the growth of society's productive capacity. Thus Marxism verges, like much of Western modernism, on technological determinism. That belief in the efficacy of technology as a cardinal agent of history was to be widely expressed in Soviet culture during the Stalin era, as exemplified by the Stakhanov movement, and the veneration of technological power in literature, art, film, and the official party ideology.

15. These innovations in the operation of technological systems also captured the imagination of V. I. Lenin, Leon Trotsky, and their Bolshevik comrades. See Antony C. Sutton, *Western Technology and Soviet Economic Development, 1917–1930* (Stanford, Calif.: Stanford University Press, 1968). For the concept of the technological system, see Jacques Ellul, *The Technological System*, trans. Joachim Neugroschel (New York: Continuum, 1980), and Lewis Mumford, *The Myth of the Machine*, vol. 1, *Technics and Human Development* (New York: Harcourt, Brace and World, 1967); for its American origins, see Alfred D. Chandler, *The Visible Hand: The Managerial Revolution in American Business* (Cambridge, Mass.: Harvard University Press, 1977); for its application to American development generally, see Thomas P. Hughes, *American Genesis* (New York: Viking Press, 1989).

16. The "mental map" of a people, as defined by image geographers, is their shared conception of their place in the world, a geographic image distorted to represent their worldview. My friend, the late Professor Rufus W. Mathewson of Columbia University, a Slavic scholar, often remarked in conversation on the many similarities between the American and Russian mental maps, each oriented to the polarity of European "civilization" and a "nature" associated with an underdeveloped frontier, but of course with the compass directions reversed, West signifying the locus of advanced civilization in the Russian mental map and wild nature in the American mental map.

17. For a useful summary of this theory (one among many) of the origins of pastoral, see David Halperin, *Before Pastoral* (New Haven, Conn.: Yale University Press, 1983), chap. 6.

18. Thomas Jefferson, *Query XIX, Notes on the State of Virginia*, ed. William Peden (Chapel Hill: University of North Carolina Press, 1955). I discuss this passage at greater length in *The Machine in the Garden*, 125.

19. Alexander Hamilton, "Report on Manufactures," in *Industrial and Commercial Correspondence of Alexander Hamilton*, ed. Arthur Harrison Cole (Chicago, 1928).

20. The idea of economic sufficiency flies in the face of prevailing (neoclassical) economic theory, for it requires the implementation of a distinction, rejected by neoclassic economists, between material wants and needs.

Appendix

THE ESSAYS in this volume are based on papers delivered at the faculty-student Workshop on Humanistic Studies of the Environment supported by the John D. and Catherine T. MacArthur Foundation at the Massachusetts Institute of Technology between 1992 and 1995. The seminar was preceded by a grant from the Rockefeller Foundation (1991–92) to consider the contributions the humanities could make to understanding current environmental problems. The MacArthur Workshop aimed at encouraging and exemplifying scholarly inquiry in the human sciences that could contribute to a better understanding of the contemporary "crisis" of the global environment. The seminar, which met monthly during the academic year, was conducted by the editors of this volume under the auspices of the Program in Science, Technology, and Society at MIT.

THE MACARTHUR WORKSHOP ON HUMANISTIC STUDIES OF THE ENVIRONMENT

Leo Marx, Jill Ker Conway, Kenneth Keniston, organizers

Program in Science, Technology, and Society (STS)
Massachusetts Institute of Technology

1991–92: Humanistic Perspectives on Atmospheric Change

September 18, 1991
>JAMES FLEMING
>Department of Physics and Astronomy, Colby College
>"Historical Perspectives on Atmospheric Change"

October 16, 1991
>GREGORY NAGY
>Department of Classics, Harvard University
>"Perspectives on the Heavens in the Ancient World"

November 20, 1991
>TIMOTHY WEISKEL
>Divinity School and Kennedy School, Harvard University
>"Key Metaphors on Environmental Problems"

December 18, 1991
> Donald Worster
> Department of History, University of Kansas
> "Drought and the Midwestern Aquifer"

January 22, 1992
> Joni Seager
> University of Vermont (Visiting Professor in Women's Studies, MIT)
> "Corporate Culture, or Who's in Charge of the Train of Progress?"

February 19, 1992
> Steven Pyne
> Department of History, Arizona State University
> "Consumed by Either Fire or Fire: A Review of the Environmental Consequences of Anthropogenic Fire"

March 18, 1992
> Joshua Cohen
> Political Science and Philosophy, MIT
> "International Equity Issues in Global Warming"

April 15, 1992
> John Richards
> Department of History, Duke University
> "Settlement Frontiers and Property Rights in Early Modern World History"

May 20, 1992
> Leo Marx
> School of Humanities and Social Science, MIT
> "Toward a History of Atmosphere Change"

1992–93: Sources of the Emerging Environmental Consciousness

September 15, 1992
> Max Oelschlaeger
> University of North Texas
> "The Idea of Wilderness"

October 13, 1992
> Elizabeth Johns
> University of Pennsylvania
> "Landscape Art"

November 17, 1992
> Leo Marx
> MIT
> "Pastoralism"

December 15, 1992
>RICHARD WHITE
>University of Washington
>"The Experience of Indigenous Peoples"

January 12, 1993
>HARRIET RITVO
>MIT
>"Human Relations with the Animal Kingdom"

February 16, 1993
>JILL KER CONWAY AND YAAKOV GARB
>MIT
>"The Experience of Women"

March 16, 1993
>STEPHEN TOULMIN
>Northwestern University
>"The Ideology of the Enlightenment"

April 13, 1993
>STEPHEN JAY GOULD
>Harvard University
>"Darwinian Evolution"

May 11, 1993
>Review, Recapitulation, Future Plans

1993–94: Modernity and the Environment

October 6, 1993
>VICTORIA DE GRAZIA
>Columbia University
>"Modern Consumerism"

November 3, 1993
>CAROL GLUCK
>Columbia University
>"Modernity in a Non-Western Context"

December 1, 1993
>ROBERT HEILBRONER
>The New School for Social Research
>"An Economist's Perspective on Modernity"

January 5, 1994
>JANET WOLFF
>University of Rochester
>"Aesthetic Modernism"

February 2, 1994

> Louis Menand
> City University of New York
> "Modernity and Literary Theory"

March 2, 1994

> Anton Struchkov
> Russian Academy of Sciences, Moscow
> "Modernity in the Former Soviet Union"

April 6, 1994

> Gyan Prakash
> Princeton University
> "Modernity in a South Asian Context"

May 4, 1994

> Review, Recapitulation, Future Plans

1994–95: The Humanities and Environmental Action

September 21, 1994

> Barbara Epstein
> History of Consciousness Board, University of California, Santa Cruz
> "Grassroots Environmental Activism: The Toxics Movement and
> Directions for Social Change"

November 2, 1994

> Stephen Kellert
> School of Forestry and Environmental Studies, Yale University
> "Psychogenetic Roots of Human-Nature Relationships"

January 4, 1995

> Terence Turner
> Anthropology Department, University of Chicago
> "The Political Struggle over Resource Use and Environmental
> Protection among the Brazilian Kayapo"

February 8, 1995

> Oleg Yanitsky
> Institute of Sociology, Russian Academy of Sciences
> "Russian Environmental Movements"

March 7, 1995

> Thomas Gladwin
> Stern School of Business, New York University
> "Envisioning Environmentally Sustainable Enterprise"

March 22, 1995
  MICHAEL SMITH
  History Department, University of California, Davis
  "Advertising and Environmental Action"

April 5, 1995
  WOLFGANG SACHS
  Wuppertal Institute for Climate, Energy and Environment
  (Germany)
  "The Political Anatomy of 'Sustainable Development'"

May 3, 1995
  BINA AGARWAL
  Institute of Economic Growth, University of Delhi (India)
  "Gender, Environment, and Collective Action"

MacArthur Workshop Participants:

Prof. Larry Buell
  English Department
  Harvard University

Prof. Jill Conway
  STS, Women's Studies
  MIT

Prof. John Dower
  History Department
  MIT

Dr. John Ehrenfeld
  Senior Research Associate
  Center for Technology, Policy, and
  Industrial Development
  MIT

Prof. Michael Fischer
  STS
  MIT

Prof. James Fleming
  Science/Technology Studies
  Colby College
  Waterville, ME

Yaakov Garb
  Fellow, Institute for Advanced
  Study
  Princeton, NJ

Slava Gerovitch
  STS
  MIT

Prof. Loren Graham
  STS
  MIT

Rebecca Herzig
  STS
  MIT

Terry Hill
  Urban Studies and Planning
  MIT

Prof. Evelyn Fox Keller
  STS
  MIT

Prof. Kenneth Keniston
  STS
  MIT

Prof. Alvin Kibel
  Literature Faculty
  MIT

Hannah Landecker
STS
MIT

Prof. Leo Marx
STS
MIT

Prof. Bruce Mazlish
History Faculty
MIT

Minakshi Menon
STS
MIT

Prof. Gregory Nagy
Classics Department
Harvard University

Russ Olwell
STS
MIT

Peter Perdue
History Faculty
MIT

Prof. Ruth Perry
Literature Faculty,
Women's Studies
MIT

James Risbey
Center for Meteorology and
Physical Oceanography
MIT

Prof. Harriet Ritvo
History Faculty
MIT

Barbara Rosenkrantz
History of Science
Harvard University

Wade Roush
STS
MIT

Robert H. (Rusty) Russell
Conservation Law Foundation
Boston, MA

Prof. Bish Sanyal
Urban Studies and Planning
MIT

Prof. Eugene Skolnikoff
Department of Political Science
MIT

Prof. Merritt Roe Smith
STS
MIT

Sam Bass Warner, Jr.
Visiting Prof., Urban Studies
and Planning
MIT

Prof. Charles Weiner
STS
MIT

Prof. Timothy Weiskel
Environmental Workshop
Harvard Divinity School

Prof. Rosalind Williams
Writing Program
MIT

Ben Williams
School of Education
Harvard University

Doug Winiarski
Harvard Divinity School

Notes on Contributors

BINA AGARWAL is a professor of economics at the Institute of Economic Growth, University of Delhi. Educated at Cambridge University and University of Delhi, she has taught at Harvard University as a visiting professor and been a fellow of the Bunting Institute (Radcliffe College), the Institute of Development Studies (Sussex), and the Science Policy Research Unit (University of Sussex). She has also been a visitor at the Institute of Advanced Study at Princeton. Her most recent book is *A Field of One's Own: Gender and Land Rights in South Asia* (Cambridge University Press, 1994), winner of the A. K. Coomaraswamy Book Prize in 1996 (Association for Asian Studies, United States); the Edgar Graham Book Prize in 1996 (Department of Development Studies, SOAS, University of London); and the K. H. Batheja Award in 1995–96 (University of Bombay). She is currently working on some aspects of gender, environment, and collective action.

JILL KER CONWAY was born in Hillston, New South Wales, Australia; graduated from the University of Sydney in 1958; and received her Ph.D. from Harvard University in 1969. From 1964 to 1975 she taught at the University of Toronto and was vice president there before serving for ten years as president of Smith College. In 1985, she and her husband, John Conway, moved to Boston, Massachusetts, where she is now a visiting scholar and professor in the Program in Science, Technology, and Society at The Massachusetts Institute of Technology.

BARBARA EPSTEIN teaches at the University of California, Santa Cruz, in the history of consciousness department. Her most recent book is *Political Protest and Cultural Revolution: Nonviolent Direct Action in the 1970s and 1980s* (University of California Press, 1991). Over the last few years she has been writing about the decline of the left in the United States, in the hope of finding ways to reverse this trend. Her research on the environmental justice movement is part of this larger project.

YAAKOV GARB, a Lady Davis Fellow at Hebrew University, Jerusalem, is currently writing a book on the politics of roads in Israel, to be filed as a thesis for a second Ph.D. in the Science, Technology, and Society Program at the Massachusetts Institute of Technology. His training is in environmental studies, and especially their social and cultural dimensions. His publications include a series of essays on episodes in the formation of environmentalist discourse in postwar America. Garb has held postdoctoral positions at the Institute for Advanced Studies at Princeton and the History of Science Program at Harvard University and served as academic director of the Arava Institute for Environmental Studies, the first regional environmental studies university program in the Middle East.

KENNETH KENISTON is Andrew Mellon Professor of Human Development in the Program in Science, Technology, and Society at the Massachusetts Institute of Technology. His current research is on the relationship of software, culture, and politics, especially in South Asia. With Jill Ker Conway and Leo Marx, he was one of the principal organizers of the workshop "The Humanities and the Environment," which led to this volume.

LEO MARX took his doctorate (in the History of American Civilization) at Harvard University in 1949. He taught in the Program in American Studies and English department at the University of Minnesota from 1949 to 1958 and at Amherst College from 1958 until 1976. His writings include *The Machine in the Garden: Technology and the Pastoral Ideal in America* (1964), *The Pilot and the Passenger* (1988), (with S. Danly) *The Railroad in American Art* (1988), and (with M. R. Smith) *Does Technology Drive History? The Dilemma of Technological Determinism* (1994); editions of work by Henry Thoreau, Nathaniel Hawthorne, and Mark Twain; as well as a number of critical essays on American writers and literary themes, especially having to do with the relations between literature and the city, the onset of industrialism, and the environment. In 1976 he was appointed the William R. Kenan, Jr., Professor of American Cultural History at The Massachusetts Institute of Technology (now Emeritus) and a member of the Program in Science, Technology, and Society. He has been president of the American Studies Association and a member of the American Academy of Arts and Sciences.

LOUIS MENAND is a professor of English at the Graduate Center of the City University of New York. He has been associate editor of *The New Republic* and literary editor of *The New Yorker* and is currently contributing editor of *The New York Review of Books*. He is the author of *Discovering Modernism: T. S. Eliot and His Context* (Oxford University Press, 1987) and the editor of *The Future of Academic Freedom* (University of Chicago Press, 1996) and *Pragmatism: A Reader* (Vintage Books, 1997).

GREGORY NAGY is the Francis Jones Professor of Classical Greek Literature and Professor of Comparative Literature at Harvard University. He served as the elected president of the American Philological Association in the academic year 1990–91. He is the author of *The Best of the Achaeans: Concepts of the Hero in Archaic Greek Poetry* (Johns Hopkins University Press, 1979), which won the Goodwin Award of Merit, American Philological Association, in 1982. Other publications include *Comparative Studies in Greek and Indic Meter* (Harvard University Press, 1974), *Greek Mythology and Poetics* (Cornell University Press, 1990), *Pindar's Homer: The Lyric Possession of an Epic Past* (Johns Hopkins University Press, 1990), *Poetry as Performance: Homer and Beyond* (Cambridge University Press, 1996), and *Homeric Questions* (University of Texas Press, 1996). His special research interests are archaic Greek literature and oral poetics, and he finds it rewarding to integrate these interests with teaching, especially in his course for Harvard's Core Curriculum, "The Concept of the Hero in Greek Civilization." He is currently the chair of Harvard's classics department.

STEPHEN J. PYNE is the author of a dozen books relating to environmental history. Best known is a suite of five volumes, Cycle of Fire, that surveys the history of fire over much of the Earth. Other works examine Antarctica, the Grand Canyon, and the history of geology. His interest in fire derives from 15 summers he spent on a forest fire crew at Grand Canyon's North Rim. He is presently a professor in the Biology Department at Arizona State University.

JOHN F. RICHARDS is a professor of history at Duke University. He received his Ph.D. in the history of South Asia from the University of California at Berkeley in 1970 and was a member of the history department of the University of Wisconsin, Madison, from 1968 to 1977 before moving to Duke. His original specialty lies in the early modern history of South Asia. His most recent work in that field is *The Mughal Empire* (Cambridge University Press, 1993). Since the late 1970s Richards has been involved in active research and publication on world environmental history. He has coedited two volumes on global deforestation in the nineteenth and twentieth centuries and has compiled a large database on the history of land use changes in South and Southeast Asia over the past century. He is currently at work on an environmental history of the early modern world.

ANTON STRUCHKOV did graduate work in the history of natural science and technology at the Russian Academy of Sciences. He is a member of the editorial board and on the staff of *Problems of the History of Natural Science and Technology*, a leading academic journal in Russia. His specialty is environmental ethics.

TERENCE TURNER is a professor of anthropology at Cornell University. He received his Ph.D. from Harvard University in 1966, for a thesis based on research among the Kayapo of Central Brazil. He has continued to work with the Kayapo until the present. His numerous writings on them cover social organization; myth; ritual; history; politics; interethnic contact; and aspects of cultural, social, political, and ideological change. He has also made ethnographic films about the Kayapo with the British Broadcasting Company and Granada Television and for the last seven years has been directing the Kayapo Video Project, in which the Kayapo have been shooting and editing videos about their own culture and relations with the Brazilians. He has published numerous theoretical papers on topics such as symbolic media; metaphor; ritual; historical consciousness; the body and subjectivity; indigenous media; documentary and representation; and anthropological applications of Marxian theoretical concepts such as value, fetishism, production, and exploitation. Turner has also been extensively involved in human rights and indigenous-support activities. He served as head of the Special Commission of the American Anthropological Association to investigate the Situation of the Brazilian Yanomami in 1991; served on the Ethics Committee of the American Anthropological Association; and was a member of the AAA Committee for Human Rights and its antecedents, the Commission for Human Rights and the Task Force on Human Rights, from 1992 to 1997.

RICHARD WHITE teaches at the University of Washington, where he got his Ph.D. in 1975. His latest book is *The Organic Machine* (Hill and Wang, 1995).

DONALD WORSTER is Hall Distinguished Professor of American History at the University of Kansas. He has published nine books on environmental history, the history of ecology, and the history of the American West. His book on the Dust Bowl of the 1930s (Oxford University Press, 1979) won the Bancroft Prize in American history, and he has held fellowships from the Guggenheim Foundation, the American Council of Learned Societies, and the National Endowment for the Humanities. He was formerly president of the American Society for Environmental History and is advisory editor for the Cambridge University Press book series Studies in Environment and History.

OLEG N. YANITSKY is chief researcher at the Institute of Sociology of the Russian Academy of Sciences. He has published extensively on environmental movements and environmental policy in Russia. In 1995 he was an academic visitor at The Massachusetts Institute of Technology (program in the Humanities and Environment). He is head of the Research Committee "Environment and Society" of Russian Society of Sociologists.

DATE DUE
